Tactile Sensing, Skill Learning, and Robotic Dexterous Manipulation

Tactile Sensing, Skill Learning, and Robotic Dexterous Manipulation

Edited by

Qiang Li
Center for Cognitive Interaction Technology, Bielefeld University
Bielefeld, Germany

Shan Luo
Department of Computer Science, University of Liverpool
Liverpool, United Kingdom

Zhaopeng Chen
Faculty of Mathematics, Informatics and Natural Science
Department Informatics, University of Hamburg
Hamburg, Germany

Chenguang Yang
Bristol Robotics Lab, University of the West of England
Bristol, United Kingdom

Jianwei Zhang
Faculty of Mathematics, Informatics and Natural Science
Department Informatics, University of Hamburg
Hamburg, Germany

ACADEMIC PRESS
An imprint of Elsevier

Academic Press is an imprint of Elsevier
125 London Wall, London EC2Y 5AS, United Kingdom
525 B Street, Suite 1650, San Diego, CA 92101, United States
50 Hampshire Street, 5th Floor, Cambridge, MA 02139, United States
The Boulevard, Langford Lane, Kidlington, Oxford OX5 1GB, United Kingdom

ISBN: 978-0-323-90445-2

For information on all Academic Press publications
visit our website at https://www.elsevier.com/books-and-journals

Publisher: Mara Conner
Acquisitions Editor: Yura R. Sonnini
Editorial Project Manager: Emily Thomson
Production Project Manager: Prasanna Kalyanaraman
Designer: Victoria Pearson

Typeset by VTeX

Contents

Contributors

Paolo Ariano, Center for Sustainable Future Technologies, Istituto Italiano di Tecnologia, Torino, Italy

Zhenshan Bing, Department of Informatics, Technical University of Munich, Munich, Germany

Nicolò Boccardo, Rehab Technologies Lab, Istituto Italiano di Tecnologia, Genova, Italy

Michele Canepa, Rehab Technologies Lab, Istituto Italiano di Tecnologia, Genova, Italy

Guanqun Cao, smARTLab, Department of Computer Science, University of Liverpool, Liverpool, United Kingdom

Simona Castellano, Centro Protesi INAIL, Vigorso di Budrio (BO), Italy

Zhaopeng Chen, University of Hamburg, Faculty of Mathematics, Informatics and Natural Science, Department Informatics, Group TAMS, Hamburg, Germany

Lorenzo De Michieli, Rehab Technologies Lab, Istituto Italiano di Tecnologia, Genova, Italy

Yitao Ding, Lab of Robotics and Human Machine Interaction, Chemnitz University of Technology, Chemnitz, Germany

Yongxiang Fan, FANUC Advanced Research Laboratory, FANUC America Corporation, Union City, CA, United States

Daniel Fernandes Gomes, smARTLab, Department of Computer Science, University of Liverpool, Liverpool, United Kingdom

Emanuele Gruppioni, Centro Protesi INAIL, Vigorso di Budrio (BO), Italy

Sami Haddadin, Chair of Robotics and Systems Intelligence, Munich Institute of Robotics and Machine Intelligence (MIRMI), Technical University of Munich (TUM), Munich, Germany
Centre for Tactile Internet with Human-in-the-Loop (CeTI), Dresden, Germany

Robert Haschke, Center for Cognitive Interaction Technology (CITEC), Bielefeld University, Bielefeld, Germany

Diego Hidalgo, Chair of Robotics and Systems Intelligence, Munich Institute of Robotics and Machine Intelligence (MIRMI), Technical University of Munich (TUM), Munich, Germany
Centre for Tactile Internet with Human-in-the-Loop (CeTI), Dresden, Germany

Jiaqi Jiang, smARTLab, Department of Computer Science, University of Liverpool, Liverpool, United Kingdom

Hannes Kisner, Lab of Robotics and Human Machine Interaction, Chemnitz University of Technology, Chemnitz, Germany

Matteo Laffranchi, Rehab Technologies Lab, Istituto Italiano di Tecnologia, Genova, Italy

Ke Li, Institute of Intelligent Medicine Research Center, Department of Biomedical Engineering, Shandong University, Jinan, Shandong, China

Qiang Li, University of Bielefeld, Bielefeld, Germany
Center for Cognitive Interaction Technology (CITEC), Bielefeld University, Bielefeld, Germany

Rui Li, School of Automation, Chongqing University, Chongqing, China

Shuang Li, Universität Hamburg, Hamburg, Germany

Tingguang Li, Tencent Robotics X, Shenzhen, China

Andrea Lince, Rehab Technologies Lab, Istituto Italiano di Tecnologia, Genova, Italy

Lorenzo Lombardi, Rehab Technologies Lab, Istituto Italiano di Tecnologia, Genova, Italy

Shan Luo, smARTLab, Department of Computer Science, University of Liverpool, Liverpool, United Kingdom
Department of Computer Science, University of Liverpool, Liverpool, United Kingdom

Andrea Marinelli, Rehab Technologies Lab, Istituto Italiano di Tecnologia, Genova, Italy

Abdeldjallil Naceri, Chair of Robotics and Systems Intelligence, Munich Institute of Robotics and Machine Intelligence (MIRMI), Technical University of Munich (TUM), Munich, Germany

Qi Qi, School of Automation, Chongqing University, Chongqing, China

Pericle Randi, Centro Protesi INAIL, Vigorso di Budrio (BO), Italy

Luigi Reale, Area sanità e salute, ISTUD Foundation, Milano, Italy

Helge Ritter, Center for Cognitive Interaction Technology (CITEC), Bielefeld University, Bielefeld, Germany

Jody A. Saglia, Rehab Technologies Lab, Istituto Italiano di Tecnologia, Genova, Italy

Marianna Semprini, Rehab Technologies Lab, Istituto Italiano di Tecnologia, Genova, Italy

Valentina Squeri, Rehab Technologies Lab, Istituto Italiano di Tecnologia, Genova, Italy

Hang Su, Dipartimento di Elettronica, Informazione e Bioingegneria, Politecnico di Milano, Milano, Italy

Antonio Succi, Rehab Technologies Lab, Istituto Italiano di Tecnologia, Genova, Italy

Ulrike Thomas, Lab of Robotics and Human Machine Interaction, Chemnitz University of Technology, Chemnitz, Germany

Simone Traverso, Rehab Technologies Lab, Istituto Italiano di Tecnologia, Genova, Italy

Ning Wang, Bristol Robotics Laboratory, University of the West of England, Bristol, United Kingdom

Chenguang Yang, Bristol Robotics Laboratory, University of the West of England, Bristol, United Kingdom

Jianwei Zhang, Universität Hamburg, Hamburg, Germany
University of Hamburg, Faculty of Mathematics, Informatics and Natural Science, Department Informatics, Group TAMS, Hamburg, Germany

Preface

Dexterous manipulation is a very challenging research topic and it is widely required in countless applications in the industrial, service, marine, space, and medical robot domains. The relevant tasks include pick-and-place tasks, peg-in-hole, advanced grasping, in-hand manipulation, physical human–robot interaction, and even complex bimanual manipulation. Since the 1990s, mathematical manipulation theories (*A Mathematical Introduction to Robotic Manipulation*, R.M. Murray, Z.X. Li, and S.S. Sastry, 1994) have been developed and we have witnessed many impressive simulations and real demonstrations of dexterous manipulation. Most of them need to assume:

1. an accurate object geometrical/physical model and a known robotic arm/hand kinematic/dynamic model,
2. the robot has the dexterous manipulation skills for the given task.

Unfortunately, as these two strong assumptions can only be feasible in a theoretical model, physics simulation, or well-customized structural environment, previous research work is biased towards motion planning and implementation. Because of the inherent uncertainty of the real world, simulation results are relatively fragile in deploying in real applications and prone to failed manipulation if the assumptions deviate from the real robot and object model. The demonstrated experiments will also fail if the structural environment is changed. Examples can be changes in kinematic/dynamic models due to wear and tear, imperfectly calibrated hand–eye ratios, and a change in the manipulated object.

In order to deal with the uncertainty from dynamic interaction and implementation, it is necessary to exploit sensors and the sensory-control loop to improve the robots' dexterous capability and robustness. Currently, the best-developed sensor feedback in robotics is vision. Vision-based perception and control have largely improved the robustness of robots in real applications. One missing aspect for vision-powered robots is their application in the context of contact. This absence is mainly because vision is not the best modality to measure and monitor contact because of occlusion issues and noisy measurements. On this point, tactile sensing is a crucial complementary modality to extract unknown contact/object information required in manipulation theories. It provides the most practical and direct information for object perception and action decisions.

Apart from sensor feedback, another unresolved issue for dexterous manipulation is how to generate the robot's motion/force trajectory – skills for the tasks. Given the diversity of the tasks, it is unpractical to hardcode all kinds of manipulation skills for robotic arms and hands. Inspired by imitation, one solution is to extract, represent, and generalize these skills from human demonstration. Then the robots use adaptive controllers to implement the learned skills on the robotic arm and hand. In recent years we have also seen researchers combine skill representation and transfer in one step – exploring and learning the dexterous controller automatically.

In this edited book, we invited the researchers working in three research directions – tactile sensing, skill learning, and adaptive control – to draw a complete picture of dexterous robotic manipulation. All of them have top-level publication records in the robotics field. We are confident that the contributed chapters can provide both scientists and engineers with an up-to-date introduction to these dynamic and developing domains and present the advanced sensors, perception, and control algorithms that will inform the important research directions and have a significant impact on our future lives. Concretely the readers can gain the following knowledge from this book:

1. tactile sensing and its applications to the property recognition and reconstruction of unknown objects;
2. human grasping and dexterous skill representation and learning;
3. the adaptive control scheme and its learning by imitation and exploration;
4. concrete applications how robots can improve their dexterity by modern tactile sensing, interactive perception, learning, and adaptive control approaches.

As editors, we believe synthesizing intelligent tactile perception, skill learning, and adaptive control is an essential path to advancing state-of-the-art dexterous robotic manipulation. We hope that readers will enjoy reading this book and find it useful for their research journey. We would like to thank all authors, and we are grateful for support from the DEXMAN project funded by the Deutsche Forschungsgemeinschaft (DFG) and the Natural Science Foundation of China (NSFC) (Project number: 410916101), support from the DFG/NSFC Transregio Collaborative Project TRR 169 "Crossmodal Learning," and support from EPSRC project "ViTac: Visual-Tactile Synergy for Handling Flexible Materials" (EP/T033517/1). We also express our appreciation to Emily Thomson and Sonnini Ruiz Yura from Elsevier for their encouragement and coordination to make this book possible.

Qiang Li
Shan Luo
Zhaopeng Chen
Chenguang Yang
Jianwei Zhang
Bielefeld
June 2021

Part I

Tactile sensing and perception

Chapter 1

GelTip tactile sensor for dexterous manipulation in clutter

Daniel Fernandes Gomes and Shan Luo
smARTLab, Department of Computer Science, University of Liverpool, Liverpool, United Kingdom

1.1 Introduction

As humans, robots need to make use of tactile sensing when performing dexterous manipulation tasks in cluttered environments such as at home and in warehouses. In such cases, the positions and shapes of objects are uncertain, and it is of critical importance to sense and adapt to the cluttered scene. With cameras, Lidars, and other remote sensors, large areas can be assessed instantly [1]. However, measurements obtained using such sensors often suffer from large uncertainties, occlusions, and variance of factors like light conditions and shadows. Thanks to the direct interaction with the object, tactile sensing can reduce the measurement uncertainties of remote sensors and it is not affected by the changes of the aforementioned surrounding conditions. Furthermore, tactile sensing gains information of the physical interactions between the objects and the robot end-effector that is often not accessible via remote sensors, e.g., incipient slip, collisions, and detailed geometry of the object. As dexterous manipulation requires precise information of the interactions with the object, especially in moments of in-contact or near-contact, it is of crucial importance to attain these accurate measurements provided by tactile sensing. For instance, failing to estimate the size of an object by 1 mm, or its surface friction coefficient, during (and also right before) a grasp might result in severely damaging the tactile sensor or dropping the object. In contrast, failing to estimate the object shape by a few centimeters will not make a big impact on the manipulation. To this end, camera vision and other remote sensors can be used to produce initial estimations of the object and plan manipulation actions, whereas tactile sensing can be used to refine such estimates and facilitate the in-hand manipulation [2,3].

The usage of tactile sensors for manipulation tasks has been studied since [4] and in the past years a wide range of tactile sensors and working principles have been studied in the literature [2,3,5], ranging from flexible electronic skins [6],

Tactile Sensing, Skill Learning, and Robotic Dexterous Manipulation
https://doi.org/10.1016/B978-0-32-390445-2.00008-8

fiber optic-based sensors [7], and capacitive tactile sensors [8] to camera-based optical tactile sensors [9,10], many of which have been employed to aid robotic grasping [11]. Electronic tactile skins and flexible capacitive tactile sensors can be adapted to different body parts of the robot that have various curvatures and geometry shapes. However, due to the necessity of dielectrics for each sensing element, they produce considerably low-resolution tactile readings. For example, a WTS tactile sensor from Weiss Robotics used in [12–14] has 14×6 taxels (tactile sensing elements). In contrast, camera-based optical tactile sensors provide higher-resolution tactile images. However, on the other side, they usually have a bulkier shape due to the requirement of hosting the camera and the gap between the camera and the tactile membrane.

Optical tactile sensors can be grouped in two main groups: marker-based and image-based, with the former being pioneered by the *TacTip* sensors [15] and the latter by the *GelSight* sensors [16]. As the name suggests, marker-based sensors exploit the tracking of markers printed on a soft domed membrane to perceive the membrane displacement and the resulting contact forces. By contrast, image-based sensors directly perceive the raw membrane with a variety of image recognition methods to recognize textures, localize contacts, and reconstruct the membrane deformations, etc. Because of the different working mechanisms, marker-based sensors measure the surface on a lower resolution grid of points, whereas image-based sensors make use of the full resolution provided by the camera. Some *GelSight* sensors have also been produced with markers printed on the sensing membrane [17], enabling marker-based and image-based methods to be used with the same sensor. Both families of sensors have been produced with either flat sensing surfaces or domed/finger-shaped surfaces.

In this chapter, we will first review existing optical tactile sensors in Section 1.2, and then we will look in detail into one example of such image-based tactile sensors, i.e., the *GelTip* [18,19], in Section 1.3. The *GelTip* is shaped as a finger, and thus it can be installed on traditional and off-the-shelf grippers to replace its fingers and enable contacts to be sensed inside and outside the grasp closure that are shown in Fig. 1.1. In Section 1.4, we will look into experiments carried out using the *GelTip* sensor that demonstrate how contacts can be localized, and more importantly, the advantages, and possibly a necessity, of leveraging all-around touch sensing in dexterous manipulation tasks in clutter. In particular, experiments carried out in a Blocks World environment show that the detected contacts on the fingers can be used to adapt planned actions during the different moments of the reach-to-grasp motion.

1.2 An overview of the tactile sensors

Compared to remote sensors like cameras, tactile sensors are designed to assess the properties of the objects via physical interactions, e.g., geometry, texture, humidity, and temperature. A large range of working principles have been actively proposed in the literature in the past decades [2,5,20]. An optical tactile sensor

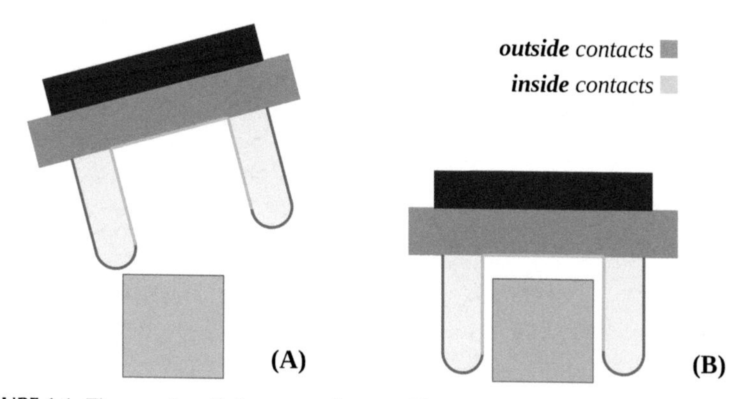

FIGURE 1.1 There are two distinct areas of contact highlighted in the robot gripper during a manipulation task: **(A)** outside contacts when the robot is probing or steering the object to be grasped; **(B)** inside contacts when the object is within the grasp closure, which can guide the grasping.

uses a camera enclosed within its shell and pointing at its tactile membrane (an opaque window membrane made of a soft material) to capture the properties of the objects from the deformations caused to its tactile membrane by the in-contact object. Such characteristics ensure that the captured tactile images are not affected by the external illumination variances. To perceive the elastomer deformations from the captured tactile images, multiple working principles have been proposed. We group such approaches in two categories: marker tracking and raw image analysis. Optical tactile sensors contrast with electronic tactile skins that usually have lower thickness and are less bulky. They are flexible and can adapt to different body parts of the robot that have various curvatures and geometry shapes. However, each sensing element of most of the tactile skins, e.g., a capacitive transducer, has the size of a few square millimeters or even centimeters, which results in a limited spatial resolution of the tactile skins. Here we do not cover such skins as these are an extensive topic on its own; however, we point the reader to two surveys that extensively cover these sensors [21,22].

1.2.1 Marker-based optical tactile sensors

The first marker-based sensor proposal can be found in [23]; however, more recently an important family of marker-based tactile sensors is the TacTip family of sensors described in [9]. Since its initial domed-shaped version [15], different morphologies have been proposed, including the TacTip-GR2 [24] of a smaller fingertip design, TacTip-M2 [25] that mimics a large thumb for in-hand linear manipulation experiments, and TacCylinder to be used in capsule endoscopy applications. Thanks to their miniaturized and adapted design, TacTip-M2 [25] and TacTip-GR2 [24] have been used as fingers (or fingertips) in robotic grippers. Although each TacTip sensor introduces some manufacturing improvements or novel surface geometries, the same working principle is shared: white pins are imprinted onto a black membrane that can then be tracked using computer vision methods.

As shown in Table 1.1, there are also other optical tactile sensors that track the movements of markers. In [26], an optical tactile sensor named FingerVision is proposed to make use of a transparent membrane, with the advantage of gaining proximity sensing. However, the usage of the transparent membrane makes the sensor lack the robustness to external illumination variance associated with touch sensing. In [27], semiopaque grids of magenta and yellow makers painted on the top and bottom surfaces of a transparent membrane are proposed, in which the mixture of the two colors is used to detect horizontal displacements of the elastomer. In [28], green fluorescent particles are randomly distributed within the soft elastomer with black opaque coating so that a higher number of markers can be tracked and used to predict the interaction with the object, according to the authors. In [29], a sensor with the same membrane construction method, four Raspberry PI cameras, and fisheye lenses has been proposed for optical tactile skins.

1.2.2 Image-based optical tactile sensors

On the other side of the spectrum, the GelSight sensors, initially proposed in [16], exploit the entire resolution of the tactile images captured by the sensor camera, instead of just tracking markers. Due to the soft opaque tactile membrane, the captured images are robust to external light variations and capture information of the touched surface's geometry structure, unlike most conventional tactile sensors that measure the touching force. Leveraging the high resolution of the captured tactile images, high-accuracy geometry reconstructions are produced in [31–36]. In [31], this sensor was used as fingers of a robotic gripper to insert a USB cable into the corresponding port effectively. However, the sensor only measures a small flat area oriented towards the grasp closure. In [37,38], simulation models of the GelSight sensors are also created.

Markers were also added to the membrane of the GelSight sensors, enabling applying the same set of methods that were explored in the TacTip sensors. There are some other sensor designs and adaptations for robotic fingers in [10,39,40]. In [10], matte aluminum powder was used for improved surface reconstruction, together with the LEDs being placed next to the elastomer and the elastomer being slightly curved on the top/external side. In [39], the GelSlim is proposed, a design wherein a mirror is placed at a shallow and oblique angle for a slimmer design. The camera was placed on the side of the tactile membrane, such that it captures the tactile image reflected onto the mirror. A stretchy textured fabric was also placed on top of the tactile membrane to prevent damages to the elastomer and to improve tactile signal strength. Recently, an even more slim design of 2 mm has been proposed [41], wherein an hexagonal prismatic shaping lens is used to ensure radially symmetrical illumination. In [40], DIGIT is also proposed with a USB "plug-and-play" port and an easily replaceable elastomer secured with a single screw mount.

TABLE 1.1 A summary of influential marker-based optical tactile sensors.

<table>
<tr><th></th><th>Sensor structure</th><th>Illumination and tactile membrane</th></tr>
<tr><td>TacTip [15]</td><td>The TacTip has a domed (finger) shape, 40 × 40 × 85 mm, and tracks 127 pins. It uses the Microsoft LifeCam HD webcam.</td><td rowspan="4">The membrane is black on the outside with white pins and filled with transparent elastomer inside. Initially the membrane was cast from VytaFlex 60 silicone rubber, the pins painted by hand, and the tip filled with optically clear silicone gel (Techsil, RTV27905); however, currently the entire sensor can be 3D printed using a multimaterial printer (Stratasys Objet 260 Connex), with the rigid parts printed in Vero White material and the compliant skin in the rubber-like TangoBlack+.</td></tr>
<tr><td>TacTip-M2 [25]</td><td>It has a thumb-like or semicylindrical shape, with TacTip-M2 32 × 102 × 95 mm, and it tracks 80 pins.</td></tr>
<tr><td>TacTip-GR2 [24]</td><td>It has a cone shape with a flat sensing membrane and is smaller than the TacTip, 40 × 40 × 44 mm, tracks 127 pins, and uses the Adafruit SPY PI camera.</td></tr>
<tr><td>TacCylinder [30]</td><td>A catadioptric mirror is used to track the 180 markers around the sensor cylindrical body.</td></tr>
<tr><td>FingerVision [26]</td><td>It uses a ELP Co. USBFHD01M-L180 camera with an 180 degree fisheye lens. It has approximately 40 × 47 × 30 mm.</td><td>The membrane is transparent, made with Silicones Inc. XP-565, with 4 mm of thickness and markers spaced by 5 mm. No internal illumination is used, as it the membrane transparent.</td></tr>
<tr><td>Subtractive color mixing [27]</td><td>N/A</td><td>Two layers of semiopaque colored markers is proposed. SortaClear 12 from Smooth-On, clear and with Ignite pigment, is used to make the inner and outer sides.</td></tr>
<tr><td>Green markers [28]</td><td>The sensor has a flat sensing surface, measures 50 × 50 × 37 mm, and is equipped with an ELP USBFHD06H RGB camera with a fisheye lens.</td><td rowspan="2">It is composed of three layers: stiff elastomer, soft elastomer with randomly distributed green fluorescent particles in it, and black opaque coating. The stiff layer is made of ELASTOSIL® RT 601 RTV-2 and is poured directly on top of the electronics, the soft layer is made of Ecoflex™ GEL (shore hardness 000-35) with the markers mixed in, and the final coat layer is made of ELASTOSIL® RT 601 RTV-2 (shore hardness 10A) black silicone. A custom board with an array of SMD white LEDs is mounted on the sensor base, around the camera.</td></tr>
<tr><td>Multicamera skin [29]</td><td>It has a flat prismatic shape of 49 × 51 × 17.45 mm. Four Pi cameras are assembled in a 2 × 2 array and fisheye lenses are used to enable its thin shape.</td></tr>
</table>

In these previous works on camera-based optical tactile sensors, multiple designs and two distinct working principles have been exploited. However, none of these sensors has the capability of sensing the entire surface of a robotic finger, i.e., both sides and the tip of the finger. As a result, they are highly constrained in object manipulation tasks, due to the fact that the contacts can only be sensed when the manipulated object is within the grasp closure [31,42,43]. To address this gap, we propose the finger-shaped sensor named GelTip that captures tactile images by a camera placed in the center of a finger-shaped tactile membrane. It has a large sensing area of approximately 75 cm^2 (vs. 4 cm^2 of the GelSight sensor) and a high resolution of 2.1 megapixels over both the sides and the tip of the finger, with a small diameter of 3 cm (vs. 4 cm of the TacTip sensor). More details of the main differences between the GelSight sensors, the TacTip sensors, and our GelTip sensor are given in Table 1.2.

With their compact design, the GelTip [18] and other GelSight [31,39–41, 46] sensors are candidate sensors to be mounted on robotic grippers. Recently, custom grippers built using the GelSight working principle have also been proposed [47,48]. Two recent works [44,45] also address the issue of the flat surface of previous GelSight sensors. However, their designs have large differences to ours. In [44], the proposed design has a tactile membrane with a surface geometry close to a quarter of a sphere. As a consequence, a great portion of contacts happening on the regions outside the grasp closure is undetectable. In [45], this issue is mitigated by the use of five endoscope microcameras looking at different regions of the finger. However, this results in a significant increase of cost for the sensor, according to the authors, approximately 3200 USD (vs. only around 100 USD for ours).

1.3 The GelTip sensor

1.3.1 Overview

As illustrated in Fig. 1.2 (A), the GelTip optical tactile sensor is shaped as a finger, and its body consists of three layers, from the inside to the outer surface: a rigid transparent body, a soft transparent membrane, and a layer of opaque elastic paint. In its base, a camera is installed, looking at inner surface of the cylinder. When an object is pressed against the tactile membrane, the elastomer distorts and indents according to the object shape. The camera can then capture the obtained imprint into a digital image for further processing. As the membrane is coated with opaque paint, the captured tactile images are immune to external illumination variances, which is characteristic of tactile sensing. To ensure that the imprint is perceptible from the camera view, LED light sources are placed adjacent to the base of the sensor, so that light rays are guided through the membrane.

TABLE 1.2 A summary of influential flat and finger-shaped GelSight sensors.

	Sensor structure	Illumination	Tactile membrane
GelSight [31]	It has a cubic design with a flat square surface. A Logitech C310 (1280 × 720) camera is placed at its base pointing at the top membrane.	Four LEDs (RGB and white) are placed at the base. The emitted light is guided by the transparent hard surfaces on the sides, so that it enters the membrane tangentially.	A soft elastomer layer is placed on top of a rigid, flat, and transparent acrylic sheet. It is painted using semispecular aluminum flake powder.
GelSight [10]	It has a close to hexagonal prism shape. The used webcam is also the Logitech C310.	Three sets of RGB LEDs are positioned (close to) tangent to the elastomer, with a 120-degree angle from each other.	A matte aluminum powder is proposed for improved surface reconstruction. Its elastomer has a flat bottom and a curved top.
GelSlim [39]	A mirror placed at a shallow oblique angle and a Raspberry Pi Spy (640 × 480) camera is used to capture the tactile image reflected by the mirror.	A single set of white LEDs is used. These are pointed at the mirror, so that the light is reflected directly onto the tactile membrane.	A stretchy and textured fabric on the tactile membrane prevents damages to the elastomer and results in improved tactile signal strength.
GelSlim v3 [41]	It is shaped similar to [10,31]; however, it is 20 mm slimmer and it has a round sensing surface.	A custom hexagonal prism is constructed to ensure radially symmetric illumination.	An elastomer with Lambertian reflectance is used, as proposed in [10].
DIGIT [40]	A prismatic design, with curved sides. An OmniVision OVM7692 (640 × 480) camera is embedded in the custom circuit board.	Three RGB LEDs are soldered directly into the circuit board, illuminating directly the tactile membrane.	The elastomer can be quickly replaced using a single screw mount.
Round fingertip [44]	It has a round membrane, close to a quarter of sphere. A single 160-degree FoV Raspberry Pi (640 × 480) is installed on its base.	Two rings of LEDs are placed on the base of the sensor, with the light being guided through the elastomer.	Both rigid and soft parts of the membrane are cast, using SLA 3D printed molds.
OmniTact [45]	It has a domed shape. Five endoscope cameras (400 × 400) are installed on a core mount and placed orthogonally to each other, pointing at the tip and sides.	RGB LEDs are soldered onto both the top and the sides of the sensor.	The elastomer gel is directly poured onto the core mount (and cameras) without any rigid surface or empty space in between.
GelTip [18]	It has a domed (finger) shape, similar to a human finger. A Microsoft Lifecam Studio webcam (1920 × 1080) is used.	Three sets of LEDs, with a 120-degree angle from each other, are placed at the sensor base, and the light is guided through the elastomer.	An acrylic test tube is used as the rigid part of the membrane. The deformable elastomer is cast using a three-part SLA/FFF 3D printed mold.

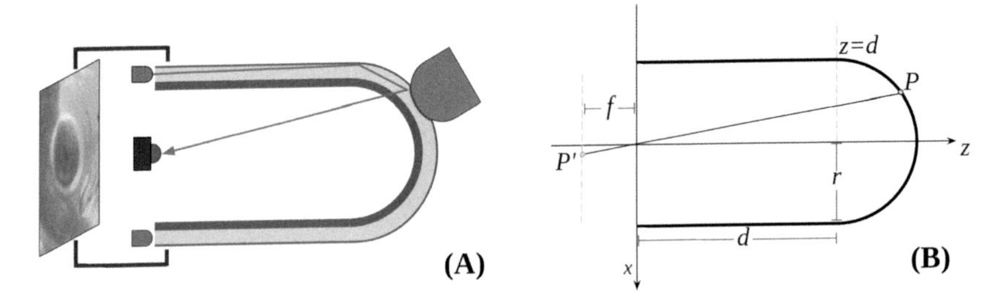

FIGURE 1.2 (**A**) The working principle of the proposed *GelTip* sensor. The three-layer tactile membrane (rigid body, elastomer, and paint coating) is shown in gray. The light rays emitted by the LEDs travel through the elastomer. As one object, shown in green, presses the soft elastomer against the rigid body, an imprint is generated. The resulting tactile image is captured by the camera sensor, placed in the core of the tactile sensor. An opaque shell, enclosing all the optical components, ensures the constant internal lighting of the elastomer surface. (**B**) Two-dimensional representation of the geometrical model of the *GelTip* sensor. The tactile membrane is modeled as a cylindrical surface and a semisphere. An optical sensor of focal length f is placed at the referential origin of the sensor, which projects a point on the surface of the sensor P into a point P' in the image plane. The sensor has a radius r and its cylindrical body has a length d.

1.3.2 The sensor projective model

For flat sensors, the relationship between the surface and the captured image can be often be easily obtained, or simply substituted by a scaling factor from single pixels to meters [10,31,42]. However, when considering highly curved sensors, it is important to study a more general projective function. In this subsection, we will look into how to derive such projective function m. As for the case of the *GelTip* sensor, m maps pixels in the image space (x', y') into points (x, y, z) on the surface of the sensor. Obtaining the protective function for other curved *GelSight* sensors should be similar, requiring only sensor-specific adaptations. The camera is assumed to be placed at the referential origin, looking in the direction of the z-axis. The sensor space takes the center of its base, which is also the center point of the camera, as the coordination origin $(0, 0, 0)$; the image space takes the center of the image as the origin $(0, 0)$. Such a projection model is necessary for, among other applications, detecting the position of contacts in the 3D sensor surface.

As illustrated in Fig. 1.2 (B), the sensor surface can be modeled as a joint semisphere and an opened cylinder, both sharing the same radius r. The cylinder surface center axis and the z-axis are collinear; therefore, the center point of the semisphere can be set to $(0, 0, d)$, where d is the distance from the center point of the base of the semisphere to the center point of the base of the sensor. The location of any point on the sensor surface (x, y, z) can be represented as follows:

$$\begin{cases} x^2 + y^2 + (z - d)^2 = r^2 & \text{for } z > d, \quad \text{(a)} \\ x^2 + y^2 = r^2 & \text{for } z <= d. \quad \text{(b)} \end{cases} \tag{1.1}$$

By making the usual thin lens assumptions, the optical sensor is modeled as an ideal pinhole camera. The projective transformation that maps a point in the world space P into a point in the tactile image P' can be defined using the general camera model [49] as

$$P' = K[\boldsymbol{R}|\boldsymbol{t}]P, \tag{1.2}$$

$$K = \begin{bmatrix} fk & 0 & c_x & 0 \\ 0 & fl & c_y & 0 \\ 0 & 0 & 1 & 0 \end{bmatrix}, \tag{1.3}$$

where $P' = [x'z, y'z, z]^T$ is an image pixel and $P = [x, y, z, 1]^T$ is a point in space, both represented in homogeneous coordinates here, $[R|t]$ is the camera's extrinsic matrix that encodes the rotation R and translation t of the camera, and K is the camera-intrinsic matrix (f is the focal length, k and l are the pixel-to-meter ratios, and c_x and c_y are the offsets in the image frame). Assuming that the used camera produces square pixels, i.e., $k = l$, fk and fl can be replaced by α, for mathematical convenience.

The orthogonal projections in the XZ and YZ of a generic projection ray can be obtained by expanding the matrix multiplication given by Eq. (1.2) and solving it w.r.t. x and y:

$$\begin{cases} x'z = \alpha x + c_x z \\ y'z = \alpha y + c_y z \\ z = z \end{cases} \Leftrightarrow \begin{cases} \alpha x = x'z - c_x z \\ \alpha y = y'z - c_y z \end{cases} \Leftrightarrow \begin{cases} x = (\frac{x' - c_x}{\alpha})z \\ y = (\frac{y' - c_y}{\alpha})z \end{cases}. \tag{1.4}$$

The desired mapping function $m : (x', y') \to (x, y, z)$ can then be obtained by constraining the z-coordinate through the intersection of the generic projection ray with the sensor surface, described in Eq. (1.5), where $\chi = x' - c_x$, $\gamma = y' - c_y$, and $\omega = \chi^2 + \gamma^2$. The discontinuity region, i.e., a circumference, is found by setting $z = d$ in Eq. (1.4):

$$m(x', y') = \begin{cases} x &= \frac{\chi}{\alpha} z, \\ y &= \frac{\gamma}{\alpha} z, \\ z &= \begin{cases} \sqrt{\frac{(r\alpha)^2}{\omega}} & \text{if } \omega < (\frac{r\alpha}{d})^2, \\ \frac{\alpha^2 2d + \sqrt{(-\alpha^2 2d)^2 - 4\omega(d^2 - r^2)\alpha^2}}{2(\omega + \alpha^2)} & otherwise. \end{cases} \end{cases} \tag{1.5}$$

The introduced sensor model is validated and visualized in Fig. 1.3. Two projection rays, corresponding to the spherical and cylindrical regions, are depicted. Each ray intersects three relevant points: the frame of reference origin,

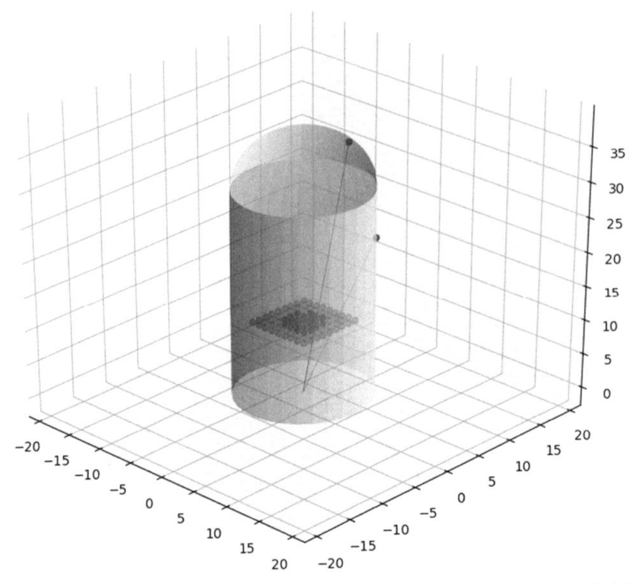

FIGURE 1.3 Two projection rays that correspond to the spherical (in red) and cylindrical (in navy blue) regions. Each ray intersects three relevant points: the frame of reference origin, a point in the sensor surface, and the corresponding projected point in the image plane.

the point in the 3D sensor surface, and the corresponding projected point in the image plane.

1.3.3 Fabrication process

As for any other optical tactile sensors, the fabrication of the GelTip sensor is the fabrication of the sensing membrane. It requires the fabrication of three parts: the bottom rigid layer, the deformable elastomer, and the coat of paint. For constructing the rigid layer, a flat sheet of transparent acrylic can be used in the case of flat sensors [17]. However, for finger-shaped sensors, a curved rigid surface is necessary. In the case of the *GelTip*, a simple off-the-shelf transparent test tube is used. These commercially available tubes are made of plastic/acrylic or glass and one disadvantage of using such test tubes, particularly the plastic ones, is that they contain small imperfections that result from the manufacturing process. An alternative approach is to print the rigid tube using a stereolithography (SLA) 3D printer and clear resin; however, proper polishing is necessary to ensure its optical transparency [44]. To fabricate the elastomer in the desired shape a mold can also be created, for instance 3D printed. Fused filament fabrication (FFF) and SLA printers yield different textured surfaces and consecutive different textured elastomers. Example 3D printed parts are shown in Fig. 1.4 (C).

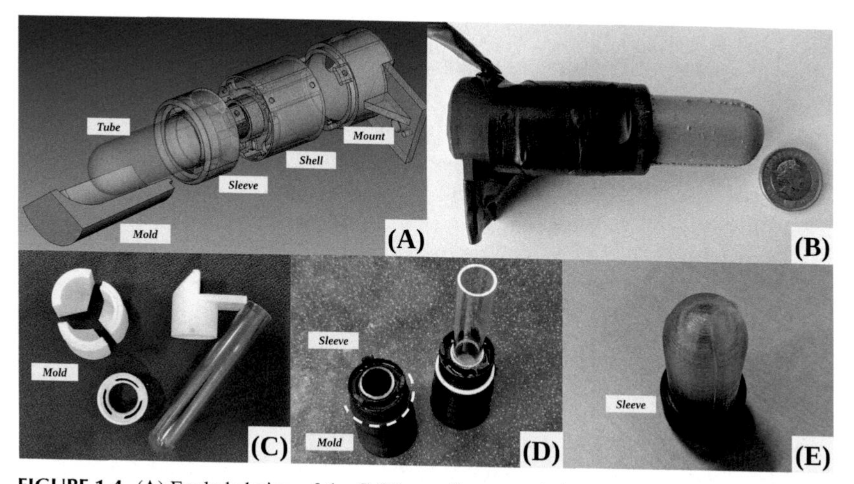

FIGURE 1.4 **(A)** Exploded view of the GelTip tactile sensor design. **(B)** A GelTip sensor, next to a British one pound coin, for relative size comparison. The sensor has a length of approximately 10 cm; its shell has a diameter of 2.8 cm; and the tactile membrane has a length of 4 cm and a diameter of 2 cm. **(C)** The three-part *mold* next to the remaining parts used in the GelTip construction. **(D)** The plastic tube is inserted into the sleeve and then mounted onto the mold; afterwards the tube is measured and trimmed and then the elastomer is poured. **(E)** The tactile membrane after being demolded and before being painted.

The soft elastomer is then created by mixing a two-component silicone, such as XP-565. Mixing these two parts in different ratios yields elastomers with different elastic properties. Additional additives can also be considered, such as Slacker for increasing the silicone tackiness. A commonly used mixture used for the *GelTip* would be 1 gram of XP-565 part-A, 22 grams of XP-565 part-B, and 22 grams of the Slacker.

For painting the transparent elastomer, off-the-shelf spray paints tend to form a rigid coat and cracks will develop in the coat when the elastomer deforms or stretches. To avoid these issues, custom paint can be fabricated and applied using a paint gun or an airbrush [10]. Pigment powder is mixed with a small portion of *part-A* and *part-B* of XP-565, with the same ratio as used in the elastomer. The paint pigment commonly consists of aluminum powder (1 μm); however, other options can also be considered. After mixing them properly, the mixture is dissolved using a silicone solvent until a watery liquid is achieved, that then can be sprayed onto the elastomer.

Finally the three sets of LEDs can be soldered: either of different colors, red, green, and blue, or all white. Since different LEDs emit different light intensities, each cluster is preceded by an independent resistor. The power source can be extracted either from the camera or from an external source, e.g., adding a secondary USB cable. At the core of any optical sensor, a camera is installed. In the case of the *GelTip*, a wide-angle lens is also considered, enabling the recording of the internal surface of the entire finger.

1.4 Evaluation

In this section, we look into two sets of experiments carried out using a *GelTip* sensor. The first set of experiments demonstrates how an image-based tactile sensor can be used to localize contacts and the second set of experiments illustrates the advantages of leveraging all-around touch sensing for dexterous manipulation. Video recordings of these experiments and the CAD models for 3D printing of the *GelTip* sensor can be found at https://danfergo.github.io/geltip/.

1.4.1 Contact localization

For the contact localization experiment, a set of seven small objects is 3D printed, each with a maximum size of $1 \times 1 \times 2$ cm^3. The objects are shown in Table 1.4. A 3D printed mount is also built and placed on top of a raised surface, ensuring that all the objects are kept in the same position throughout the experiment. *GelTip* sensors are installed on a robotic actuator, i.e., the 6-DoF Universal Robots UR5 arm with a Robotiq 2F-85 gripper. The actuator rotates and translates to tap objects at multiple known positions of one of its fingers' surface, as illustrated in Fig. 1.5. The actuator starts with its fingers pointing downwards, i.e., orientation 0, being visually aligned with the cone object. Contacts are then registered, firstly on the sensor tip, by rotating the sensor, and then on the side, by translating the sensor. In Fig. 1.5 (B) markings show the location of such contacts and in Fig. 1.5 (C) the necessary $(\Delta x, \Delta z)$ translation to obtain contacts on the finger skin is also shown. To use the projection model described in Section 1.3, five parameters must be known: r, d, c_x, c_y, and α. The first two are extracted from the dimensions of the sensor design; however, the latter three are the intrinsic parameters of the camera, which need to be calibrated. To this end, we obtain such parameters from a known pair of corresponding (x', y') and (x, y, z) points. We set the actuator to tap the object in the 15 mm translation position. The center of the sensor tip (c_x, c_y) and the contacted point are manually annotated in the image space. The α parameter can then be derived by fitting the known information into Eq. (1.4). After detecting the contact in the image space and projecting it into (x, y, z) coordinates, the Euclidean distances between the predicted and the true contact positions are computed. For each of the seven objects, a total of eight contacts are recorded, i.e., four rotations (θ), $0, \pi/6, \pi/4, \pi/3$, and four translations (τ), 0 mm, 5 mm, 10 mm, and 15 mm.

The resulting localization errors between the observed and true localization, expressed in millimeters, are summarized in Table 1.3. Overall, the variance of the localization errors is large: in some contacts the obtained errors are lower than 1 mm, while in others they are over 1 cm. On the other hand, the localization error, for each object or position, is correlated with its variance. The largest localization errors happen on objects with large or rounded tops, i.e., sphere, edge, and slab; contrariwise, the lowest errors are observed for objects with sharp tops, i.e., cone, tube, and cylinder. In terms of the localization errors at different positions, contacts happening near the sensor tip, i.e., the rotations,

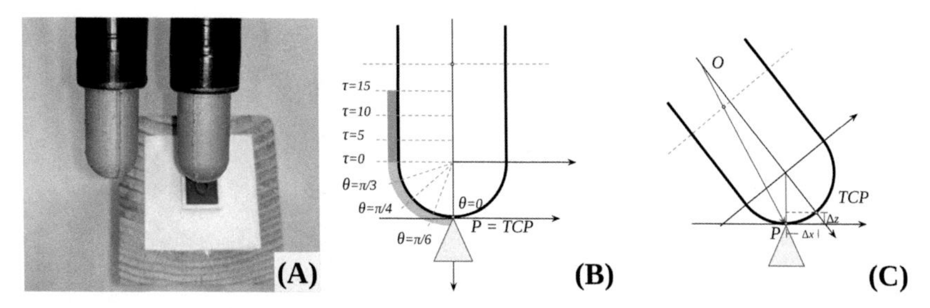

FIGURE 1.5 **(A)** Two GelTip sensors are installed on a robotic actuator and a 3D printed mount that holds a small 3D printed shape (a cylinder here) placed on top of a wooden block. The actuator moves in small increments and collects tactile images annotated with the known contact positions. **(B, C)** Illustration of the motion of the sensor during the data collection. The sensor starts pointing downwards, as shown in **(B)**. To obtain contacts on the sensor surface, while moving, the sensor is also translated by $(\Delta x, \Delta z)$, as shown in **(C)**. A total of eight contacts are collected per object: four rotations (θ) on the sensor tip and four translations (τ) on the sensor side, as highlighted in **(B)**.

TABLE 1.3 Contact errors per position, expressed in millimeters.

Rotations				Translations			
0	$\pi/6$	$\pi/4$	$\pi/3$	0	5	10	15
4.71	2.01	1.04	6.96	7.87	8.03	7.55	4.86
±0.75	±0.90	±0.46	±4.82	±5.08	±1.92	±5.00	±8.41

TABLE 1.4 Contact errors per object, expressed in millimeters.

Cone	Sphere	Irregular	Cylinder	Edge	Tube	Slab
3.63	6.79	5.61	4.57	7.47	3.33	6.27
±3.26	±5.38	±4.08	±4.30	±6.29	±1.90	±8.17

present lower errors than contacts happening on the sensor side, i.e., translations. In particular, contacts happening at $\pi/4$ and $\pi/6$ have the lowest errors.

From these experiments, we find three main challenges using a finger-shaped sensor, such as the *GelTip*, for contact localization: (1) weak imprints, created by light contacts, may not be captured by the localization algorithm; (2) forces perpendicular to the main axis of the sensor may flex the sensor tip, resulting in localization errors; and (3) imperfections in the sensor modeling and calibration further contribute to these localization inaccuracies. Examples of captured tactile images and corresponding predictions for the smallest (i.e., $< Cone, \theta = 0 >$) and largest localization errors (i.e., $< Slab, \tau = 15 \text{ mm} >$) are shown in Fig. 1.6. In the first case, due to the bright imprint provided by the sharp cone top, the algorithm successfully locates the contact. In the second

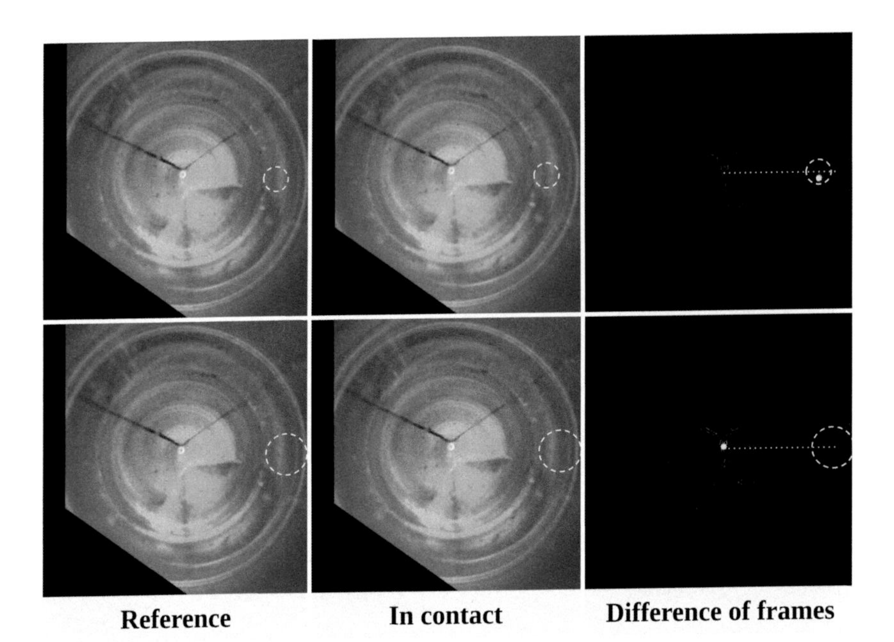

FIGURE 1.6 Reference and in-contact frames for two evaluated contacts. The expected contact region is highlighted with a dash circumference, the predicted contact position with a yellow circle, and the axis where the contacts occur with dotted lines. The top row shows the contact for the $< Cone, \theta = 0 >$ that results in the smallest error, and the bottom row shows the $< Slab, \tau = 15\text{ mm} >$ that results in the largest error.

case, due to the imperceptible contact imprint the algorithm incorrectly predicts the contact in the sensor tip.

1.4.2 Touch-guided grasping in a Blocks World environment

In the task of grasping objects, the initial motion of the gripper is often planned using remote sensing, e.g., camera vision or Lidar. However, remote sensing suffers from occlusions and inaccurate predictions about geometry and distances. In such cases, the final grasp and regrasp control policies have to rely on inaccurate information of where and when contacts occur. In contrast, touch sensing offers accurate feedback about such contacts.

We can clearly verify the importance of considering touch sensing by studying two different policies in a simple grasp experiment, i.e., ***random grasp*** (Rg) and ***random grasp + touch-informed regrasp*** (RgTr), in a simple grasp experiment. In this experiment, the robot is presented with a 4×4 board with one wooden block placed in each row in unknown columns (to the robot). The robot attempts to grasp each block, and if it fails to grasp one block after five attempts, it is considered a failure, and it skips to the next one, as shown in Fig. 1.7. Here, the random grasp in Rg and RgTr mimics the inaccuracies from remote sensing-

FIGURE 1.7 The Blocks World experimental setup. In each experiment, the robot actuator moves row by row, attempting to grasp each block. The experiment shows that, even with the initial uncertainty, the robot grasps all the blocks successfully, using the all-around touch feedback.

TABLE 1.5 The table summarizes the percentage of failing to grasp blocks (failure rate) and the average number of attempts and collisions per block in all the grasping attempts (4 x 5). It can be noted that the *random Grasp + touch-informed regrasp* policy outperforms *random grasp* in all three metrics, i.e., obtains a lower failure rate, requires fewer attempts, and causes fewer collisions.

Policies	Failure rate	Number of attempts per block	Number of collisions per block
Control	0%	1	0
Random grasp	20%	3.30	1.45
Random grasp + touch-informed regrasp	0%	1.85	0.55

based grasping, by sampling the block position randomly. The touch-informed regrasp in RgTr mimics an adaptation carried out by touch sensing by sensing possible collisions and moving the gripper towards the column in which the contact is detected. A ***control policy*** (C) can also be implemented for reference. In this case, the agent always knows the position of each block and consequently always moves directly towards it. The results of this experiment are summarized in Table 1.5, after executing these policies five times. As can be seen, in all the measured metrics, RgTr is a more successful policy than Rg. For instance, Rg fails to grasp 20% of the blocks, i.e., on average one block is left on the board at the end of each run. In contrast, with the RgTr policy all the blocks are grasped, resulting in a failure rate of 0%. Similarly, both the average number of attempts and the average number of collisions per block are lower with the RgTr policy than with the Rg policy, i.e., 1.85 and 0.55 vs. 3.30 and 1.45. This difference in performance is justified by the fact that in the case of RgTr, once a collision occurs the regrasp policy ensures that the grasp attempt is successful. If the grasp position is sampled randomly, there will be a success chance of $1/4$ for

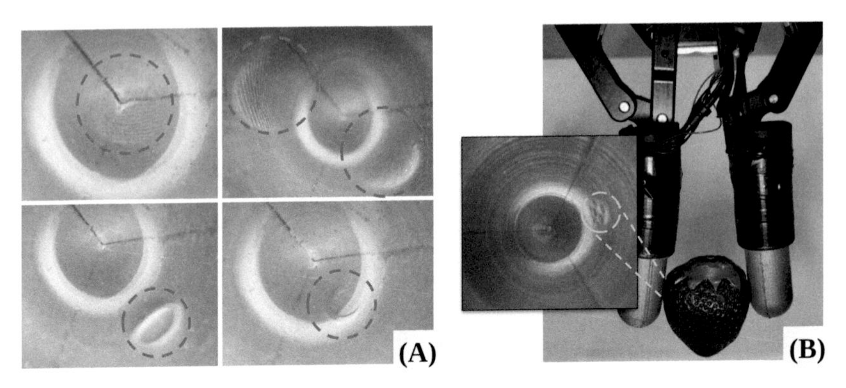

FIGURE 1.8 **(A)** Tactile images captured using our proposed GelTip sensor. From left to right, top to bottom: a fingerprint pressed against the tip of the sensor, two fingerprints on the sides, an open-cylinder shape being pressed against an side of the sensor, and the same object being pressed against the corner of the tip, all highlighted in red circles. **(B)** A plastic strawberry being grasped by a parallel gripper equipped with two GelTip sensors, with the corresponding imprint highlighted in the obtained tactile image (in grayscale).

each grasping attempt. In contrast, with the touch feedback enabled, this chance jumps to 2.5/4 on average. As a consequence, the RgTr policy finds a successful grasp more quickly and thus grasps more blocks within the maximum five attempts. This experiment shows that sensing contacts outside the grasp closure offers an important feature to improve the success chance of a given grasp attempt.

1.5 Conclusions and discussion

In this chapter, we have reviewed the tactile sensors for robot grasping and manipulation, highlighting our proposed GelTip sensor, which can detect contacts around the robot finger. As illustrated in Fig. 1.8, it can capture fine ridges of human fingerprints and the fine texture of a plastic strawberry. The grasping experiments in the *Blocks World* environment show the potential of all-around finger sensing in facilitating dynamic manipulation tasks. In our future research, we will introduce imprinted markers to the GelTip sensor to track the force fields. The use of the GelTip sensor in the manipulation tasks, such as grasping in cluttered environments, will also be of our interest.

Compared to the GelSight sensors [10,31], due to the sensor design of a finger shape, the light distribution throughout the sensor internal surface is no longer homogeneous. Specifically, a brightly illuminated ring can be observed near the discontinuity region (see Fig. 1.2 (B)). Shadows can also be observed in the bottom-left sample of Fig. 1.8 (A) when contacts of large pressure are applied, due to the placement of the camera and light sources. It may pose a challenge to geometry reconstruction using the Poisson reconstruction method [10,31,44], which builds a fixed mapping of pixel intensities to surface orientations and requires carefully placed RGB LEDs. In future research, convolutional neural networks could be used for geometry reconstruction of the GelTip sensor.

Acknowledgment

This work was supported by the EPSRC project "ViTac: Visual-Tactile Synergy for Handling Flexible Materials" (EP/T033517/1).

References

[1] H. Peel, S. Luo, A. Cohn, R. Fuentes, Localisation of a mobile robot for bridge bearing inspection, Automation in Construction 94 (2018) 244–256.

[2] S. Luo, J. Bimbo, R. Dahiya, H. Liu, Robotic tactile perception of object properties: a review, Mechatronics 48 (2017) 54–67.

[3] S. Luo, N.F. Lepora, U. Martinez-Hernandez, J. Bimbo, H. Liu, Vitac: integrating vision and touch for multimodal and cross-modal perception, Frontiers in Robotics and AI 8 (2021).

[4] P. Dario, D. De Rossi, Tactile sensors and the gripping challenge: increasing the performance of sensors over a wide range of force is a first step toward robotry that can hold and manipulate objects as humans do, IEEE Spectrum 22 (8) (1985) 46–53.

[5] R.S. Dahiya, P. Mittendorfer, M. Valle, G. Cheng, V.J. Lumelsky, Directions toward effective utilization of tactile skin: a review, IEEE Sensors Journal 13 (11) (2013) 4121–4138.

[6] M. Kaltenbrunner, T. Sekitani, J. Reeder, T. Yokota, K. Kuribara, T. Tokuhara, M. Drack, R. Schwödiauer, I. Graz, S. Bauer-Gogonea, et al., An ultra-lightweight design for imperceptible plastic electronics, Nature 499 (7459) (2013) 458–463.

[7] H. Xie, H. Liu, S. Luo, L.D. Seneviratne, K. Althoefer, Fiber optics tactile array probe for tissue palpation during minimally invasive surgery, in: 2013 IEEE/RSJ International Conference on Intelligent Robots and Systems, IEEE, 2013, pp. 2539–2544.

[8] P. Maiolino, M. Maggiali, G. Cannata, G. Metta, L. Natale, A flexible and robust large scale capacitive tactile system for robots, IEEE Sensors Journal 13 (10) (2013) 3910–3917.

[9] B. Ward-Cherrier, N. Pestell, L. Cramphorn, B. Winstone, M.E. Giannaccini, J. Rossiter, N.F. Lepora, The TacTip family: soft optical tactile sensors with 3D-printed biomimetic morphologies, Soft Robotics 5 (2) (2018) 216–227.

[10] W. Yuan, S. Dong, E.H. Adelson, GelSight: high-resolution robot tactile sensors for estimating geometry and force, Sensors (Basel, Switzerland) 17 (12) (11 2017).

[11] Z. Kappassov, J.-A. Corrales, V. Perdereau, Tactile sensing in dexterous robot hands, Robotics and Autonomous Systems 74 (2015) 195–220.

[12] S. Luo, W. Mou, K. Althoefer, H. Liu, Novel tactile-sift descriptor for object shape recognition, IEEE Sensors Journal 15 (9) (2015) 5001–5009.

[13] S. Luo, W. Mou, K. Althoefer, H. Liu, Iterative closest labeled point for tactile object shape recognition, in: IEEE/RSJ International Conference on Intelligent Robots and Systems (IROS), 2016, pp. 3137–3142.

[14] S. Luo, W. Mou, K. Althoefer, H. Liu, iclap: shape recognition by combining proprioception and touch sensing, Autonomous Robots 43 (4) (2019) 993–1004.

[15] C. Chorley, C. Melhuish, T. Pipe, J. Rossiter, Development of a tactile sensor based on biologically inspired edge encoding, in: International Conference on Advanced Robotics (ICAR), 2009.

[16] M.K. Johnson, E.H. Adelson, Retrographic sensing for the measurement of surface texture and shape Retrographic sensing for the measurement of surface texture and shape, in: IEEE Conference on Computer Vision and Pattern Recognition, 2009.

[17] S. Dong, W. Yuan, E.H. Adelson, Improved gelsight tactile sensor for measuring geometry and slip, in: 2017 IEEE/RSJ International Conference on Intelligent Robots and Systems (IROS), 2017, pp. 137–144.

[18] D.F. Gomes, Z. Lin, S. Luo, Blocks world of touch: exploiting the advantages of all-around finger sensing in robot grasping, Frontiers in Robotics and AI 7 (2020) 127.
[19] D.F. Gomes, Z. Lin, S. Luo, Geltip: a finger-shaped optical tactile sensor for robotic manipulation, in: 2020 IEEE/RSJ International Conference on Intelligent Robots and Systems (IROS), IEEE, 2020, pp. 9903–9909.
[20] R.S. Dahiya, G. Metta, M. Valle, G. Sandini, Tactile sensing—from humans to humanoids, IEEE Transactions on Robotics 26 (1) (2009) 1–20.
[21] H. Yousef, M. Boukallel, K. Althoefer, Tactile sensing for dexterous in-hand manipulation in robotics - a review, Sensors and Actuators, A: Physical 167 (2) (2011) 171–187.
[22] B. Shih, D. Shah, J. Li, T.G. Thuruthel, Y.-L. Park, F. Iida, B. Zhenan, R. Kramer-Bottiglio, M.T. Tolley, Electronic skins and machine learning for intelligent soft robots, Science Robotics 41 (5) (2020).
[23] K. Vlack, K. Kamiyama, T. Mizota, H. Kajimoto, N. Kawakami, S. Tachi, Gelforce: a traction field tactile sensor for rich human-computer interaction, in: IEEE Conference on Robotics and Automation, 2004. TExCRA Technical Exhibition Based, 2004, pp. 11–12.
[24] B. Ward-Cherrier, N. Rojas, N.F. Lepora, Model-free precise in-hand manipulation with a 3d-printed tactile gripper, IEEE Robotics and Automation Letters 2 (4) (2017) 2056–2063.
[25] B. Ward-Cherrier, L. Cramphorn, N.F. Lepora, Tactile manipulation with a tacthumb integrated on the open-hand m2 gripper, IEEE Robotics and Automation Letters 1 (1) (2016) 169–175.
[26] A. Yamaguchi, C.G. Atkeson, Combining finger vision and optical tactile sensing: reducing and handling errors while cutting vegetables, in: IEEE-RAS International Conference on Humanoid Robots, 2016, pp. 1045–1051.
[27] X. Lin, M. Wiertlewski, Sensing the frictional state of a robotic skin via subtractive color mixing, IEEE Robotics and Automation Letters 4 (3) (2019) 2386–2392.
[28] C. Sferrazza, R. D'Andrea, Design, motivation and evaluation of a full-resolution optical tactile sensor, Sensors 19 (4) (2019).
[29] C. Trueeb, C. Sferrazza, R. D'Andrea, Towards vision-based robotic skins: a data-driven, multi-camera tactile sensor, in: 2020 3rd IEEE International Conference on Soft Robotics (RoboSoft), 2020, pp. 333–338.
[30] B. Winstone, C. Melhuish, T. Pipe, M. Callaway, S. Dogramadzi, Toward bio-inspired tactile sensing capsule endoscopy for detection of submucosal tumors, IEEE Sensors Journal 17 (3) (2017) 848–857.
[31] R. Li, R. Platt Jr, W. Yuan, A. Pas, N. Roscup, M.A. Srinivasan, E.H. Adelson, Localization and manipulation of small parts using GelSight tactile sensing, in: IEEE International Conference on Intelligent Robots and Systems, 2014.
[32] S. Luo, W. Yuan, E. Adelson, A.G. Cohn, R. Fuentes, Vitac: feature sharing between vision and tactile sensing for cloth texture recognition, in: IEEE International Conference on Robotics and Automation (ICRA), IEEE, 2018, pp. 2722–2727.
[33] J.-T. Lee, D. Bollegala, S. Luo, "Touching to see" and "seeing to feel": robotic cross-modal sensory data generation for visual-tactile perception, in: 2019 International Conference on Robotics and Automation (ICRA), IEEE, 2019, pp. 4276–4282.
[34] G. Cao, Y. Zhou, D. Bollegala, S. Luo, Spatio-temporal attention model for tactile texture recognition, 2020.
[35] C. Lu, J. Wang, S. Luo, Surface following using deep reinforcement learning and a gelsight-tactile sensor, arXiv preprint, arXiv:1912.00745, 2019.
[36] J. Jiang, G. Cao, D.F. Gomes, S. Luo, Vision-guided active tactile perception for crack detection and reconstruction, in: The 29th Mediterranean Conference on Control and Automation (MED 2021), 2021.
[37] D.F. Gomes, A. Wilson, S. Luo, Gelsight simulation for sim2real learning, in: ICRA ViTac Workshop, 2019.
[38] D.F. Gomes, P. Paoletti, S. Luo, Generation of gelsight tactile images for sim2real learning, IEEE Robotics and Automation Letters 6 (2) (2021) 4177–4184.

[39] E. Donlon, S. Dong, M. Liu, J. Li, E. Adelson, A. Rodriguez, Gelslim: a high-resolution, compact, robust, and calibrated tactile-sensing finger, in: IEEE/RSJ International Conference on Intelligent Robots and Systems (IROS), IEEE, 2018, pp. 1927–1934.

[40] M. Lambeta, P.-W. Chou, S. Tian, B. Yang, B. Maloon, V.R. Most, D. Stroud, R. Santos, A. Byagowi, G. Kammerer, et al., Digit: a novel design for a low-cost compact high-resolution tactile sensor with application to in-hand manipulation, IEEE Robotics and Automation Letters 5 (3) (2020) 3838–3845.

[41] I. Taylor, S. Dong, A. Rodriguez, Gelslim3. 0: high-resolution measurement of shape, force and slip in a compact tactile-sensing finger, arXiv preprint, arXiv:2103.12269, 2021.

[42] S. Dong, D. Ma, E. Donlon, A. Rodriguez, Maintaining grasps within slipping bounds by monitoring incipient slip, in: IEEE International Conference on Robotics and Automation (ICRA), 2019, pp. 3818–3824.

[43] R. Calandra, A. Owens, D. Jayaraman, J. Lin, W. Yuan, J. Malik, E.H. Adelson, S. Levine, More than a feeling: learning to grasp and regrasp using vision and touch, IEEE Robotics and Automation Letters 3 (4) (2018) 3300–3307.

[44] B. Romero, F. Veiga, E. Adelson, Soft, round, high resolution tactile fingertip sensors for dexterous robotic manipulation, in: IEEE International Conference on Robotics and Automation, 2020.

[45] A. Padmanabha, F. Ebert, S. Tian, R. Calandra, C. Finn, S. Levine, Omnitact: a multi-directional high resolution touch sensor, in: IEEE International Conference on Robotics and Automation, 2020.

[46] G. Cao, J. Jiang, C. Lu, D.F. Gomes, S. Luo, Touchroller: a rolling optical tactile sensor for rapid assessment of large surfaces, arXiv preprint, arXiv:2103.00595, 2021.

[47] A. Wilson, S. Wang, B. Romero, E. Adelson, Design of a fully actuated robotic hand with multiple gelsight tactile sensors, arXiv preprint, arXiv:2002.02474, 2020.

[48] Y. She, S. Wang, S. Dong, N. Sunil, A. Rodriguez, E. Adelson, Cable manipulation with a tactile-reactive gripper, arXiv preprint, arXiv:1910.02860, 2019.

[49] R. Szeliski, Computer Vision Algorithms and Applications, Springer Science & Business Media, 2010.

Chapter 2

Robotic perception of object properties using tactile sensing

Jiaqi Jiang and Shan Luo
smARTLab, Department of Computer Science, University of Liverpool, Liverpool, United Kingdom

2.1 Introduction

Humans explore the environment in their close vicinity using rich sensory information such as the visual information obtained from our eyes and the tactile feeling through physical interaction with cutaneous receptors. Among these sensing modalities, the sense of touch is a key source of information for predicting object properties. From touch sensing, diverse sensory information can be obtained, such as pressure, vibration, pain, and temperature.

For robots, tactile sensing is also an irreplaceable modality. Compared to vision and other sensing modalities, tactile sensing is superior at processing material characteristics and detailed shapes of objects [1–3]. By analyzing the data from tactile sensors, the properties of the object in contact can be extracted such as the pressure distribution, vibrations, and surface textures. Much of such information is inaccessible for remote sensors like cameras due to the occlusion of the end-effectors, which makes tactile sensing highly important in grasping and manipulation. Furthermore, tactile sensing is not subject to factors like light conditions that the performance of vision heavily depends on.

To enable robots to have the sense of touch as humans, in the past decades researchers in the field of robotics have endeavored to develop tactile sensing solutions [4–11] for the perception of object properties. In this chapter, we review the recent development in robotic perception of object properties with tactile sensing. Three types of object properties that are important for robot grasping tasks are investigated: the material, shape, and pose of the object. The grasping stability prediction using tactile sensing is also discussed, which is vital in object grasping.

Through the survey, we find that many robotic tasks, especially dexterous grasping and manipulation, can only be performed with vision and tactile sensing together. To demonstrate the coordination of visual and tactile perception, our recent work in vision-guided tactile crack perception [12] is introduced. In this work, camera vision is first used to quickly search the candidate crack regions, and then a high-resolution optical tactile sensor is applied against these

Tactile Sensing, Skill Learning, and Robotic Dexterous Manipulation
https://doi.org/10.1016/B978-0-32-390445-2.00009-X

candidate regions. Finally, a refined crack shape is reconstructed from the obtained tactile images. In this task, vision shows a large field of view and is able to detect the potential crack areas. On the other hand, tactile sensing can enhance the fineness of the reconstructed crack shape thanks to the characteristics of being less susceptible to light and noise. To evaluate the proposed method, we collected a dataset by investigating 10 mock-up structures that were manufactured using 3D printing. The extensive experimental results demonstrate that compared to only using vision itself, our proposed approach achieves a significant reduction of mean distance error from 0.82 mm to 0.24 mm for crack reconstruction. Furthermore, the proposed method is more than 10 times faster than passive tactile perception in terms of tactile data collection time.

The remainder of this chapter is organized as follows. The works on material properties recognition using tactile sensing are discussed in Section 2.2, followed by discussions of object shape estimation and object pose estimation via touch and grasping stability prediction in Section 2.3, Section 2.4, and Section 2.5, respectively. Vision-guided tactile perception for crack reconstruction is introduced in Section 2.6. At the end of this chapter, a conclusion is given along with the discussion on open issues and potential future directions.

2.2 Material properties recognition using tactile sensing

To effectively grasp and manipulate objects, robots need to recognize the materials of the objects that they are interacting with. The material properties of an object include roughness, softness, and hardness, etc. By recognizing the object's material properties, the robot can infer a reasonable grasping force and further achieve effective interaction with an object, e.g., grasping a fragile glass cup without damaging it. Furthermore, the understanding of material properties enables robots to facilitate the task. For example, in a task of laundry sorting, by recognizing the materials of different garments with tactile sensing (and also vision), the robot will be able to separate the items, select appropriate laundry options, and prevent potential damage to the clothes. In general, there are two types of approaches for material recognition using tactile sensing: texture-based and stiffness-based.

Texture-based tactile material recognition. Surface texture information is usually closely related to the friction coefficients, roughness, and microstructure patterns of objects. Among early studies, handcrafted features such as force variation [13,14] were widely used. In [13], the vibrations detected by the accelerometer were first transferred to the frequency domain by fast Fourier transform (FFT), and then two machine learning algorithms, i.e., support vector machine (SVM) and k-nearest neighbors (k-NN), were used to infer the surface texture class. In [14], the surface physical properties were determined by the application of a dynamic friction model with a finger-shaped force/torque sensor sliding along the object surface at varying speeds. In [15], a microelectromechanical system (MEMS)-based tactile sensor was developed and used to discriminate the roughness of object surfaces based on the estimated vibratory patterns during the sliding motion. In [16], a MEMS-based tactile array sensor was employed to distinguish simple textures using maximum likelihood (ML) estimation. In [17], a joint kernel sparse coding method was developed, and this method can explicitly take the intrinsic relations between tactile sensors of BarrettHand into account.

Instead of using electronic tactile sensors that were used in the above works, some studies classified surface textures with optical tactile sensors [18,6]. In such sensors, cameras are used to capture the deformation of a silicone layer that is placed on the top of the camera. Thanks to the use of a camera in the optical tactile sensors, detailed textures of the object can be captured. In [19], the camera-based GelSight sensor [20] was used to obtain tactile images by pressing the sensor against different objects. An algorithm named multiscale local binary pattern (MLBP) was proposed to classify different surface textures. Similarly, another camera-based tactile sensor, TacTip [21], was used to analyze the object textures in [22].

Neural networks have recently been used as feature learners to infer material properties. In [23], the raw 24,000-dimensional sensor signal was fed to a convolutional neural network and a classification accuracy of up to 97.3% was achieved. In [24], convolutional and recurrent neural networks were applied to learn feature representations from raw data acquired with the hybrid touch and sliding movements. On the other hand, the use of deep neural networks may result in high computational costs and make it challenging to deploy the learned models in real-time operations.

To enhance the efficiency of the texture classification, a spiking neural network was used to can achieve a high classification accuracy comparable to that of artificial neural network (ANN) with a faster inference in [25]. In [26], a spatiotemporal attention model was proposed for the material recognition by using the tactile texture sequence, where the spatial attention was used to pay more attention to the dominant features for each tactile frame, and temporal attention was used to model the relevance of different regions in the whole contact event. As a result, redundant features can be suppressed, and the salient features can be extracted effectively for the material recognition.

Stiffness based tactile material recognition. Object stiffness is another cue that can be used for material property recognition [27]. By using a BioTac [28] sensor, the object compliance, i.e., the reciprocal of stiffness, can be estimated either using the contact angle of the fingertip [29] or investigating the BioTac electrode data [30]. In [31], a piezoresistive tactile sensor was used, and the objects were classified by applying a KNN classifier as well as the dynamic time warping algorithm [32], which can calculate the distance between time series. The camera-based GelSight sensor was also applied to object stiffness estimation [33]. In [33], object hardness was estimated based on a numerical model that compared the contact area changes and contact surface geometry as well as the normal force value during pressing of the GelSight sensor.

TABLE 2.1 A summary of material recognition methods with touch sensing.

Method type	Texture-based		Stiffness-based
Motions	Sliding, tapping, scratching	Pressing	Squeezing, knocking, pressing
Data types	Acceleration, force vibration, acoustic data	Tactile images	Force variances, tactile images
Advantages	Low cost and limited computational expenses	Microstructure patterns of object textures can be captured in one tactile image	Limited computational costs; complementary to other information
Disadvantages	Interaction actions like scratching may damage objects	Low resolution of tactile sensors may cause difficulties for processing	Interaction actions like knocking may damage objects
References	[13], [14]	[19], [22]	[31], [34]

The object materials can be recognized by using different tactile sensors and different sensing cues. As summarized in Table 2.1, different motions are used in texture-based and stiffness-based material recognition methods: data of acceleration and force vibration and acoustic data can be obtained via the motions of sliding, tapping, and scratching to reveal the surface textures, whereas data of force variances can be acquired via the motions of squeezing, knocking, and pressing to reveal the stiffness of the object. When an optical tactile sensor is used, cues of both object surface textures and stiffness can be obtained by analyzing the tactile images. The motions employed and resultant data types also result in pros and cons of each method: the use of cameras in the optical tactile sensors can capture the detailed microstructure patterns of object textures in one high-resolution tactile image, whereas it also comes along with high computational cost compared to the use of other tactile sensors.

2.3 Object shape estimation using tactile sensing

The ability to identify or reconstruct the shape of objects is crucial for robots to perform multiple tasks such as grasping and in-hand manipulation. By gathering the shape information of objects, robots can better plan and execute grasping strategies and trajectories. There are two main approaches to achieve shape estimation, i.e., shape recognition and shape exploration, and we will discuss them below.

Tactile shape recognition. Due to the limitation of the low resolution, in most early works tactile sensors were used to collect contact points and generate point clouds to represent object shapes [35]. The mosaics of tactile measure-

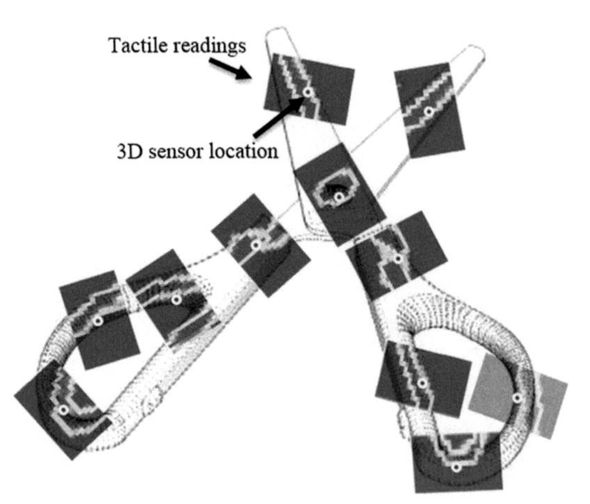

FIGURE 2.1 An illustration of haptic data collection from a pair of scissors in [39,40]. When the tactile sensor moves on the object, two types of cues can be obtained via tactile sensing: At each data point, a tactile reading that represents the local features of the object (shown in rectangle blocks in the figure) and the 3D sensor locations that show how the contact points are distributed in the 3D space (shown in red dots with white edges at the block centers in the figure) are collected.

ments [36] were used to reconstruct object global geometry. Such approaches are time consuming, especially when a large object surface needs to be explored.

Instead of collecting contact points or local measurements, the object shapes can be represented by extracting features from tactile readings [37]. In [38], a new tactile-SIFT descriptor was proposed to extract features in the tactile image and a classification accuracy in shape recognition of 91.33% was achieved. In [39,40], a novel method named Iterative Closest Labeled Point (iCLAP) was introduced that fundamentally linked kinesthetic cues (contact points) and tactile features extracted from local patterns and showed superior object recognition performance compared to methods using either contact points or the local tactile features only. As shown in Fig. 2.1, when a tactile sensor interacts with an object, two types of cues are obtained: local features of the object that indicate the local shapes of the object and the distribution of the contact points that gives a global view of the object in the 3D space.

In the above methods, either local contacts or local features are gathered to train a model first, and then the trained model is used to recognize the objects in the test set by matching the new observations with the ones in the training. To this end, a database of the object shapes needs to be built. Such methods show a good performance in recognizing object shapes that are known in the database. However, when the object shape is not available in the database, tactile shape exploration is needed.

Tactile shape exploration. To reduce the dependence on prior knowledge such as object shapes, researchers have proposed several approaches that reconstruct object shape through shape exploration [41], particularly in the paradigm of active shape exploration. In [42], a probabilistic approach based on Bayes' rule was proposed to explore the object's surface through rotating the tactile sensor and moving the sensor tangentially over the object surface. To improve the generalization performance of shape exploration, a deep convolutional neural network was designed in [4] to replace the probabilistic model in [42]. Nevertheless, the performance of those methods heavily depends on the localization error of the end-effector and requires a long exploration time. Hence, some other studies use Gaussian process implicit surface (GPIS) [43] as object shape representations [44–47] to reduce the impact of localization uncertainty and accelerate the exploration.

In [44,45], active touch probing was performed at discrete query points. To avoid inefficient touch-and-retract motions, an active learning framework was proposed in [46] based on optimal query paths to address the efficiency issue of tactile shape exploration. Furthermore, Driess et al. [47] extended the previous work [46] and used multiple end-effectors at the same time to achieve a more efficient exploration.

However, the objects used in the above works were fixed on the table to avoid the movement of objects, which makes it hard to be applied in real grasping tasks. To address the uncertainties of movable objects, some studies attempted to recover a movable object's shape by using a series of contacts [48,49]. Inspired by the paradigm of simultaneous localization and mapping (SLAM) in mobile robotics, an approach that uses contact measurements and planar pushing mechanics as constraints in batch optimization was proposed in [48]. Due to the usage of ordered control points as object shape representations, the method proposed in [48] is not suitable for exploring objects of arbitrary shapes, and will fail easily when an incorrect data association exists. To address this issue, a recent work [49] expanded the method in [48] by combining efficient GPIS regression with factor graph optimization over geometric and physics-based constraints. The results showed a promising localization and shape estimation performance in both simulated and real-world settings. The localization and exploration of the contact on the object can also be facilitated by matching the tactile features with a visual map of the object. In [41], the visual-tactile localization problem is also treated as a probabilistic estimation problem for the first time, i.e., a SLAM problem, that was solved in a framework of recursive Bayesian filtering.

As summarized in Table 2.2, most works in tactile shape recognition focus on the shape estimation of fixed objects, which simplifies the task by ignoring the object's dynamics models. Only a few works can estimate the shape of a movable object, whereas they could only be applied in two-dimensional shape estimation. As for the shape representations, there are three different ones: shape class, ordered points, and GPIS. The different object states and shape representations also result in pros and cons of each method: those methods using shape classes as the representations are only applicable to specific scenarios such as assembly automation, whereas they are able to perform the perception much

TABLE 2.2 A summary of the methods for tactile object shape estimation.

Method Type	Shape recognition	Shape exploration		
Object state	Fixed	Fixed	Fixed	Movable
Shape representation	Shape class	Ordered points	GPIS	Ordered points or GPIS
Advantages	Low computational cost	High accuracy	Suitable for 3D exploration	Suitable for movable objects
Disadvantages	Need some prior shape information, not suitable for unseen objects	Long exploration time and performance will drop when exploring 3D objects	The computational effort is not suitable for real-time tasks	Long exploration time and can only reconstruct 2D shapes
References	[36], [38], [39]	[42], [4]	[46], [47]	[48], [49]

faster than those methods using points and GPIS as shape representations. Tactile object shape exploration methods are less dependent on priors and therefore more suitable for the estimation of unseen objects, objects of arbitrary shapes, and even movable objects.

2.4 Object pose estimation using tactile sensing

For dexterous manipulation, it is essential to estimate the pose of the object in hand accurately. A small estimation error of object pose could result in misplacing robotic fingers on the object and even damaging the object. Currently, pose estimation is an important topic in the field of computer vision. However, during object manipulation, the gripper or the arm of the robot may partially or completely occlude the object from vision, which makes visual pose estimation not stable anymore. To address this problem, tactile sensing can be used as a complementary modality for determining the pose of an object in grasping and manipulation.

Particle filtering has been extensively used in vision-based robot localization problems. In the past years, it has also found its applications in tactile object pose estimation [50–53]. In [50], particle filters were applied to estimate a tube's pose, in which some parts of the robot's fingers were not in contact with the tube. In [51], a particle filtering approach using proprioceptive and tactile measurements was proposed to localize the small objects, e.g., button or grommet, embedded in a flexible material such as thin plastic. In [52] an approach named memory unscented particle filter (MUPF) was proposed that localizes objects recursively in real-time. In [53], a particle filter-based method was proposed to estimate in-hand object poses by bringing the object into contact with a surface.

In some other works, the pose of the in-hand object was estimated by matching geometry with a pressure sensor [54] or a camera-based tactile sensor [55,56]. In [54], the in-hand object pose was estimated by matching principal component analysis (PCA)-based features extracted from tactile pressure readings to the object's geometric features. In [55], a feature-based height map registration method was proposed to localize the small objects, e.g., a USB cable. Bauza et al. [56] used CNNs to compute heightmaps and a coarse-to-fine strategy for localizing objects. With this approach, each object has to be explored individually to determine the tactile map of its global shape.

In summary, particle filtering and feature matching have been used to achieve tactile object pose estimation. However, the pose estimation problems are still not fully solved yet. For example, most particle filter-based approaches are sensitive to parameters of models and need a great number of iterations to infer the estimated results. In contrast, feature-based matching methods for pose estimation have to obtain a tactile map for each object and extensively explore the object, which is time consuming.

2.5 Grasping stability prediction using tactile sensing

Apart from the material, shape, and pose of the object, it is also important to predict the grasping stability with tactile sensing of objects while they are being grasped or manipulated. With the prediction of grasping stability, a robot can find a stable grasping pose and prevent objects from falling. The prediction methods can be summarized into two types: slip detection and grasping stability regression. We will discuss them as follows.

Tactile slip detection. Some researchers simplify the grasping stability prediction as a slip detection task, when the object is being lifted. In [57], a force/torque sensor and a matrix-based tactile sensor were used to detect both translational and rotational slip with the prior of the contact surface's frictional coefficient. In [58], a fabric sensor of woven electroconductive yarns was used to detect the slip by detecting the change of resistance that depends on the yarn stretch. Although the above method showed good performance in grasp stability prediction, the need for prior knowledge like the frictional coefficient in [57] and the initial stretch of yarn in [58] pose a restriction on their applications.

To predict the slip between the object and the hand without the priors, a three-layer neural network was used in [59] to predict the slip, which achieves a classification accuracy of 80%. However, as the data in [59] were squeezed into a one-dimensional array, the spatial distribution inherent in the tactile sensor was lost. To exploit the spatial relationships between different electrodes on the BioTac sensor, in [60] graph neural networks (GNNs) were used, which yielded a validation accuracy of 92.7% in predicting the grasp stability of novel objects.

Tactile images obtained from optical-based tactile sensors have also been used to detect the slip in recent years. In [61], a two-fingered gripper composed of two GelSight sensors was used to detect slip events by measuring the relative displacement between object texture and markers. In [62,63], two other different optical tactile sensors, i.e., TacTip [64] and T-MO [65], were employed for detecting slip. The results in [62,63] showed that the slip is detected effectively employing an SVM classifier with the movement of pins as features.

Tactile grasping stability regression. Different from slip detection that is used to prevent in-hand objects from dropping, grasping stability regression is normally used for picking up objects with a robot arm. To assess the grasp stability, multiple works [66–68] used the image moments of tactile readings as features in their classifiers. In [66,67], a kernel-logistic regression model was presented. In the model, pose and touch are the conditional factors for grasping probability. To reduce the modeling effort and computational load, a wrench-based reasoning method was proposed in [68] which achieved comparable performance with [66,67]. In [69], CNNs were combined with long short-term memory networks (LSTMs) for predicting the grasp stability.

Though promising results have been achieved, the above works can only estimate the stability of an ongoing grasp with tactile readings and are not capable of regrasping the object when a failure of grasp happens. The tactile readings can also be used for computing the action level information [70]. Inspired by the actions that humans can adjust hand posture to perform dexterous grasping by using only tactile information, some researchers have investigated assessing current grasp and selecting grasp adjustments to produce a stable new grasp simultaneously. In [71], the grasp outcome estimated with a grasp stability predictor was used as a reward signal that supervises and provides feedback to the regrasping reinforcement learning algorithm. In [72], an action-conditional model using tactile readings of a GelSight sensor as input was proposed to predict the outcome of a candidate grasp adjustment. In contrast to [71], the action-conditional model presented in [72] learned the actions entirely from raw inputs. Similar to [72], a regrasping approach was presented in [73] using a camera-based tactile sensor. It used CNNs to predict grasp quality and made local adjustments through simulating rigid-body transformations of tactile readings.

As summarized in Table 2.3, slip detection and grasping stability regression are used for different application scenarios. Slip detection methods are normally used for preventing in-hand or lifted objects from dropping, whereas grasping stability regression methods are usually used to select the best grasp proposal for picking up ungrasped objects. Moreover, tactile grasping stability regression has a wider application prospect compared to tactile slip detection. For example, the grasp outcome from tactile grasping stability regression can be used to adjust grasp proposals and secure the object grasp.

2.6 Vision-guided tactile perception for crack reconstruction

By reviewing the above research on tactile object perception, it is found that many robotic tasks can only be accomplished with vision and tactile sensing

TABLE 2.3 A summary of methods for tactile grasping stability prediction.

Method type	Slip detection	Grasping stability regression
Object states	In-hand or lifted	Ungrasped
Application scenarios	To prevent in-hand or lifted objects from dropping	To pick up objects and prevent them from dropping
Advantages	Low-cost computation, generalized to novel objects	Can be used for ungrasped objects and to adjust hand posture to produce a stable new grasp based on the ongoing grasp
Disadvantages	Not suitable for ungrasped objects and may need the prior knowledge of frictional coefficients	The low resolution of tactile sensors may cause difficulties for processing and it may need the prior knowledge of object models
References	[57], [59], [63]	[67], [69], [72], [73]

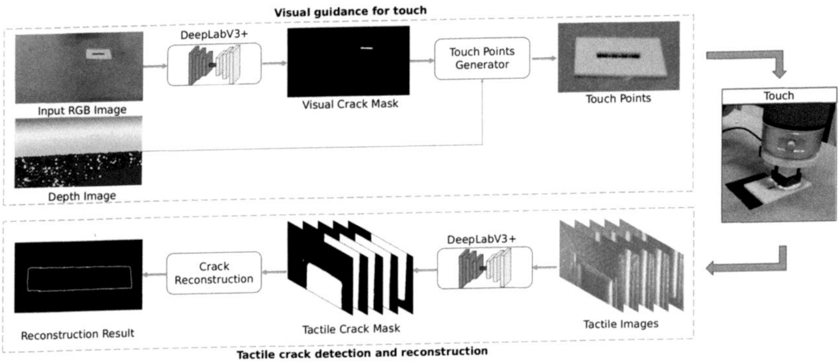

FIGURE 2.2 An overview of our vision-guided active tactile crack detection and reconstruction method. **Top row (from left to right):** The Deeplabv3+ model is used to segment the cracks in the visual image. Given the visual crack mask and the depth image, a set of contact points are generated to guide the collection of tactile images. **Bottom row (from right to left):** Another deep convolutional network is used to segment the crack in the collected tactile images. Given the detected tactile crack mask, the crack shape is reconstructed based on the geometrical model of the GelSight sensor and the coordinate transforming relation between the tactile sensor coordinate and the world coordinate.

together. To illustrate the coordination between visual and tactile perception, we present our work on vision-guided tactile perception for crack detection and reconstruction, with an overview of the framework illustrated in Fig. 2.2.

As for crack detection tasks, skilled inspectors typically first examine the surface to search for areas that have color or shape characteristics similar to cracks, and then use hammers or ultrasonic devices to double-check those areas instead of traversing all regions. Inspired by those observations, in this work camera vision is first used for a quick search of the candidate crack regions; a high-resolution optical tactile sensor is then applied against these candidate regions and a refined crack shape is reconstructed from the obtained tactile images.

To evaluate our proposed method, we collected a dataset by investigating 10 mock-up structures that were manufactured using 3D printing. The extensive experimental results demonstrate that compared to only using vision itself, our proposed approach achieves a significant reduction of mean distance error from 0.82 mm to 0.24 mm for crack reconstruction. Furthermore, the proposed method is more than 10 times faster than passive tactile perception in terms of tactile data collection time.

2.6.1 Visual guidance for touch sensing

A deep semantic segmentation network is first utilized to predict pixel-wise masks of cracks. To guide the touch sensing, the contact points with crack skeletons are then generated. Details of these two steps are given below.

Visual crack segmentation. We treat visual crack segmentation as a semantic segmentation problem that classifies pixels of an input image into one of two categories: (a) background and (b) cracks. In order to generate these segmentations, we apply the DeepLabv3+ model [74], which is a state-of-the-art deep learning model for semantic image segmentation. Considering that the number of pixels in the background is much larger than that of the cracks, the network may easily converge to the status that treats all the pixels as background. To address this issue, we use the original images as input instead of resizing it to a smaller size as done in previous works [75] and use a weighted cross-entropy loss with crack pixels weighted $10\times$ more than background pixels. In addition, we set the output stride value to 8 since smaller values give finer details in the output mask. To train the model, we started with a pretrained model for semantic segmentation based on the COCO dataset [76] and used the following hyperparameters: SGD optimizer with a constant learning rate of $1e-6$, a momentum of 0.9, and a weight decay of $5e-4$.

Contact points generation. Based on the predicted pixel-wise crack mask in the color image, we can extract the crack mask skeleton with a pattern thinning method [77]. To represent the topology of the crack pattern, two types of keypoints (i.e., end points and branch points) and minimal edges are defined as follows:

- end points: if they have less than two neighbors;
- branch points: if they have more than two neighbors;
- minimal edge E_{ij}: if there is a continuous path between two keypoints p_i and p_j and all points on the path are neither end points nor branch points.

For every minimal edge E_{ij} which consists of a number of ordered points, the keypoint p_i is initially selected as the current contact point $p_{current}$. Then we iteratively choose the next contact point p_k using the following formula:

$$\begin{aligned} &\max_k D[p_{current}, p_k] \\ &\text{s.t. } D[p_{current}, p_k] < d, \end{aligned} \tag{2.1}$$

where $D[p_{current}, p_k]$ is the distance between two points in the world coordinate system. The hyperparameter d is the threshold of the distance between two points that is related to the coverage and speed of the tactile exploration and perception. A smaller d will increase the coverage while reducing the perception speed. In our case, d is empirically set to four-fifths of the tactile sensor's view length. As shown in Fig. 2.2, the end point pixels and the generated contact points are tagged with red dots and green dots, respectively. For each contact point p_i, the yaw angle of the end-effector is parallel to the vector $< p_i, p_n >$, where p_n is the nearest contact point to p_i, so that the end-effector can contact the surface perpendicularly.

2.6.2 Guided tactile crack perception

Tactile crack detection. To address the issue of false positives in the visual crack detection caused by light changes and shadows, we apply tactile information to refine the vision-based detection results and reconstruct crack shapes in 3D space. At first, we control a robot arm with an embedded camera-based GelSight tactile sensor to collect tactile images autonomously at the generated contact points in Section 2.6.1.

The GelSight sensor is a camera-based optical tactile sensor that can capture fine details of the object surface. As shown in Fig. 2.3, a webcam under an elastomer captures the deformation of the elastomer as it interacts with the object. The sensor has a flat surface and a view range of 14 mm × 10.5 mm and can capture tactile images at a frequency of 30 Hz [26].

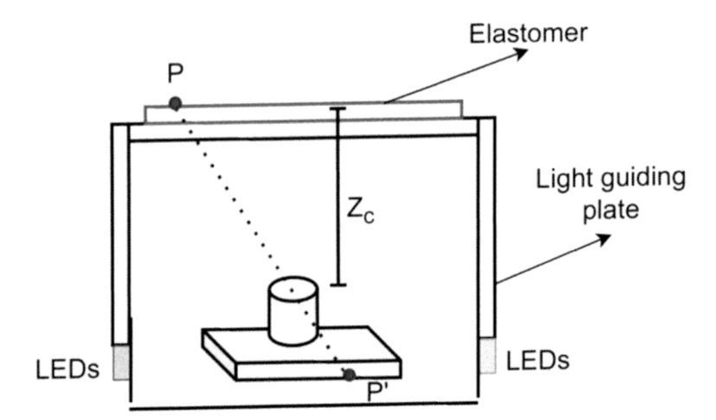

FIGURE 2.3 The geometrical model of the GelSight sensor. The webcam at the bottom captures the deformations of the elastomer and the LEDs project light to illuminate the space.

After collecting the tactile images, another Deeplabv3+ convolutional network [74] with ResNet-101 [78] is then used to predict the crack masks in tactile images. Since the number of background pixels is similar to that of the crack pixels in tactile images, we use the vanilla cross-entropy loss instead of the weighted cross-entropy loss in Section 2.6.1. Using the predicted masks in tactile images, we can double-check the visual segmentation results. In case there is more than one tactile image with a predicted crack area smaller than a predetermined threshold, all minimal edges will be classified as false positives and deleted. The threshold is empirically set as one-fiftieth of the total number of pixels in the tactile image.

Tactile crack reconstruction. For crack perception, it is crucial to estimate the shape and size of the crack when assessing its potential risk to the building and infrastructure. Due to the limitation of the depth camera's accuracy, current vision-based reconstruction techniques cannot reconstruct small cracks precisely. Therefore, we use the tactile images obtained from the GelSight sensor for the reconstruction of cracks, whose spatial resolution is approximately 20 to 30 μm.

Using the detected boundaries of pixel-wise masks in the tactile images, we first predict the location of cracks on the surface of the GelSight sensor. To simplify the problem, the webcam is modeled as a pinhole camera, and the surface of the GelSight sensor is treated as a flat plane perpendicular to the camera's z-axis. Hence, the transformation between a contact point $P = [X_c, Y_c, Z_c]^T$ in the tactile sensor coordinates (the tactile sensor takes the center of the webcam as the origin) to the pixel $P' = [u, v]^T$ in the tactile image coordinates can be calculated:

$$Z_c \begin{bmatrix} u \\ v \\ 1 \end{bmatrix} = K \begin{bmatrix} X_c \\ Y_c \\ Z_c \\ 1 \end{bmatrix}, \tag{2.2}$$

where Z_c is the distance from the optical center to the elastomer surface of the tactile sensor; K is the matrix of the intrinsic parameters of the webcam and can be represented as

$$K = \begin{bmatrix} \frac{f}{d_x} & 0 & u_0 & 0 \\ 0 & \frac{f}{d_y} & v_0 & 0 \\ 0 & 0 & 1 & 0 \end{bmatrix}, \tag{2.3}$$

where f is the focal length of the webcam, d_x and d_y denote the pixel size, and (u_0, v_0) is the center point of the tactile image.

After obtaining the position P in the tactile sensor coordinates, we can calculate its position P_W in the world coordinate system:

$$P_W = T_E^W T_C^E P, \tag{2.4}$$

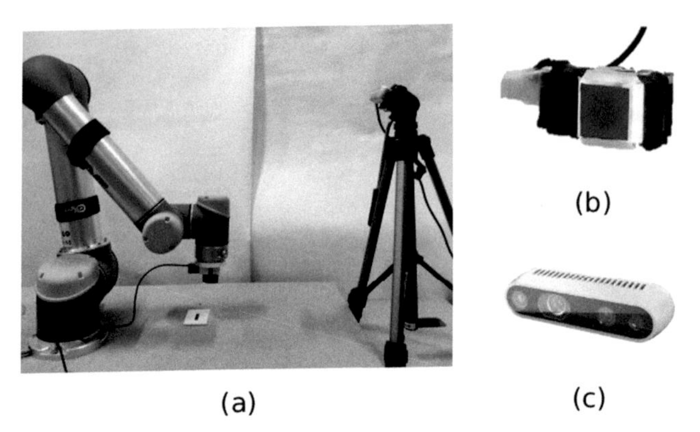

FIGURE 2.4 a) The UR5 robot setup with GelSight sensor and Realsense D435i. b) GelSight sensor. c) Realsense D435i.

where T_C^E and T_E^W are the transformation matrices from the tactile sensor coordinates to the end-effector coordinate system and from the end-effector coordinate system to the world coordinate system, respectively.

2.6.3 Experimental setup

In this subsection, we introduce the robot setup used for data collection and experiments with an overview shown in Fig. 2.4.

3D printed structures with cracks. Following part of the data acquisition protocol in [79], a set of 10 structures with cracks of different widths (holes in the structures) are manufactured with PLA plastic. To test the robustness of the proposed method, several structures are painted with fake cracks on the surfaces. The samples are shown in Fig. 2.5.

Visual data collection. For training the visual model, we put those 3D printed structures on a table and took 10 images of each one containing only real cracks. Then we took three images of each object separately for the test. All the images were taken with a RealSense D435i camera.

Tactile data collection. The tactile data collection setup consists of two parts: a 6-DoF Universal Robots UR5 collaborative robot arm and a GelSight sensor mounted on a 3D printed end-effector. To collect the tactile data autonomously and repeatedly, we built data collection software based on Robot Operating System (ROS). Using the software, a robot arm can be controlled to move across a structure following the predetermined initial position, step length, and steps in x- and y-axes. In each position, we rotate the sensor with different angles about the axis perpendicular to the surface in order to generalize the dataset. When the pressure reaches a threshold, the robot arm will stop moving and the GelSight will record a tactile image. In this way, we can obtain high-quality tactile data and avoid stopping the robot arm for unnecessarily protective reasons. In total, 544 valid tactile images were collected and split into training

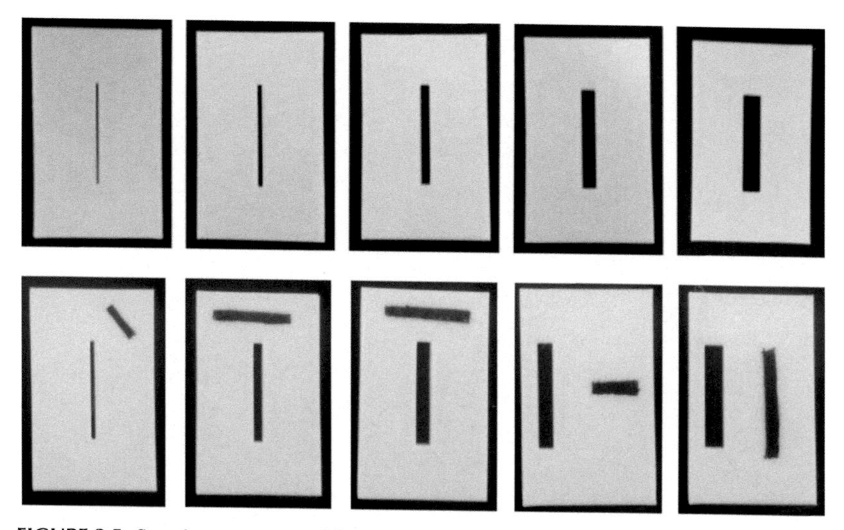

FIGURE 2.5 Sample structures used for collecting the visual and tactile dataset. **Top row**: Printed structures with real cracks (holes). **Bottom row**: Printed pad with fake cracks (painted black blocks).

FIGURE 2.6 Visualization of the tactile images and their annotations for different cracks.

and test sets (370 and 174 for each, respectively), with some samples and their annotations shown in Fig. 2.6.

2.6.4 Experimental results

Detection results. Our proposed methods are evaluated using standard evaluation metrics of pixel accuracy (pixAcc) and Intersection over Union (IoU). The results of segmentation based solely on visual information and based on both visual and tactile information are summarized in Table 2.4. Due to the presence

TABLE 2.4 Crack detection accuracy.

Modalities	Fake Painting	pixAcc	IoU
Vision	×	0.899	0.504
Vision	✓	0.866	0.376
Vision-tactile	✓	**0.909**	**0.636**

TABLE 2.5 Reconstruction accuracy.

Method	MeanD (mm)	SD (mm)	MaxD (mm)	time (s)
Vision	0.82	0.92	4.87	**1**
Aligned-vision	0.55	0.53	3.78	**1**
Passive-tactile	**0.20**	0.17	0.99	400
Active-tactile (ours)	0.24	**0.16**	**0.82**	35

of fake painting cracks, both pixel accuracy and IoU of visual semantic segmentation have dropped significantly from 0.899 to 0.866 and from 0.504 to 0.376, respectively. In addition, the results show that the crack detection can be effectively improved after tactile information is used to find the fake paintings and refine the visual results. Due to the fact that cracks on the RGB images only take up fewer pixels compared to the background, the pixel accuracy does not change dramatically as IoU.

Reconstruction results. To evaluate the accuracy of our proposed crack reconstruction method, we used the mean, maximum, and standard deviation (SD) of the shortest distances between the actual crack shape and the reconstructed crack location. There are four methods used for comparison. The vision method uses point clouds obtained from visual detection and depth information to represent cracks. Aligned vision reduces the impact of depth information accuracy on reconstruction by projecting the point cloud to the table surface. Passive tactile methods collect tactile images by traversing the entire surface of a 3D printed object.

Table 2.5 shows that our approach significantly improves the mean distance error from 0.55 mm to 0.24 mm for crack reconstruction compared to the aligned vision method. Two examples of reconstructed crack profiles are shown in Fig. 2.7, which shows that reconstructed crack profiles with tactile data are much closer to the ground truth compared to the vision-based method. Furthermore, our proposed vision-guided tactile perception is 10 times faster than passive tactile perception in terms of the time it takes to collect tactile data without affecting the accuracy of crack reconstruction.

2.7 Conclusion and discussion

In this chapter, we first briefly illustrated the irreplaceability of tactile sensing. Then we reviewed the state-of-the-art tactile perception of object properties

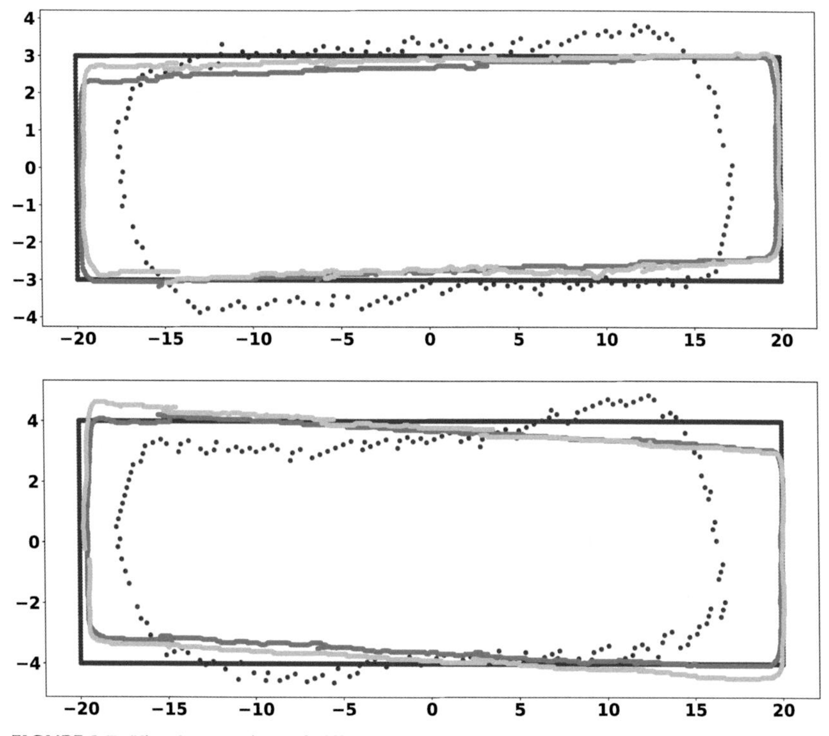

FIGURE 2.7 Visual comparison of different reconstruction methods. Blue, red, yellow, and green curves represent the ground truth of the crack profile, aligned-vision, passive-tactile, and our method for crack reconstruction, respectively.

from four perspectives: material properties, object shape, object pose, and grasping stability. A recent trend shows that deep learning, especially CNNs, has been widely applied in robotic tactile perception and optical tactile sensors play a greater role than traditional force tactile sensors thanks to the quick development of both optical sensors and the computer vision field. Moreover, we introduce a novel vision-guided tactile perception for crack detection and reconstruction. The cooperation between those two modalities addresses the false positives in visual detection results. The experimental results show that our proposed method can reconstruct crack shapes accurately and efficiently. It has the potential to enable robots to inspect and repair the concrete infrastructure.

As for object property perception using tactile sensing, however, there still exist multiple open issues expected to be investigated: (1) how to achieve real-time perception with optical tactile sensors; (2) how to acquire the costly data with labeling and ground truth; and (3) how to close the perception-action loop with tactile sensing. Those open issues and future directions are discussed below.

Real-time tactile perception. Due to the development of optical tactile sensor technologies and the domination of deep learning in the computer vision field, CNNs are investigated to achieve better performance at the expense of high computational load. To this end, lighter networks obtained via methods like network pruning can be also considered in those tasks that require real-time processing in the future.

Data acquirement and labeling. It is challenging to obtain the ground truth for tactile perception tasks due to the high cost and difficulties in labeling or measuring tactile properties of objects. Most of the recent literature still uses a limited number of human annotators to label data which may bring bias to the ground truths. To this end, unsupervised learning and sim-to-real transfer learning [80,81] will be of benefit in the future research of object perception using tactile sensing.

Close the perception-action loop with tactile sensing. In the physical interactions with an object, humans can not only infer object properties but also adjust the hand posture to better feel features that are relevant to the current task. However, most of the studies in the field still focus on research on a dataset collected in advance, whereas the actions taken to collect the data are not considered in the design of the algorithms. To close the perception-action loop, it will be an interesting direction to generate synergies between perception and actions when robots perform complex tasks with tactile sensing [82].

References

[1] R.S. Dahiya, M. Valle, Robotic Tactile Sensing: Technologies and System, Springer Science & Business Media, 2012.

[2] S.J. Lederman, R.L. Klatzky, Haptic perception: a tutorial, Attention, Perception, & Psychophysics 71 (7) (2009) 1439–1459.

[3] S. Luo, J. Bimbo, R. Dahiya, H. Liu, Robotic tactile perception of object properties: a review, Mechatronics 48 (2017) 54–67.

[4] N.F. Lepora, A. Church, C. De Kerckhove, R. Hadsell, J. Lloyd, From pixels to percepts: highly robust edge perception and contour following using deep learning and an optical biomimetic tactile sensor, IEEE Robotics and Automation Letters 4 (2) (2019) 2101–2107.

[5] S. Luo, L. Zhu, K. Althoefer, H. Liu, Knock-knock: acoustic object recognition by using stacked denoising autoencoders, Neurocomputing 267 (2017) 18–24.

[6] S. Luo, W. Yuan, E. Adelson, A.G. Cohn, R. Fuentes, Vitac: feature sharing between vision and tactile sensing for cloth texture recognition, in: 2018 IEEE International Conference on Robotics and Automation (ICRA), IEEE, 2018, pp. 2722–2727.

[7] D.F. Gomes, Z. Lin, S. Luo, Geltip: a finger-shaped optical tactile sensor for robotic manipulation, in: 2020 IEEE/RSJ International Conference on Intelligent Robots and Systems (IROS), IEEE, 2020, pp. 9903–9909.

[8] D.F. Gomes, Z. Lin, S. Luo, Blocks world of touch: exploiting the advantages of all-around finger sensing in robot grasping, Frontiers in Robotics and AI 7 (2020).

[9] G. Cao, J. Jiang, C. Lu, D.F. Gomes, S. Luo, Touchroller: a rolling optical tactile sensor for rapid assessment of large surfaces, arXiv preprint, arXiv:2103.00595, 2021.

[10] H. Xie, H. Liu, S. Luo, L.D. Seneviratne, K. Althoefer, Fiber optics tactile array probe for tissue palpation during minimally invasive surgery, in: 2013 IEEE/RSJ International Conference on Intelligent Robots and Systems, IEEE, 2013, pp. 2539–2544.

[11] S. Luo, N.F. Lepora, U. Martinez-Hernandez, J. Bimbo, H. Liu Vitac, Integrating vision and touch for multimodal and cross-modal perception, Frontiers in Robotics and AI 8 (2021).
[12] J. Jiang, G. Cao, D.F. Gomes, S. Luo, Vision-guided active tactile perception for crack detection and reconstruction, in: The 29th Mediterranean Conference on Control and Automation (MED 2021), 2021.
[13] J. Sinapov, V. Sukhoy, R. Sahai, A. Stoytchev, Vibrotactile recognition and categorization of surfaces by a humanoid robot, IEEE Transactions on Robotics 27 (3) (2011) 488–497.
[14] H. Liu, X. Song, J. Bimbo, L. Seneviratne, K. Althoefer, Surface material recognition through haptic exploration using an intelligent contact sensing finger, in: 2012 IEEE/RSJ International Conference on Intelligent Robots and Systems, IEEE, 2012, pp. 52–57.
[15] C.M. Oddo, M. Controzzi, L. Beccai, C. Cipriani, M.C. Carrozza, Roughness encoding for discrimination of surfaces in artificial active-touch, IEEE Transactions on Robotics 27 (3) (2011) 522–533.
[16] S.-H. Kim, J. Engel, C. Liu, D.L. Jones, Texture classification using a polymer-based mems tactile sensor, Journal of Micromechanics and Microengineering 15 (5) (2005) 912.
[17] H. Liu, D. Guo, F. Sun, Object recognition using tactile measurements: kernel sparse coding methods, IEEE Transactions on Instrumentation and Measurement 65 (3) (2016) 656–665.
[18] J.-T. Lee, D. Bollegala, S. Luo, "Touching to see" and "seeing to feel": robotic cross-modal sensory data generation for visual-tactile perception, in: 2019 International Conference on Robotics and Automation (ICRA), IEEE, 2019, pp. 4276–4282.
[19] R. Li, E.H. Adelson, Sensing and recognizing surface textures using a gelsight sensor, in: Proceedings of the IEEE Conference on Computer Vision and Pattern Recognition, 2013, pp. 1241–1247.
[20] M.K. Johnson, F. Cole, A. Raj, E.H. Adelson, Microgeometry capture using an elastomeric sensor, ACM Transactions on Graphics (TOG) 30 (4) (2011) 1–8.
[21] C. Chorley, C. Melhuish, T. Pipe, J. Rossiter, Development of a tactile sensor based on biologically inspired edge encoding, in: 2009 International Conference on Advanced Robotics, 2009, pp. 1–6.
[22] B. Winstone, G. Griffiths, T. Pipe, C. Melhuish, J. Rossiter, Tactip-tactile fingertip device, texture analysis through optical tracking of skin features, in: Conference on Biomimetic and Biohybrid Systems, Springer, 2013, pp. 323–334.
[23] S.S. Baishya, B. Bäuml, Robust material classification with a tactile skin using deep learning, in: 2016 IEEE/RSJ International Conference on Intelligent Robots and Systems (IROS), IEEE, 2016, pp. 8–15.
[24] T. Taunyazov, H.F. Koh, Y. Wu, C. Cai, H. Soh, Towards effective tactile identification of textures using a hybrid touch approach, in: 2019 International Conference on Robotics and Automation (ICRA), IEEE, 2019, pp. 4269–4275.
[25] T. Taunyazov, Y. Chua, R. Gao, H. Soh, Y. Wu, Fast texture classification using tactile neural coding and spiking neural network.
[26] G. Cao, Y. Zhou, D. Bollegala, S. Luo, Spatio-temporal attention model for tactile texture recognition, in: IEEE/RSJ International Conference on Intelligent Robots and Systems (IROS), 2020.
[27] T. Nanayakkara, A. Jiang, M.d.R.A. Fernandez, H. Liu, K. Althoefer, J. Bimbo, Stable grip control on soft objects with time-varying stiffness, IEEE Transactions on Robotics 32 (3) (2016) 626–637.
[28] N. Wettels, V.J. Santos, R.S. Johansson, G.E. Loeb, Biomimetic tactile sensor array, Advanced Robotics 22 (8) (2008) 829–849.
[29] D. Xu, G.E. Loeb, J.A. Fishel, Tactile identification of objects using Bayesian exploration, in: 2013 IEEE International Conference on Robotics and Automation, IEEE, 2013, pp. 3056–3061.
[30] Z. Su, J.A. Fishel, T. Yamamoto, G.E. Loeb, Use of tactile feedback to control exploratory movements to characterize object compliance, Frontiers in Neurorobotics 6 (2012) 7.

[31] A. Drimus, G. Kootstra, A. Bilberg, D. Kragic, Design of a flexible tactile sensor for classification of rigid and deformable objects, Robotics and Autonomous Systems 62 (1) (2014) 3–15.
[32] H. Sakoe, S. Chiba, Dynamic programming algorithm optimization for spoken word recognition, IEEE Transactions on Acoustics, Speech, and Signal Processing 26 (1) (1978) 43–49.
[33] W. Yuan, M.A. Srinivasan, E.H. Adelson, Estimating object hardness with a gelsight touch sensor, in: 2016 IEEE/RSJ International Conference on Intelligent Robots and Systems (IROS), IEEE, 2016, pp. 208–215.
[34] M. Kaboli, P. Mittendorfer, V. Hügel, G. Cheng, Humanoids learn object properties from robust tactile feature descriptors via multi-modal artificial skin, in: 2014 IEEE-RAS International Conference on Humanoid Robots, IEEE, 2014, pp. 187–192.
[35] P.K. Allen, K.S. Roberts, Haptic object recognition using a multi-fingered dextrous hand, 1988.
[36] Z. Pezzementi, C. Reyda, G.D. Hager, Object mapping, recognition, and localization from tactile geometry, in: 2011 IEEE International Conference on Robotics and Automation, IEEE, 2011, pp. 5942–5948.
[37] S. Luo, X. Liu, K. Althoefer, H. Liu, Tactile object recognition with semi-supervised learning, in: International Conference on Intelligent Robotics and Applications, Springer, 2015, pp. 15–26.
[38] S. Luo, W. Mou, K. Althoefer, H. Liu, Novel tactile-SIFT descriptor for object shape recognition, IEEE Sensors Journal 15 (9) (2015) 5001–5009.
[39] S. Luo, W. Mou, K. Althoefer, H. Liu, Iterative closest labeled point for tactile object shape recognition, in: 2016 IEEE/RSJ International Conference on Intelligent Robots and Systems (IROS), IEEE, 2016, pp. 3137–3142.
[40] S. Luo, W. Mou, K. Althoefer, H. Liu, iCLAP: shape recognition by combining proprioception and touch sensing, Autonomous Robots 43 (4) (2019) 993–1004.
[41] S. Luo, W. Mou, K. Althoefer, H. Liu, Localizing the object contact through matching tactile features with visual map, in: 2015 IEEE International Conference on Robotics and Automation (ICRA), IEEE, 2015, pp. 3903–3908.
[42] N.F. Lepora, K. Aquilina, L. Cramphorn, Exploratory tactile servoing with active touch, IEEE Robotics and Automation Letters 2 (2) (2017) 1156–1163.
[43] O. Williams, A. Fitzgibbon, Gaussian process implicit surfaces, in: Gaussian Proc. in Practice, 2007, pp. 1–4.
[44] S. El-Khoury, M. Li, A. Billard, On the generation of a variety of grasps, Robotics and Autonomous Systems 61 (12) (2013) 1335–1349.
[45] N. Jamali, C. Ciliberto, L. Rosasco, L. Natale, Active perception: building objects' models using tactile exploration, in: 2016 IEEE-RAS 16th International Conference on Humanoid Robots (Humanoids), IEEE, 2016, pp. 179–185.
[46] D. Driess, P. Englert, M. Toussaint, Active learning with query paths for tactile object shape exploration, in: 2017 IEEE/RSJ International Conference on Intelligent Robots and Systems (IROS), IEEE, 2017, pp. 65–72.
[47] D. Driess, D. Hennes, M. Toussaint, Active multi-contact continuous tactile exploration with Gaussian process differential entropy, in: 2019 International Conference on Robotics and Automation (ICRA), IEEE, 2019, pp. 7844–7850.
[48] K.-T. Yu, J. Leonard, A. Rodriguez, Shape and pose recovery from planar pushing, in: 2015 IEEE/RSJ International Conference on Intelligent Robots and Systems (IROS), IEEE, 2015, pp. 1208–1215.
[49] S. Suresh, M. Bauza, K.-T. Yu, J.G. Mangelson, A. Rodriguez, M. Kaess, Tactile slam: real-time inference of shape and pose from planar pushing, in: 2021 International Conference on Robotics and Automation (ICRA), IEEE, 2021.
[50] C. Corcoran, R. Platt, A measurement model for tracking hand-object state during dexterous manipulation, in: 2010 IEEE International Conference on Robotics and Automation, IEEE, 2010, pp. 4302–4308.

[51] R. Platt, F. Permenter, J. Pfeiffer, Using Bayesian filtering to localize flexible materials during manipulation, IEEE Transactions on Robotics 27 (3) (2011) 586–598.

[52] G. Vezzani, U. Pattacini, G. Battistelli, L. Chisci, L. Natale, Memory unscented particle filter for 6-dof tactile localization, IEEE Transactions on Robotics 33 (5) (2017) 1139–1155.

[53] F. von Drigalski, S. Taniguchi, R. Lee, T. Matsubara, M. Hamaya, K. Tanaka, Y. Ijiri, Contact-based in-hand pose estimation using Bayesian state estimation and particle filtering, in: 2020 IEEE International Conference on Robotics and Automation (ICRA), IEEE, 2020, pp. 7294–7299.

[54] J. Bimbo, S. Luo, K. Althoefer, H. Liu, In-hand object pose estimation using covariance-based tactile to geometry matching, IEEE Robotics and Automation Letters 1 (1) (2016) 570–577.

[55] R. Li, R. Platt, W. Yuan, A. ten Pas, N. Roscup, M.A. Srinivasan, E. Adelson, Localization and manipulation of small parts using gelsight tactile sensing, in: 2014 IEEE/RSJ International Conference on Intelligent Robots and Systems, IEEE, 2014, pp. 3988–3993.

[56] M. Bauza, O. Canal, A. Rodriguez, Tactile mapping and localization from high-resolution tactile imprints, in: 2019 International Conference on Robotics and Automation (ICRA), IEEE, 2019, pp. 3811–3817.

[57] C. Melchiorri, Slip detection and control using tactile and force sensors, IEEE/ASME Transactions on Mechatronics 5 (3) (2000) 235–243.

[58] D. Kondo, S. Okada, T. Araki, E. Fujita, M. Makikawa, S. Hirai, et al., Development of a low-profile sensor using electro-conductive yarns in recognition of slippage, in: 2011 IEEE/RSJ International Conference on Intelligent Robots and Systems, IEEE, 2011, pp. 1946–1953.

[59] Z. Su, K. Hausman, Y. Chebotar, A. Molchanov, G.E. Loeb, G.S. Sukhatme, S. Schaal, Force estimation and slip detection/classification for grip control using a biomimetic tactile sensor, in: 2015 IEEE-RAS 15th International Conference on Humanoid Robots (Humanoids), IEEE, 2015, pp. 297–303.

[60] A. Garcia-Garcia, B.S. Zapata-Impata, S. Orts-Escolano, P. Gil, J. Garcia-Rodriguez, Tactilegcn: a graph convolutional network for predicting grasp stability with tactile sensors, in: 2019 International Joint Conference on Neural Networks (IJCNN), IEEE, 2019, pp. 1–8.

[61] S. Dong, W. Yuan, E.H. Adelson, Improved gelsight tactile sensor for measuring geometry and slip, in: 2017 IEEE/RSJ International Conference on Intelligent Robots and Systems (IROS), IEEE, 2017, pp. 137–144.

[62] J.W. James, N. Pestell, N.F. Lepora, Slip detection with a biomimetic tactile sensor, IEEE Robotics and Automation Letters 3 (4) (2018) 3340–3346.

[63] J.W. James, N.F. Lepora, Slip detection for grasp stabilization with a multifingered tactile robot hand, IEEE Transactions on Robotics (2020).

[64] B. Ward-Cherrier, N. Pestell, L. Cramphorn, B. Winstone, M.E. Giannaccini, J. Rossiter, N.F. Lepora, The tactip family: soft optical tactile sensors with 3d-printed biomimetic morphologies, Soft Robotics 5 (2) (2018) 216–227.

[65] J.W. James, A. Church, L. Cramphorn, N.F. Lepora, Tactile model o: fabrication and testing of a 3d-printed, three-fingered tactile robot hand, Soft Robotics (2020).

[66] Y. Bekiroglu, R. Detry, D. Kragic, Learning tactile characterizations of object-and pose-specific grasps, in: 2011 IEEE/RSJ International Conference on Intelligent Robots and Systems, IEEE, 2011, pp. 1554–1560.

[67] Y. Bekiroglu, D. Song, L. Wang, D. Kragic, A probabilistic framework for task-oriented grasp stability assessment, in: 2013 IEEE International Conference on Robotics and Automation, IEEE, 2013, pp. 3040–3047.

[68] R. Krug, A.J. Lilienthal, D. Kragic, Y. Bekiroglu, Analytic grasp success prediction with tactile feedback, in: 2016 IEEE International Conference on Robotics and Automation (ICRA), IEEE, 2016, pp. 165–171.

[69] J. Li, S. Dong, E. Adelson, Slip detection with combined tactile and visual information, in: 2018 IEEE International Conference on Robotics and Automation (ICRA), IEEE, 2018, pp. 7772–7777.

[70] Q. Li, O. Kroemer, Z. Su, F.F. Veiga, M. Kaboli, H.J. Ritter, A review of tactile information: perception and action through touch, IEEE Transactions on Robotics 36 (6) (2020) 1619–1634.
[71] Y. Chebotar, K. Hausman, Z. Su, G.S. Sukhatme, S. Schaal, in: 2016 IEEE/RSJ International Conference on Intelligent Robots and Systems (IROS), IEEE, 2016, pp. 1960–1966.
[72] R. Calandra, A. Owens, D. Jayaraman, J. Lin, W. Yuan, J. Malik, E.H. Adelson, S. Levine, More than a feeling: learning to grasp and regrasp using vision and touch, IEEE Robotics and Automation Letters 3 (4) (2018) 3300–3307.
[73] F.R. Hogan, M. Bauza, O. Canal, E. Donlon, A. Rodriguez, Tactile regrasp: grasp adjustments via simulated tactile transformations, in: 2018 IEEE/RSJ International Conference on Intelligent Robots and Systems (IROS), IEEE, 2018, pp. 2963–2970.
[74] L.-C. Chen, Y. Zhu, G. Papandreou, F. Schroff, H. Adam, Encoder-decoder with atrous separable convolution for semantic image segmentation, in: Proceedings of European Conference on Computer Vision (ECCV), 2018.
[75] L. Pauly, S. Luo, D. Hogg, R. Fuentes, H. Peel, Deeper networks for pavement crack detection, in: Proceedings of the 34th ISARC, IAARC, 2017, pp. 479–485.
[76] T.-Y. Lin, M. Maire, S. Belongie, J. Hays, P. Perona, D. Ramanan, P. Dollár, C.L. Zitnick, Microsoft COCO: common objects in context, in: Proceedings of European Conference on Computer Vision (ECCV), Springer, 2014, pp. 740–755.
[77] T. Zhang, C.Y. Suen, A fast parallel algorithm for thinning digital patterns, Communications of the ACM 27 (3) (1984) 236–239.
[78] K. He, X. Zhang, S. Ren, J. Sun, Deep residual learning for image recognition, in: Proceedings of the IEEE Conference on Computer Vision and Pattern Recognition, 2016, pp. 770–778.
[79] F. Palermo, J. Konstantinova, K. Althoefer, S. Poslad, I. Farkhatdinov, Implementing tactile and proximity sensing for crack detection, in: 2020 IEEE International Conference on Robotics and Automation (ICRA), IEEE, 2020, pp. 632–637.
[80] D.F. Gomes, P. Paoletti, S. Luo, Generation of gelsight tactile images for sim2real learning, IEEE Robotics and Automation Letters 6 (2) (2021) 4177–4184.
[81] D.F. Gomes, A. Wilson, S. Luo, Gelsight simulation for sim2real learning, in: ICRA ViTac Workshop, 2019.
[82] C. Lu, J. Wang, S. Luo, Surface following using deep reinforcement learning and a gelsight-tactile sensor, arXiv preprint, arXiv:1912.00745, 2019.

Chapter 3

Multimodal perception for dexterous manipulation

Guanqun Cao and Shan Luo
smARTLab, Department of Computer Science, University of Liverpool, Liverpool, United Kingdom

3.1 Introduction

Humans interact and perceive the world with various senses, including vision, touch, sound, taste, and smell. Multiple feelings can be combined to have a synergistic effect which is superior to the sum of the results by using each sense separately. Even in daily tasks like cleaning our teeth, both visual and tactile senses are used together subconsciously: we use eyes to recognize and locate the toothbrush; when the toothbrush is placed in the mouth, tactile feedback is used to complete the blind spot cleaning due to the occluded views.

For robots, vision and touch are also two main sensing modalities for perception. Vision is often achieved with a camera by capturing the image from a distance, which can provide the robot with the appearance, shape, and color properties of the object. The visual modality has been extensively used and investigated in many applications such as recognition [1], object detection [2], and pose estimation [3]. However, it is still challenging to complete the manipulation task using the visual modality only. For example, it is difficult for robots to recognize transparent objects whose color is similar to that of their surroundings due to reflection. In addition to vision, tactile information can also be used by robots, enabling the understanding of the world from another dimension [4,5]. The tactile information is achieved by direct interaction between tactile sensors [6–12] and contacted objects, which can provide the robot with the object's surface properties, including the textures [13–15], local geometry [16–19], friction, hardness, and pose of the object [20], even under occlusions. Furthermore, objects that cannot be detected by vision, like transparent objects, are easily recognized through touch. Recently, researchers have achieved considerable progress in robotic object perception using tactile sensing, such as object recognition [21–23], slip detection [24,25], shape perception [26,27], and pose estimation [28]. As vision and touch can provide complementary information of objects, it is desirable for robots to combine vision and touch for multimodal perception in tasks like dexterous manipulation [29–31].

Tactile Sensing, Skill Learning, and Robotic Dexterous Manipulation
https://doi.org/10.1016/B978-0-32-390445-2.00010-6

The human brain employs a multisensory model to perceive the world in daily experience [32]. It is indicated that the same features from various sensory modalities share a common subspace. Through the common subspace, it is possible to perform a unitary presentation from multiple modalities or transfer the feature from one modality to another. For instance, we are able to imagine the taste of food by smelling it or tell the smell by the taste, which is called *synesthesia*. Inspired by this, in this chapter, our recent work on cross-modal sensory data generation for the translation between vision and touch [15] is introduced, through which we can generate realistic pseudodata to make up the inaccessible real data.

Apart from learning object properties across vision and touch, we have also investigated the multiple modalities in attention-based touch perception, i.e., both temporal and spatial modalities in touch sensing. We propose a spatiotemporal attention model for tactile texture recognition. It not only pays attention to salient spatial features but also models the correlation of each location through the time. The experimental results on fabric texture recognition show that our selective spatiotemporal attention model can largely improve the recognition performance by up to 18.8%, compared to the nonattention-based models.

The rest of chapter is organized as follows. An overview of multimodal perception is discussed including cross-modal learning in Section 3.2 and the spatiotemporal attention mechanism for tactile texture recognition in Section 3.3. Conclusions and future work are summarized in Section 3.4.

3.2 Visual-tactile cross-modal generation

Cross-modal generation not only learns a subspace between different modalities but also generates new data across them, i.e., mapping from one modality space to the other. It generates novel images that are unseen or data in other modalities that are not easy to access. In this chapter, we present our recent work on visual-tactile cross-modal generation [15].

3.2.1 "Touching to see" and "seeing to feel"

Specifically, as shown in Fig. 3.1, a framework of cross-modal sensory data generation for visual tactile perception is proposed [15], i.e., visual images and tactile textures of fabrics are generated from data of the other modality. Therefore, it is possible to generate a visual image using tactile images or tactile feelings by seeing the objects, which is called "touching to see" and "seeing to feel." The generated results are photorealistic compared with the real data, and the results can be used to expand the dataset to improve the performance of recognition.

As shown in the overall architecture in Fig. 3.2, the auxiliary information y denotes data of input modality while x represents the target modality. The generator takes y to generate the realistic pseudodata that are close to the distribution

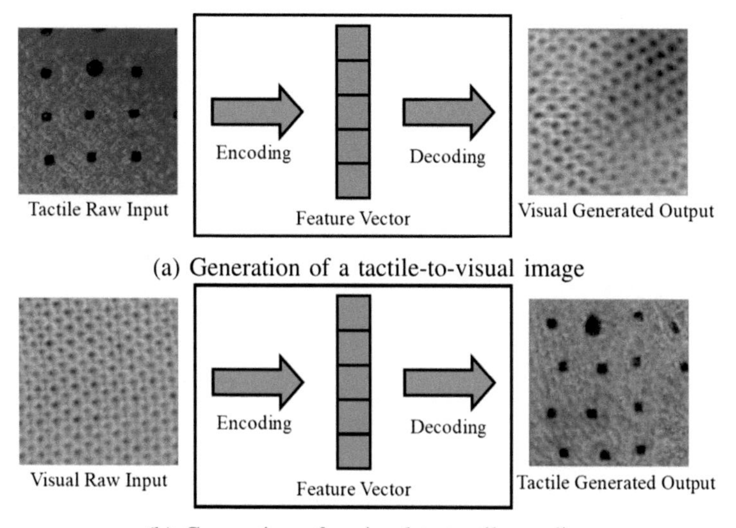

FIGURE 3.1 Cross-modal sensory data generation between vision and touch.

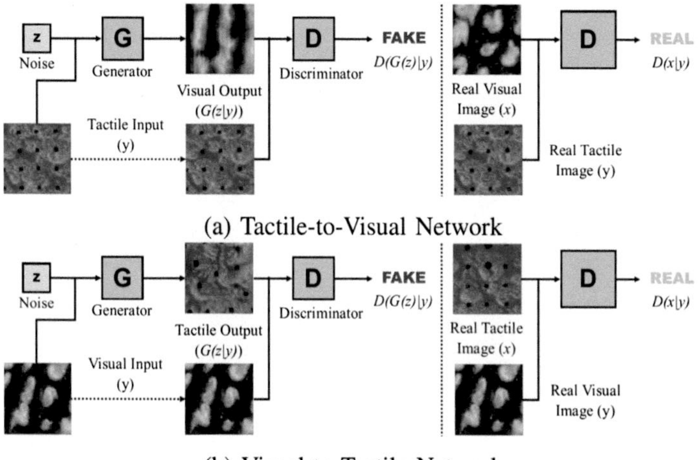

FIGURE 3.2 Architecture of the proposed method including visual-to-tactile network and tactile-to-visual network.

of x, which is used to fool the fixed discriminator. The discriminator is trained to determine if the input image is from real data or generated images with a fixed generator. The framework includes the translations both from vision to touch and from touch to vision. As a result, the mapping between vision and tactile modalities can be learned through this framework. The objective function can be represented as follows:

$$\min_G \max_D V(D, G) = \mathbb{E}_{x \sim p_{\text{data}}(x)}[\log D(x \mid y)] + \mathbb{E}_{z \sim p_z(z)}[\log(1 - D(G(z \mid y)))], \tag{3.1}$$

where D and G represent the generator and discriminator, respectively, x and y denote the data from the target domain and the input domain, and z represents the noise used in the generation.

3.2.2 Experimental results

We use the ViTac Cloth dataset [13] in our experiments for a texture recognition task, which consists of both visual images and corresponding tactile images collected by a digital camera and a GelSight sensor [6], respectively, from 100 pieces of fabrics. More detailed information can be found in [13]. Extensive experiments have been conducted to measure the performance of the cross-modal generation model under different conditions. (1) ***Cloth properties:*** Different cloth types own unique properties including the weaving patterns, colors, and textures. To this end, selected materials are used to test the performance of the model with different materials. (2) ***Network parameters:*** The internal parameters in the network have different impacts on the results of generated data. Therefore, the internal parameters are altered such as increasing the batch size, adding L1 loss into the objective function, and increasing the number of iterations, the number of training images, and the resolution. (3) ***Region of interest:*** As the tactile data are collected from different locations of fabric, the region of interest (ROI) is selected to reduce the views of tactile modality, which allows the training image to be more consistent. (4) ***Paired dataset:*** The generated images are projected back to the original input domain, enforcing a consistent cycle training, in which the input in one domain is transferred to another domain and then back to the original domain (i.e., cycle training), and the output is expected to be close to the input (i.e., consistent cycle training).

Moreover, a color structural similarity (Color-SSIM) index is proposed to measure the similarity between real images and generated images, which can be represented as

$$\text{Color-SSIM}(x, y) = \frac{\left(2\mu_x^{(i)}\mu_y^{(i)} + C_1\right)\left(2\sigma_{xy}^{(i)} + C_2\right)}{\left(\left(\mu_x^{(i)}\right)^2 + \left(\mu_y^{(i)}\right)^2 + C_1\right)\left(\left(\sigma_x^{(i)}\right)^2 + \left(\sigma_y^{(i)}\right)^2 + C_2\right)}, \tag{3.2}$$

where μ_x and μ_y represent the means of pixel values in image x and y, respectively, σ_x and σ_y denote the variances, σ_{xy} is the covariance of x and y, i represents the index of channels in the image, and C_1 and C_2 are scalars with a small value.

TABLE 3.1 The similarity between visual-to-tactile images and real tactile images is measured by Color-SSIM. The metric is implemented on different conditions and the resulting average is calculated.

Visual-to-tactile	Color-SSIM
Cloth types	0.89717
Network parameters	0.90896
ROI, with noise	0.92059
ROI, no noise	0.91188
Paired dataset	0.90471

TABLE 3.2 The similarity between tactile-to-visual images and real visual images is measured by Color-SSIM. The metric is implemented on different conditions and the resulting average is calculated.

Tactile-to-visual	Color-SSIM
Cloth types	0.77279
Network parameters	0.89595
Paired dataset	0.91338

From Table 3.1, it can be found that most methods give a Color-SSIM value of around 0.9 as the generated tactile results own the key characteristic of the real tactile data. However, more realistic data can be achieved with further adjusting the parameters or training methods. Table 3.2 demonstrates a lower score compared with the results of Table 3.1. One possible reason is that the pattern and color of visual images are difficult to infer from the tactile information due to different sensing principles. In general, as can be seen from Fig. 3.3, the proposed methods output realistic results from both the visual-to-tactile and the tactile-to-visual perspective.

The generated images are also evaluated on the classification tasks. The network is trained under two different settings: 1) only real data are fed into the network; 2) both real data and generated data are fed into the network for training. In the test, we calculate the recognition accuracy of real test data using the networks trained in the above two cases respectively. Table 3.3 shows that the generated data can effectively improve the performance in the early iterations. The generated results have great potential to expand the limited dataset and enable the learning more efficiently.

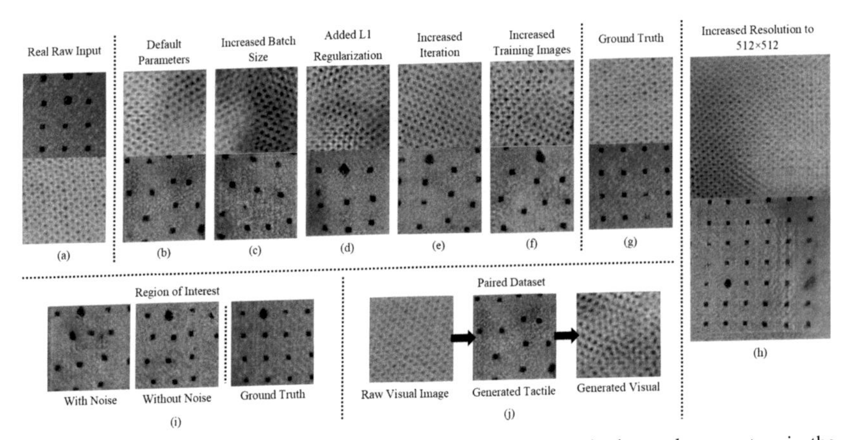

FIGURE 3.3 The visualization of generated results while altering the internal parameters in the network. (a) Input images. (b) Outputs with default parameters. (c) Increase batch size in each iteration. (d) Add L1 loss to the objective function. (e) Increase the iteration number. (f) Increase the number of training images. (g) Ground truth. (h) Image resolution changes from 256×256 to 512×512. (i) Select ROI. (j) The output is projected back to the original domain.

TABLE 3.3 Classification accuracy results by using real data and using both real data and translated data in visual classification and tactile classification.

Iteration	Visual images		Tactile images	
	Real images	Real/gen.	Real images	Real/gen.
1	0.9623	0.9710	0.7776	0.8506
2	0.9789	0.9770	0.8249	0.8465
3	0.9802	0.9853	0.8902	0.8699
4	0.9830	0.9839	0.8952	0.8736
5	0.9858	0.9835	0.9118	0.8911
6	0.9821	0.9867	0.9007	0.8989
7	0.9871	0.9867	0.9127	0.9067
8	0.9858	0.9867	0.9141	0.9044
9	0.9876	0.9894	0.9099	0.8989
10	0.9881	0.9894	0.9131	0.9058

3.3 Spatiotemporal attention model for tactile texture perception

The attention mechanism was first proposed in the natural language processing (NLP) field [33]. It was soon applied in the visual perception field [34–36]. The attention mechanism assigns higher weights to address the salient features and lower weights to suppress the redundant features, which enables learning effectively.

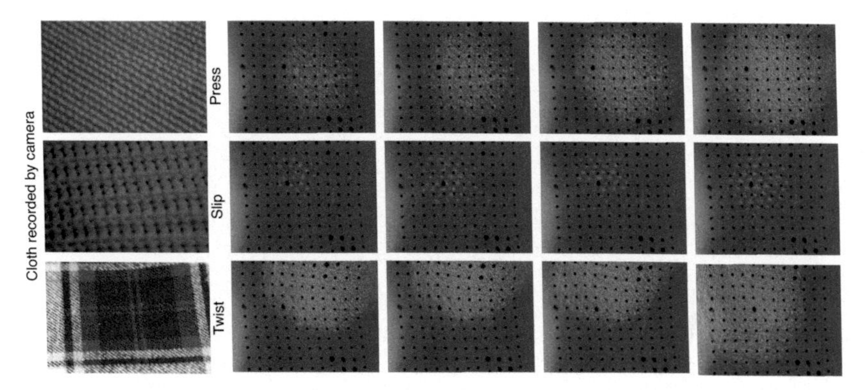

FIGURE 3.4 ***Sample tactile sequences.*** Leftmost column: Visual images recorded by a digital camera from fabrics. Right four columns: tactile sequences collected by the GelSight sensor with different interaction with fabric including pressing (first row), slipping (second row), and twisting (third row).

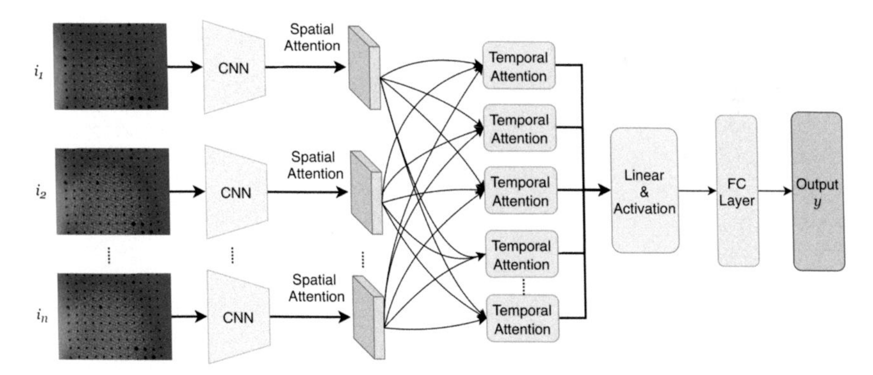

FIGURE 3.5 ***The proposed STAM framework for tactile texture recognition.*** The proposed methods consists of three parts: (1) a CNN module to extract a high-dimensional feature map, (2) a spatial attention module to emphasize informative features in each frame, and (3) a temporal attention module to model the temporal correlation in a whole sequence. Finally, the model outputs a predicted category by using fully connection layers.

3.3.1 Spatiotemporal attention model

Recently, we introduced a spatiotemporal attention model (STAM) for tactile texture recognition [14]. The textures are shown in Fig. 3.4. The tactile textures are collected by a GelSight sensor [6] by different interactions with the fabrics including pressing, slipping, and twisting with 100 categories of fabrics. In the spatiotemporal attention model, the spatial attention module is used to emphasize the crucial area in each frame of texture sequence, whereas temporal attention is used to select the important features considering the whole sequence regardless of the distance through the time dimension.

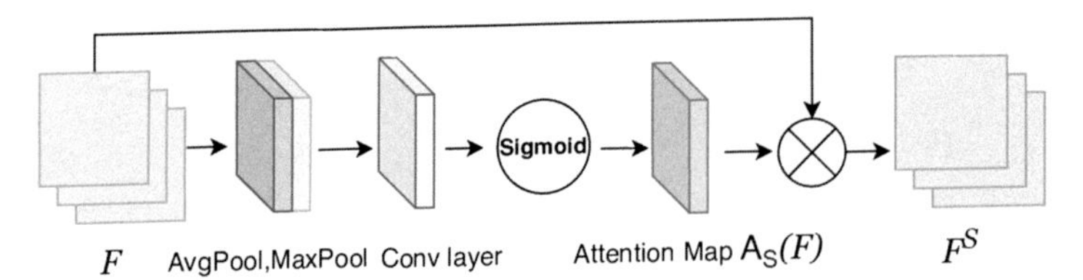

FIGURE 3.6 ***Spatial attention module.*** Average-pooling and max-pooling are used to learn the feature efficiently. A convolutional layer is used to calculate the attention map.

The overall architecture of the spatiotemporal attention model is illustrated in Fig. 3.5. The texture sequence is firstly fed into a pretrained convolutional neural network (CNN) to obtain a feature map $\boldsymbol{F} \in \mathbb{R}^{h \times w \times c}$, and then the spatial attention and temporal attention are used to highlight the salient features from space dimension and time dimension, respectively. At last, the network outputs the predicted category of the fabric y.

3.3.2 Spatial attention

Spatial attention aims to assign higher weights to the salient features in each frame. As shown in Fig. 3.6, max-pooling and average-pooling are applied to the feature map $\boldsymbol{F}$ to describe the features. Then a convolutional layer with a kernel of 7×7 with a sigmoid function is used to calculate the attention map that describes which location is more informative. The attention map $\boldsymbol{A_S(F)}$ can be calculated as

$$\begin{aligned} A_S(F) &= \sigma(f^{7\times 7}([MaxPool(F); AvgPool(F)])) \\ &= \sigma(f^{7\times 7}([F^S_{max}; F^S_{avg}])), \end{aligned} \tag{3.3}$$

where σ represents the activation function. The output feature map can be calculated as

$$F^S = A_S(F) \otimes F, \tag{3.4}$$

where $\otimes$ denotes element-wise multiplication.

3.3.3 Temporal attention

Different from spatial attention considering a single frame, temporal attention is used to compute the relevance of all locations in a sequence. As shown in Fig. 3.7, each frame in the sequence is concatenated and denoted as $\boldsymbol{F}^{\boldsymbol{S(n)}}$, and then $\boldsymbol{F}^{\boldsymbol{S(n)}}$ is turned into two different feature spaces using $1 \times 1 \times 1$ convolutional layers, which can be represented as

$$q(F^{S(n)}) = W_q F^{S(n)}, \tag{3.5}$$

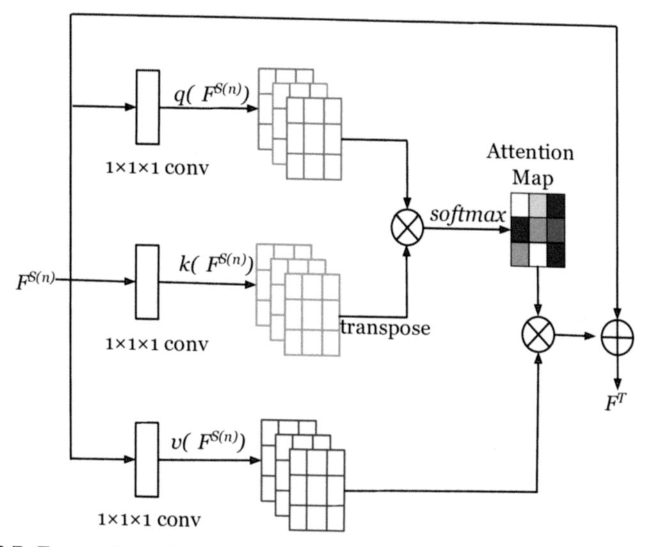

FIGURE 3.7 ***Temporal attention module.*** $1 \times 1 \times 1$ convolutions is used to turn $F^{S(n)}$ into different feature spaces. $\oplus$ denotes element-wise addition.

$$k(F^{S(n)}) = W_k F^{S(n)}, \tag{3.6}$$

where $q(F^{S(n)})$ and $k(F^{S(n)}) \in \mathbb{R}^{m \times c}$ $(m = n \times h \times w)$ are used to calculate the temporal attention map that describes the correlation of all the regions regardless of time. The attention map can be computed as follows:

$$A_T(F^{S(n)})_{j,i} = \frac{\exp(s_{ij})}{\sum_{i=1}^{m} \exp(s_{ij})}, \tag{3.7}$$

$$\text{in which} \quad s_{ij} = q\left(F_i^{S(n)}\right) k\left(F_j^{S(n)}\right)^T \tag{3.8}$$

where $A_T(F^{S(n)})_{j,i}$ illustrates the extent of $F_i^{S(n)}$ correlating with $F_j^{S(n)}$. The temporal attention outputs the feature map:

$$F^T = (F_1^T, F_2^T, ..., F_j^T, ..., F_m^T) \tag{3.9}$$

$$\text{in which} \quad F_j^T = \sum_{i=1}^{m} A_T(F^{S(n)})_{j,i} v\left(F_i^{S(n)}\right) + F_j^{S(n)}, \tag{3.10}$$

$$v\left(F^{S(n)}\right) = W_v F^{S(n)}, \tag{3.11}$$

where W_v denotes a trainable matrix. Moreover, as illustrated in Fig. 3.5, multiple temporal attention modules are applied, which allows joint learning from various feature spaces. At last, the outputs of the temporal attention module are connected with a fully connected layer to predict the category of the fabric.

TABLE 3.4 Texture recognition results with different models and different lengths of input sequences.

Models	$n=2$	$n=3$	$n=4$	$n=5$	$n=6$	$n=7$
CNNs	67.23%	72.04%	75.26%	78.06%	79.56%	81.29%
CNNs + spatial attention	72.12%	73.97%	78.60%	80.43%	80.43%	80.86%
STAM	**76.50%**	**79.35%**	**80.00%**	**80.64%**	**81.72%**	**81.93%**

TABLE 3.5 Texture recognition results while the dataset includes some noisy data collected before contact.

Models	$n=2$	$n=3$	$n=4$	$n=5$	$n=6$	$n=7$
CNNs	53.20%	58.20%	59.60%	61.23%	64.60%	69.40%
CNNs + spatial attention	55.40%	60.80%	62.60%	62.80%	65.40%	71.00%
STAM	**72.00%**	**72.20%**	**75.80%**	**76.61%**	**80.80%**	**80.20%**

3.3.4 Experimental results

The experiments consist of three parts. (1) An ablation study is performed, i.e., only the spatial attention module is used at first, and then both the spatial attention and temporal attention modules are applied, to understand the effectiveness of the designed method step by step. (2) It is also investigated how the length of a sequence affects the recognition performance and the processing time. Therefore, different lengths of input sequences are used and compared in our experiment. (3) Due to the uncertainty in the dynamic environment, some tactile images are collected before the contact happened. These tactile images are taken as noisy data to test the robustness of the model.

As shown in Table 3.4 and Table 3.5, the proposed method has achieved the best recognition results in all conditions with different lengths of sequences used. Compared with using CNNs only, there is a 4.45% average improvement in recognition accuracy using the proposed STAM method. In a further step, after introducing the noisy data, the accuracy increases by 15.23% on average compared with the baselines. Furthermore, with a longer length of sequences applied, the accuracy of recognizing the fabrics shows an upward trend.

From Table 3.4, it can be learned that the lengths of sequences lead to different impacts on the performance of different models. For example, when the length of sequences is changed from 7 to 2, the recognition accuracy of the proposed STAM method only decreases by 5.43%, whereas those of the CNN method and the spatial attention method decrease by 14.06% and 8.74%, respectively. It can be found that by using the spatial attention methods, the accuracy improves slightly in most of the cases compared with the baselines. By using both spatial attention and temporal attention modules, the performance has a further improvement, achieving the highest accuracy in all cases. The results in-

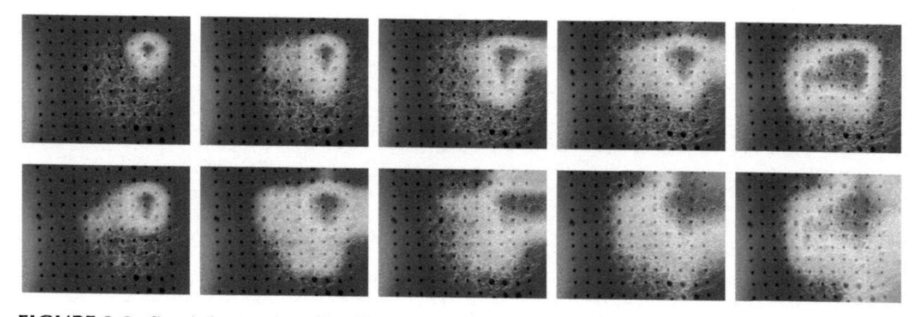

FIGURE 3.8 ***Spatial attention distribution.*** The first and second rows represent the same tactile sequence, but are activated by the nonattention method and the spatial attention method with Grad-CAM, respectively. It can be found that more contact regions are activated while using the spatial attention method.

dicate that the proposed STAM method has better ability to extract the obvious features compared to the baseline approaches, especially when limited data are available.

From Table 3.5, it can be found that the introduction of noisy data leads to a lower accuracy on a whole for the texture recognition task. However, the proposed STAM method sustains a similar recognition result with a small decrease compared with Table 3.4, while the other two methods have an obvious drop in performance. Specifically, when the length of sequence is 6, the accuracy of the baseline drops by 14.96% and the recognition accuracy of the proposed STAM method falls by only 0.92%. The results demonstrate that the proposed spatiotemporal attention method is robust to the noise and has the ability to select salient features.

To summarize, two strengths of the proposed STAM model can be listed. (1) By using the attention mechanism, the informative features can be selected and noise can be restrained. (2) The learning of the feature is more efficient and effective as the proposed STAM method has a better performance when limited data are available, i.e., the length of the sequence is short.

3.3.5 Attention distribution visualization

In this section, the attention distributions are visualized for spatial attention and temporal attention to illustrate how they work in the recognition tasks.

Spatial attention distribution. To visualize the spatial attention distribution, the gradient class activation map (Grad-CAM) [37] is performed on both spatial attention-based methods and the methods using CNNs only. In the Grad-CAM, the gradients are used to express the saliency of regions according to the classification results. The Grad-CAM allows us to understand the informative location in different methods. As demonstrated in Fig. 3.8, larger contact regions are highlighted by the spatial attention method compared with the baseline method, which indicates that more informative locations are utilized by the spatial attention.

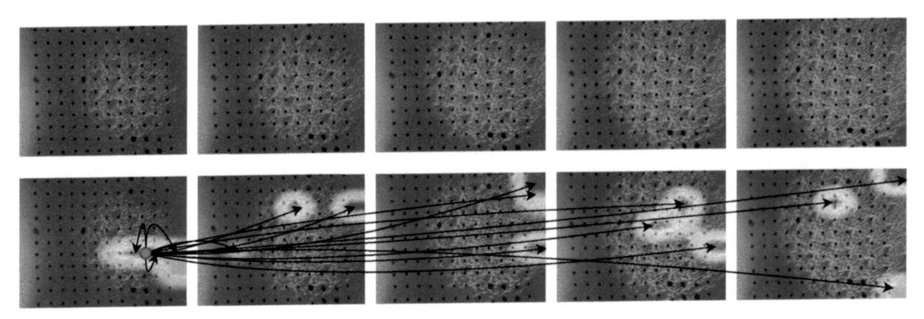

FIGURE 3.9 ***Temporal attention distribution.*** Upper row: The sequence of tactile textures in order. Lower row: The temporal attention distribution, where the green dot refers to a randomly selected region and the highlighted regions pointed by arrows are top three related regions in each frame. It can be observed that the most highlighted regions are the latest locations that are in contact with the fabric.

Temporal attention distribution. To understand the correlation between each location across the time dimension, the attention map of the temporal attention is also visualized. Due to the application of multiple temporal attention modules, the values of the attention maps are averaged. As shown in Fig. 3.9, a green dot is randomly selected in the first frame, and then three most related locations, pointed by arrows in the figure, are highlighted according to the attention maps in the other frames. As shown in the figure, the most highlighted regions are located in the latest contact area in all the second, third, fourth, and fifth frames, showing that the temporal attention is able to make use of the latest information in the time dimension for the recognition task.

3.4 Conclusion and discussion

In this chapter, we discuss the multimodal perception for dexterous manipulation including visual-tactile cross-modal learning and attention-based spatiotemporal tactile texture recognition. As vision and touch sensing provide complementary information about objects, it is natural to have the robots make use of multiple modalities to perceive the world as humans. As a result, the robots are able to observe and interact with the objects from different dimensions, enhancing their understanding of the physical world. Specifically, we discuss a cross-modal sensory data generation framework to make up inaccessible data and a spatiotemporal attention model for tactile texture recognition, which can learn the informative features from both spatial and temporal modalities.

In future works, the selective attention mechanism can be used to facilitate dexterous manipulation tasks and there are a few promising applications of the attention mechanism to be explored. Instead of feeding data of different modalities into the network together, attention methods can be used to learn selective visual and tactile features across the two modalities. In the current works, most of the tasks are isolated tasks and do not contribute to each other. To enable the robot to handle different tasks together, a lifelong learning framework can be developed to leverage the prior knowledge so as to improve the efficiency of the learning.

Acknowledgment

This work was supported by the EPSRC project "ViTac: Visual-Tactile Synergy for Handling Flexible Materials" (EP/T033517/1).

References

[1] K. He, X. Zhang, S. Ren, J. Sun, Deep residual learning for image recognition, in: Proceedings of the IEEE Conference on Computer Vision and Pattern Recognition, 2016, pp. 770–778.

[2] J. Redmon, A. Farhadi, Yolo9000: better, faster, stronger, in: Proceedings of the IEEE Conference on Computer Vision and Pattern Recognition, 2017, pp. 7263–7271.

[3] E. Brachmann, A. Krull, F. Michel, S. Gumhold, J. Shotton, C. Rother, Learning 6d object pose estimation using 3d object coordinates, in: European Conference on Computer Vision, Springer, 2014, pp. 536–551.

[4] S. Luo, N.F. Lepora, U. Martinez-Hernandez, J. Bimbo, H. Liu, Vitac: integrating vision and touch for multimodal and cross-modal perception, Frontiers in Robotics and AI 8 (2021).

[5] S. Luo, J. Bimbo, R. Dahiya, H. Liu, Robotic tactile perception of object properties: a review, Mechatronics 48 (2017) 54–67.

[6] W. Yuan, S. Dong, E. Adelson, Gelsight: high-resolution robot tactile sensors for estimating geometry and force, Sensors (2017).

[7] D.F. Gomes, Z. Lin, S. Luo, Geltip: a finger-shaped optical tactile sensor for robotic manipulation, in: 2020 IEEE/RSJ International Conference on Intelligent Robots and Systems (IROS), IEEE, 2020, pp. 9903–9909.

[8] D.F. Gomes, Z. Lin, S. Luo, Blocks world of touch: exploiting the advantages of all-around finger sensing in robot grasping, Frontiers in Robotics and AI 7 (2020).

[9] G. Cao, J. Jiang, C. Lu, D.F. Gomes, S. Luo, Touchroller: a rolling optical tactile sensor for rapid assessment of large surfaces, arXiv preprint, arXiv:2103.00595, 2021.

[10] D.F. Gomes, P. Paoletti, S. Luo, Generation of gelsight tactile images for sim2real learning, arXiv preprint, arXiv:2101.07169, 2021.

[11] D.F. Gomes, A. Wilson, S. Luo, Gelsight simulation for sim2real learning, in: ICRA ViTac Workshop, 2019.

[12] H. Xie, H. Liu, S. Luo, L.D. Seneviratne, K. Althoefer, Fiber optics tactile array probe for tissue palpation during minimally invasive surgery, in: 2013 IEEE/RSJ International Conference on Intelligent Robots and Systems, IEEE, 2013, pp. 2539–2544.

[13] S. Luo, W. Yuan, E. Adelson, A.G. Cohn, R. Fuentes, Vitac: feature sharing between vision and tactile sensing for cloth texture recognition, in: 2018 IEEE International Conference on Robotics and Automation (ICRA), IEEE, 2018, pp. 2722–2727.

[14] G. Cao, Y. Zhou, D. Bollegala, S. Luo, Spatio-temporal attention model for tactile texture recognition, in: 2020 IEEE/RSJ International Conference on Intelligent Robots and Systems (IROS), 2020, pp. 9896–9902.

[15] J.-T. Lee, D. Bollegala, S. Luo, "Touching to see" and "seeing to feel": robotic cross-modal sensory data generation for visual-tactile perception, in: 2019 International Conference on Robotics and Automation (ICRA), IEEE, 2019, pp. 4276–4282.

[16] S. Luo, W. Mou, K. Althoefer, H. Liu, Novel tactile-sift descriptor for object shape recognition, IEEE Sensors Journal (2015).

[17] S. Luo, X. Liu, K. Althoefer, H. Liu, Tactile object recognition with semi-supervised learning, in: ICIRA, 2015.

[18] S. Luo, W. Mou, K. Althoefer, H. Liu, Iterative closest labeled point for tactile object shape recognition, in: IROS, 2016.
[19] S. Luo, W. Mou, K. Althoefer, H. Liu, iCLAP: shape recognition by combining proprioception and touch sensing, Autonomous Robots (2019).
[20] J. Bimbo, S. Luo, K. Althoefer, H. Liu, In-hand object pose estimation using covariance-based tactile to geometry matching, IEEE Robotics and Automation Letters 1 (1) (2016) 570–577.
[21] N. Gorges, S.E. Navarro, D. Göger, H. Wörn, Haptic object recognition using passive joints and haptic key features, in: 2010 IEEE International Conference on Robotics and Automation, IEEE, 2010, pp. 2349–2355.
[22] S.-H. Kim, J. Engel, C. Liu, D.L. Jones, Texture classification using a polymer-based mems tactile sensor, Journal of Micromechanics and Microengineering 15 (5) (2005) 912.
[23] R. Li, E.H. Adelson, Sensing and recognizing surface textures using a gelsight sensor, in: Proceedings of the IEEE Conference on Computer Vision and Pattern Recognition, 2013, pp. 1241–1247.
[24] J. Li, S. Dong, E. Adelson, Slip detection with combined tactile and visual information, in: 2018 IEEE International Conference on Robotics and Automation (ICRA), IEEE, 2018, pp. 7772–7777.
[25] J.W. James, N. Pestell, N.F. Lepora, Slip detection with a biomimetic tactile sensor, IEEE Robotics and Automation Letters 3 (4) (2018) 3340–3346.
[26] A. Aggarwal, P. Kampmann, J. Lemburg, F. Kirchner, Haptic object recognition in underwater and deep-sea environments, Journal of Field Robotics 32 (1) (2015) 167–185.
[27] M. Meier, M. Schopfer, R. Haschke, H. Ritter, A probabilistic approach to tactile shape reconstruction, IEEE Transactions on Robotics 27 (3) (2011) 630–635.
[28] A. Petrovskaya, O. Khatib, S. Thrun, A.Y. Ng, Bayesian estimation for autonomous object manipulation based on tactile sensors, in: Proceedings 2006 IEEE International Conference on Robotics and Automation, 2006, ICRA 2006, IEEE, 2006, pp. 707–714.
[29] O. Kroemer, C.H. Lampert, J. Peters, Learning dynamic tactile sensing with robust vision-based training, IEEE Transactions on Robotics 27 (3) (2011) 545–557.
[30] T. Taunyazov, W. Sng, H.H. See, B. Lim, J. Kuan, A.F. Ansari, B.C. Tee, H. Soh, Event-driven visual-tactile sensing and learning for robots, perception 4 5.
[31] J. Ilonen, J. Bohg, V. Kyrki, Fusing visual and tactile sensing for 3-d object reconstruction while grasping, in: 2013 IEEE International Conference on Robotics and Automation, IEEE, 2013, pp. 3547–3554.
[32] S. Lacey, C. Campbell, K. Sathian, Vision and touch: multiple or multisensory representations of objects?, Perception 36 (10) (2007) 1513–1521.
[33] D. Bahdanau, K. Cho, Y. Bengio, Neural machine translation by jointly learning to align and translate, arXiv preprint, arXiv:1409.0473, 2014.
[34] V. Mnih, N. Heess, A. Graves, K. Kavukcuoglu, Recurrent models of visual attention, arXiv preprint, arXiv:1406.6247, 2014.
[35] J. Hu, L. Shen, G. Sun, Squeeze-and-excitation networks, in: Proceedings of the IEEE Conference on Computer Vision and Pattern Recognition, 2018, pp. 7132–7141.
[36] S. Woo, J. Park, J.-Y. Lee, I.S. Kweon, Cbam: convolutional block attention module, in: Proceedings of the European Conference on Computer Vision (ECCV), 2018, pp. 3–19.
[37] R.R. Selvaraju, M. Cogswell, A. Das, R. Vedantam, D. Parikh, D. Batra, Grad-cam: visual explanations from deep networks via gradient-based localization, in: ICCV, 2017.

Chapter 4

Capacitive material detection with machine learning for robotic grasping applications

Hannes Kisner, Yitao Ding, and Ulrike Thomas
Lab of Robotics and Human Machine Interaction, Chemnitz University of Technology, Chemnitz, Germany

4.1 Introduction

4.1.1 Motivation

Consider a scenario in which a robot has the task of grasping an object. Before the execution of the actual grasping motion, the robot must perform several computation steps. First, the robot has to recognize the object with sensors, and then the object's pose in the robot's frame must be estimated. With the given pose, a grasp planner calculates feasible grasping motions and poses by taking into account the object's geometry to establish friction- and form-fit connections between gripper and object. During the grasp execution, sensor information, such as from tactile or visual perception, provides necessary feedback to maintain a stable grasp.

Apparently, the grasp process involves many different engineering and research disciplines from perception and recognition to motion generation, where advancements in each discipline can contribute to successful robotic grasping. This book chapter focuses on the perception domain and presents how a new perception modality can provide new information to improve grasping. Conventional perception methods rely mostly on visual perception to recognize objects during precontact. When grasped, tactile information can be used for object recognition, haptic exploration, pose estimation, and contact-force regulation. However, besides the sensing challenges with cameras and tactile sensors, what if objects are similar in shape, color, and texture? For example, many objects have paint coatings and other thin layers and underlying materials that are not visible to the robot. Objects with similar appearance but different materials are difficult to distinguish by vision and touch alone and require information from another sensing modality.

Tactile Sensing, Skill Learning, and Robotic Dexterous Manipulation
https://doi.org/10.1016/B978-0-32-390445-2.00011-8

4.1.2 Concept

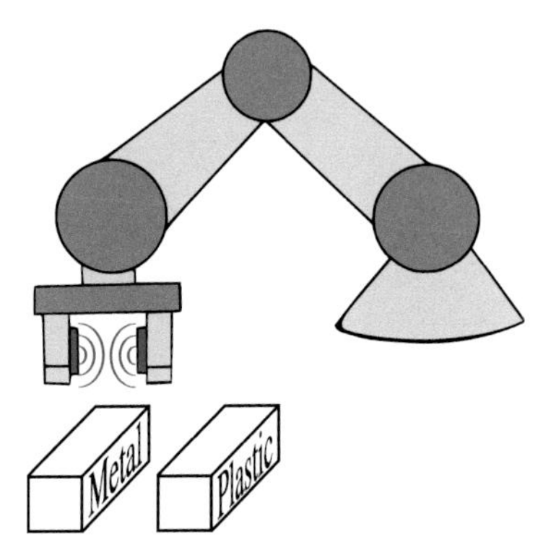

FIGURE 4.1 Capacitance sensors in the gripper for material detection.

We proposed the use of electrical capacitance sensors to infer material information in our prior work [1,2] (Fig. 4.1). The sensors can penetrate most outside layers without mechanical destruction to perform measurements on the underlying materials. The material information is useful not only to distinguish objects, but also for the proper application of grasping forces by estimating the weights or limiting maximum grasping forces and deformation according to material properties. For example, the sensors are able to detect the filling state of a cup filled with liquid, which is useful information to provide sufficient grasping force without deforming the cup or slippage. The capacitance sensors measure electrical capacitance between the sensor and the object at different excitation frequencies, resulting in a capacitance spectrum unique to each material. Even though electrical impedance measurement methods exist, they require probes with direct electrical contact to the materials, which is often not given with surface layers. Instead of focusing on the sensing hardware, the challenge lies in the processing and classification of sensor information. In this work we will discuss the use of capacitance spectra in more detail. Many different classification methods exist (see Section 4.1.3), but not all are suited to classify the capacitance spectra. In our previous work, we gave an overview of different machine learning methods for classification on a small range of different material classes. Now we focus on some of those material classes but at the finer, more granular level. For example, instead of differentiating between metal, wood, and plastics, we determine which kind of metal, wood, or plastics the object is made of. Our focus lies in metals and plastics in this work. We demonstrate the use of machine learning as well as an image classifier based on deep learning methods. The results are accurate and robust enough and can be integrated in further grasp parameterizations.

This work serves as a guideline for researchers who are interested in implementing their own capacitive material detection system by providing all the necessary steps but also by pointing out interesting and important remarks.

4.1.3 Related work

There are many contributions to material recognition in robotics, but many other research areas also show transferable results to robotics applications. Most works make use of capacitive sensing and tactile perception to generate the information needed for classification algorithms. The available algorithms can be divided into traditional and machine learning classification methods.

As for traditional classification methods, Novak et al. [3] and Kirchner et al. [4] demonstrated early works on detecting materials based on their permittivity using capacitive proximity sensors. The detection of material for capacitive proximity perception is relevant because accurate knowledge of permittivity allows more precise determination of distances. We determined the materials at different frequencies based on the generated impedance spectra in our previous work [2]. The applied classification algorithm matches the measured value with previously recorded values via a simple 1-nearest neighbor method. Another interesting technology related to material classification is X-ray spectrometry [5], which exploits the unique reemitted light spectrum from X-ray excitation.

In the machine learning domain for material detection, Alagi et al. [6] used an artificial neural network for classification based on capacitive proximity information and motion information. Chin et al. [7] used capacitance sensors in combination with tactile information for garbage sorting with artificial neural networks and demonstrated a high success rate. Xie et al. [8] used tactile information, such as from sliding motions, to detect materials and the material properties of objects. Similar to this work, they evaluated different machine learning classification algorithms in terms of performance and accuracy. The classification algorithm based on artificial neural networks showed promising detection rates in the range of 95%. Kaboli et al. [9] showed how to apply transfer learning on tactile information, such as texture, stiffness, and thermal conductivity, to learn new objects for object detection with a discrimination accuracy of 72%. Pastor et al. [10] and Gandarias et al. [11] applied convolutional neural networks (CNNs) on tactile images for material classification. Alameh et al. [12] used recurrent neural networks on tactile time-series information to determine different touch modalities, such as sliding or rolling. Besides tactile material classification, Helwan et al. [13] showed promising results on detecting breast cancer with electrical impedance spectroscopy combined with artificial neural networks.

4.2 Basic knowledge

4.2.1 Capacitance perception

4.2.1.1 Sensing hardware

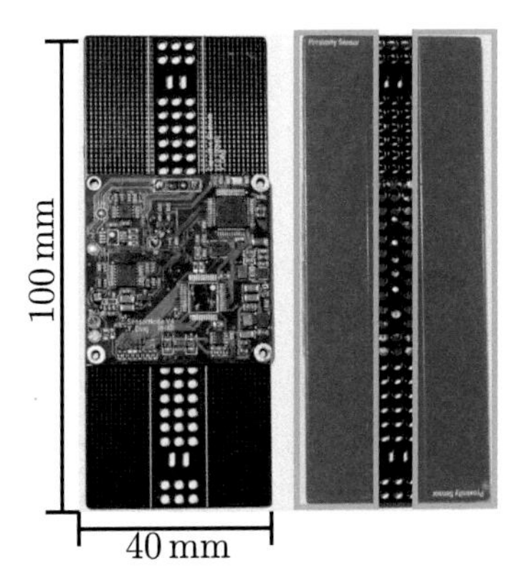

FIGURE 4.2 Capacitive proximity sensor with two differentially driven electrodes on the back.

Originally the capacitance sensors were designed for proximity perception (Fig. 4.2), which allows the robot to detect obstacles in the close-range region. The capacitance sensors have capacitive electrodes, which create electrostatic fields between the electrodes and nearby objects. Depending on the material properties, the objects disturb the electric fields through amplification or attenuation through their permittivity index. This disturbance results in a measurable change in capacitance between the electrodes of the sensor.

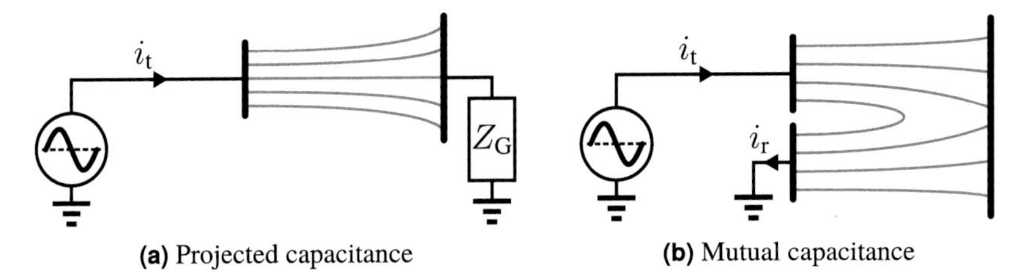

FIGURE 4.3 The two main sensing modes for capacitive proximity perception.

In general, there are two main capacitive proximity sensing modes (Fig. 4.3): projected capacitance and mutual capacitance. In projected capacitance, the sensor forms a capacitor together with the target object and relies on the object's proper grounding to close the equivalent circuit. The sensor requires only one electrode in this configuration, and as long as the object provides a proper grounding, this sensing mode has the longest proximity sensing range. However, in our material detection applications, the sensors are placed very close to the target objects. Thus the range is not an issue, rather than robust detection. For mutual capacitance, the sensor itself forms a capacitor with two electrodes, and the target object serves as a dielectric between the electrodes.

The design objectives of proximity sensing and material detection are mostly similar and differ only slightly with respect to electrode geometry selection. Capacitive proximity sensing focuses mostly on high sensitivity to achieve a far range with limited electrode area. Range is not an issue for capacitive material detection as the electrodes are directly applied on the object's surface. Nevertheless, many materials with low permittivity, such as plastics, are still hard to detect due to their low capacitive coupling and require high sensitivity to capture the material differences. In contrast to proximity sensing, the electrode geometry is optimized to focus mostly on the close range area. Fig. 4.4 illustrates the electrical circuit of the capacitance sensor. A signal generator provides excitation signals of variable frequency. The voltage drop across the resistor R_{m} determines the current i_{c} through the capacitor. We also measure the source signal to compensate for noise from the signal generator, to achieve the highest sensitivity with a coherence/correlation-based signal processing algorithm. We use differentially driven electrodes, which allow robust detection regardless of the grounding state of the target. Furthermore, this configuration eliminates electromagnetic interference as the signals cancel each other out at long distance. The back of the sensing electrodes has shielding electrodes, which are driven from a low-impedance source with the same signal as the sensing electrodes. The shieldings allow the sensor to measure only in the forward direction, while the sensor remains insensitive to objects in the back.

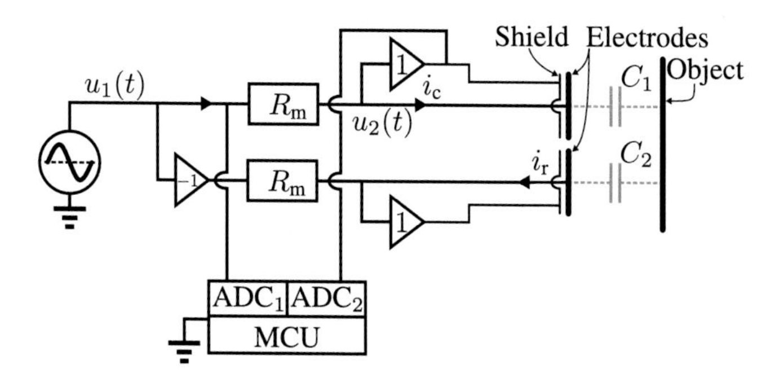

FIGURE 4.4 Electrical circuit of the capacitance sensor.

Another important aspect is the electrode design. The penetration depth and the sensitivity of capacitance measurements depend on the material itself and the shape of the sensing electrodes. It makes sense to choose different electrode designs according to the final application. Given a fixed total area, electrodes with the same spacing as their width provide a good trade-off between range and sensitivity (Fig. 4.5). However, in grasping applications, where the electrodes are located within the gripper and make direct contact with the materials, range

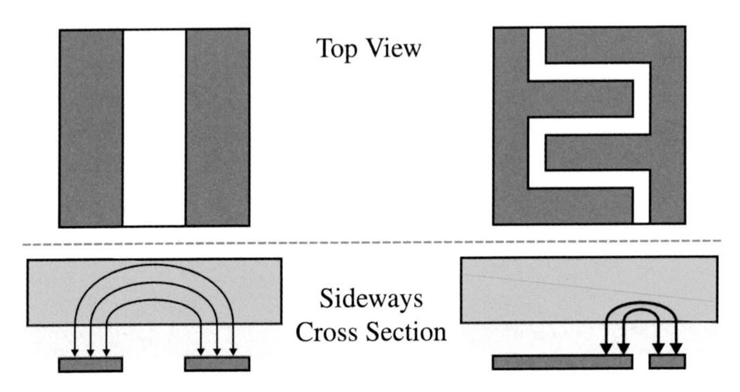

FIGURE 4.5 Electrode designs with different penetration depths.

is not an issue. Therefore, it makes sense to choose a comb structure with close spacing to generate strong electric fields, ideally suited to measure materials with low permittivity. The close electrode distance limits the penetration depths and allows for measurement at the material surface region.

4.2.1.2 Signal processing

In order to determine the capacitance, the electrode current must be determined by measuring the voltage across the measurement resistor R_{m}. Since the excitation signal is a sine wave, the challenge lies in the efficient determination of the signal amplitudes. Given a signal of the form

$$u(t) = \hat{u}\cos(2\pi f t + \varphi), \tag{4.1}$$

common methods rely on the use of discrete Fourier transform (DFT) algorithms to extract the amplitude $\hat{u}$ and phase φ from the time-series data using convolution with a reference signal. However, in a microcontroller, computing the reference signals for varying frequencies is time consuming. We use a coherence-based and auto-correlation approach to determine the signal parameters $\hat{u}$ and φ. We measure the source signal u_1 additionally to the electrode signal u_2 and use it as reference signal. Through mixing both signals with themselves, we gain the phase and amplitude of u_2 in reference to the source signal u_1:

$$\hat{u}_1\cos(2\pi f t + \varphi) + \hat{u}_2\cos(2\pi f t) \tag{4.2}$$

$$= \frac{\hat{u}_1\hat{u}_2}{2}\left[\cos(\varphi) + \cos(4\pi f t + \varphi)\right]. \tag{4.3}$$

As we can see, the first term holds information of the phase φ while the second term is a signal with double the frequency (image frequency). The higher frequency can be eliminated through low pass filtering. In our case, we use the inner product formulation to gain back amplitudes and phase:

$$\langle x, y\rangle = \frac{1}{N}\sum_{n=1}^{N} x_n y_n, \tag{4.4}$$

$$\frac{\hat{u}_1\hat{u}_2}{2}\cos(\varphi) = \langle u_1, u_2\rangle, \tag{4.5}$$

$$\frac{\hat{u}_1^2}{2} = \langle u_1, u_1\rangle, \tag{4.6}$$

$$\frac{\hat{u}_2^2}{2} = \langle u_2, u_2\rangle, \tag{4.7}$$

$$\hat{u}_1 = \sqrt{2\cdot\langle u_1, u_1\rangle}, \tag{4.8}$$

$$\hat{u}_2 = \sqrt{2\cdot\langle u_2, u_2\rangle}, \tag{4.9}$$

$$\varphi = \arccos\left(\frac{2}{\hat{u}_1\hat{u}_2}\langle u_1, u_2\rangle\right). \tag{4.10}$$

The benefit of the coherence-based method is that disturbances in the source signal can be compensated as only the ratio between u_1 and u_2 is necessary to compute the impedance between the electrodes.

4.2.1.3 Capacitance spectroscopy

The impedance spectrum consists of individual impedance measurements scanned through different excitation frequencies. The impedance at the capacitive electrodes Z_C for a single excitation frequency is

$$Z_C = R_\text{m}\left(\frac{\hat{u}_2 e^{j\varphi}}{\hat{u}_1 - \hat{u}_2 e^{j\varphi}}\right). \tag{4.11}$$

The impedance spectrum holds material-specific information, as many materials have frequency-dependent impedances. In our system, we use frequencies from 21 to 1000 kHz with a resolution of 1 kHz. However, as Z_C results mostly from capacitive coupling between the target object and the sensor, the impedance spectrum will be highly frequency-dependent, where small changes of the impedance by material properties remain unnoticeable in the large dynamic range caused by the inverse proportional frequency dependency. Therefore we use the capacitance spectrum, which is a better representation of the material properties:

$$C = \frac{1}{j2\pi f Z_C}. \tag{4.12}$$

For example, dielectric materials with partial resistive conductance have imaginary capacitance, which is inversely proportional to frequency. On the other hand, pure real capacitances represent nonconductive materials. The capacitance is a better reflection of the material's individual properties.

4.2.2 Classification for material detection

The capacitive sensor provides a capacitance spectrum depending on the measured material. The next step is evaluating the acquired information and classifying the correspondences between the capacitance spectrum and the material. Classifiers can be differentiated into different systems based on the supervision category and the training data used. We aim to classify predefined materials based on a set of real and imaginary capacitance values. Thus, we have labeled data that lead to supervised classifiers. In the following, we discuss five different classification methods in detail, introduce their concept, and apply them to our material classification experiment.

4.2.2.1 *k*-Nearest neighbors

k-Nearest neighbors (*k*-NN) is one of the simplest classifiers. The algorithm uses its predefined classes and sets of labeled training datasets to find a best fit. A set of features represents a class and can be distinguished by distance metrics, e.g., Euclidean distances. Also, distances, such as Manhattan or Minkowski distance, can be used to optimize the results if the structure of the input data is better suited. However, since our sensed data are a set of real and imaginary 1D values it is suited for the Euclidean distance metric. Furthermore, *k*-NN is nonparametric except for the maximum number of neighbors k that should be considered during classification. Thus, this algorithm is simple to use, has low computational effort, and does not require training beforehand.

4.2.2.2 Support vector machines

The capacitance spectra have, depending on the frequency interval, large feature spaces (each frequency can be considered as one feature). Support vector machines (SVMs) are more accurate in such cases. SVMs try to fit two parallel hyperplanes that distinguish the feature points with the maximum distance between the planes. A set of training data points represents each hyperplane to increase memory efficiency. Various kernel functions can represent different distance metrics, e.g., linear, radial, and sigmoid functions. The shape of the resulting hyperplanes highly depends on the used kernel. Thus the results can be optimized per dataset. However, these adjustments increase flexibility while decreasing simplicity and generalization to other datasets, making SVMs dependent on the used parameters. SVMs perform best with datasets with clear margins between the features but have problems with noise or overlapping features of target classes. Additionally, large datasets lead to inefficient training times. The most common usage of SVMs is a binary classification. Determining materials is a multiclass classification. Thus we have to use either the concept of *one-vs.-one* or *one-vs.-rest*, which result in different performances.

4.2.2.3 *Random forest classifier*

Training SVMs with a large amount of training data and possibly noisy input data may lead to long training times and overfitting. A random forest (RF) classifier overcomes these problems. RF is an ensemble of individual decision trees. Each tree predicts a class, and the tree with the highest probability is selected. The number of decision trees can be predefined. Each tree samples a random subset of the input data during training. The meaning of results and the high number of decision trees can prevent overfitting. The parallel training of trees is faster compared to SVM training. However, the prediction time increases by the number of decision trees. RFs evaluate importance of input features during training to identify the most significant features in the dataset.

4.2.2.4 *Feedforward neural networks*

Feedforward neural networks (FFNNs) can represent more complex classification functions. An FFNN includes an input, output, and several hidden layers. The number of hidden layers represents how deep a network is. All layers include interconnected nodes. The connections are constructed by weights and their predefined activation functions. The network structure is not fixed and can be adjusted in various ways, which makes it very flexible and applicable to any classification task. The weights are initialized randomly or set to a specific number. An optimizer adjusts the weights during training using a loss function. The loss function represents the difference between ground truth data and the output of the network for a given input dataset. The optimizer aims to decrease the loss function. The network performance depends on the used structure, the optimization strategy, and the quality of the training dataset. More complex and deeper FFNNs can learn more complex mapping functions but require more training data to adjust all network weights correctly. Thus the computational effort increases with network complexity as well as with training data size.

4.2.2.5 *Convolutional neural networks*

CNNs are a special type of deep neural network. CNNs focus on image classification tasks as they can handle a matrix representation of input data. CNNs are implemented in various dimensions. 1D-CNNs are able to recognize patterns in 1D signals, e.g., in time-series [14] or frequency-series data. The capacitance spectrum is a set of ordered 1D signals (real and imaginary values), a 1D representation of data. 1D-CNNs can learn from the values of the features and also from the order of the features, which can lead to more accurate classifications. However, state-of-the-art CNN approaches focus on their use as 2D-CNNs in image classification tasks. There are powerful image recognition systems, e.g., *ResNet* [15]. The high performance and availability of these systems leads to the idea of representing the capacitance spectrum data as images, which is the same concept as visualizing data in graphs. Our experience is that humans are able to classify the material data using only data diagrams. Consequently, a 2D-CNN should do the same when we transform data from 1D to 2D. We describe this process, which is applied to our data, in Section 4.3.1 in more detail. The drawback of using CNNs is the higher computational effort and a large amount of required training data. However, they are more robust to sensor noise and can learn the classification function by themselves, especially when the necessary metric to distinguish data is very complex and the feature space is high-dimensional.

4.3 Methods

4.3.1 Data preparation

Our sensor measures the capacitance between the sensor and different materials and generates a set of feature vectors. Section 4.2.1 describes the functionality and structure of the capacitance sensor in more detail. In the following, we discuss how to prepare and forward the data to be used in classification tasks, first starting with the raw data and then encoding raw data into 2D images.

4.3.1.1 *Raw data*

The capacitance sensor sweeps the excitation frequency from $[21, 1000]$ kHz with a resolution of 1 kHz, leading to 980 data points (Fig. 4.6). Each data point is comprised of real and imaginary parts of the capacitance. We prepare three different datasets to evaluate the performances of the described classifiers. The first dataset $Set_{Re:Im}$ is a 1D feature vector with real and imaginary values stacked horizontally. We split the capacitance spectrum for the second dataset Set_{Re} and the third dataset Set_{Im} to show the importance of combining the real and the imaginary part. Additionally, we prepare a dataset $Set_{abs(Re:Im)}$ using the absolute value of the capacitance spectra to investigate if the raw capacitance spectra are more precise.

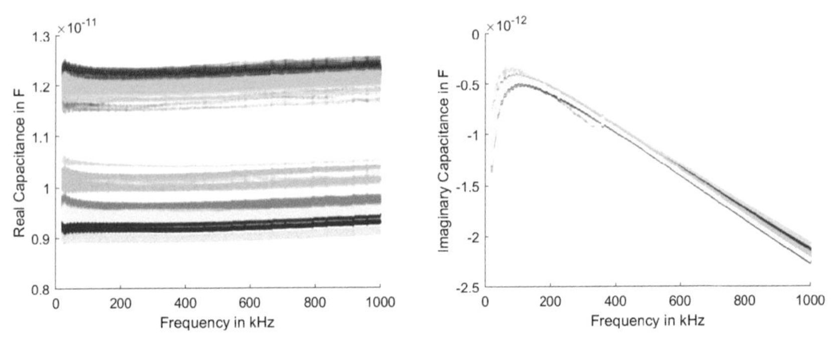

FIGURE 4.6 Raw capacitance spectrum captured for 11 different materials during motion. Many spectra overlap and are difficult to distinguish.

4.3.1.2 Image generation

In addition to raw data classifications, we also use image classifiers to classify materials. Therefore, we encode two 1D feature vectors into an image. CNN image classifiers predict class labels for 2D images.[1] Image features are automatically learned during the network training phase. The classifiers try to find and align the learned features in the input image in the inference phase. A 1D feature vector cannot be directly applied to an image classifier. Therefore, we developed a pipeline projecting the values of the feature vector towards an image plane. In doing so, we first have to define the constraints. We use a state-of-the-art CNN image classifier, e.g., *ResNet* [15], *InceptionV3* [16], and *VGG* [17]. The measured data, or in our case the capacitance spectra, can be plotted using an out-of-the-box data visualizer (see Fig. 4.7 (A)). We found during data analysis that humans are able to make rough classification predictions based on these plots. The authors of [18] demonstrate the conversion from time-series data to images that can be identified by image classifiers. We recreate the graphs by using the constraints given by the CNN image classifier. In addition, the diagrams could be extended with further information. For example, features or information could be converted into an image file. This would provide the classifier with further distinguishing criteria.

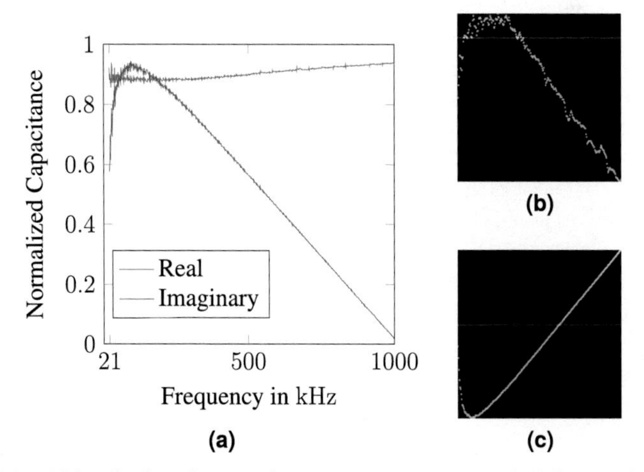

FIGURE 4.7 (a) Visualization of an capacitance spectrum using a standard diagram visualizer. (b) and (c) Image generation results for image-based convolutional neural networks.

Since the input data are RGB images, we can encode our data into three color channels. *ResNet* and *InceptionV3* need images with a fixed resolution of 299 × 299 pixels and the *VGG* variation *VGG16* needs 224 × 224 pixels. We cannot encode the absolute feature values directly into the images with the predefined resolution. There are CNN applications that resize any image size to

[1] There are various configurations and also nD-classifiers available. However, these are out of the scope of this work.

the needed resolution. Image resizing may lead to unwanted stretching of data or compression artifacts, leading to bad performance. Therefore, we encode the scale of the values into the three color channels. There are various color models available. We use the RGB model as all channels are equal in calculating the resulting color.

We have two values per feature, one real and one imaginary value. We encode the real value into the red channel and the imaginary value into the blue channel. Each channel ranges from 0% to 100%. We scale each feature using the minimum and maximum values in the dataset. Therefore, the minimum represents 0% in the color channel.

Our initial image pixels represent zeros which result in black regions. Adding small feature values respectively small color values would lead to very dark resulting image colors which are not clearly visible. We use the green color channel to increase the brightness of the pixels. Therefore, each pixel includes 100% green and a percentage of the real (red) or imaginary (blue) value. In addition to the brightness respectively color value, we encode the capacitance spectra in pixel positions. Images correspond to a 2D matrix where each position/pixel implies a row and a column. We use the columns to represent frequencies (from low to high frequency), and the row index is set to the percentage of the real or imaginary values. Each index is scaled to the image resolution. However, in previous work, we explored that adding more features into the images results in more precise classifications. For this purpose, we use horizontal lines, which additionally represent each part of the capacitance spectrum's mean value.

Red color encodes real capacitance values. Therefore, we draw the horizontal line using the red color channel on full brightness. The mean of the real capacitance values per measurement in relation to the overall dataset corresponds to the relative position in the image. The resulting pixel position of the line results from the image resolution.

We determine the horizontal line for the imaginary part similar to the real part but using the blue color channel. Fig. 4.7 (B) visualizes an output image of the proposed pipeline. Each frequency-wise capacitance spectra measurement results in one image. Each material class includes about 250 images generated from premeasured data. Note the visual similarities between Fig. 4.7 (A) and Fig. 4.7 (B).

At last, we split each dataset in three parts, as is common for the evaluation of machine learning methods. The first part contains 70% of the data as training data. We halve the remaining data for testing and evaluation purposes.

4.3.2 Classifier configurations

Every classifier has its parametrization. The following section describes some adjustable parameters in more detail and how we used them to optimize the mean average precision (mAP). mAP measures the classification results with respect to the ground truth data. Therefore, we use mAP to evaluate the performances of each classifier.

The k-NN is the simplest classifier to parametrize; k defines the number of nearest neighbors that should be considered for classification. A small k decreases the accuracy for noisy input data, whereas a large k considers classes that are further away from the correct class. We use k-NN in two different approaches. We have the capacitance spectra given as our feature vector. The first approach k-NN_1 considers the whole spectrum to make classification decisions. However, we explored the input data beforehand and did not detect a strong connection between the materials and the capacitance per frequency. Thus, we set up k-NN_2 to classify the material frequency-wise and set the resulting class to the one with the highest matches. We set $k = 5$ to get a trade-off between robustness and accuracy.

SVM classifiers have multiple adjustable parameters. We use scikit-learn for implementation purposes [19]. Thus, we can use a set of default values that may or may not influence the mAP score. However, the defined kernel most strongly influences the result. We investigate to use a linear kernel in SVM_L and a kernel with radial basis SVM_{RBF}. Each SVM is based on the *one-vs.-rest* method which is the default parameter in the scikit-learn implementation.

The most important parameter for RF-based classifiers is the number of decision trees. The strength of an RF is the large number of individual decision trees, but many trees increase the decision time. We set up two RF configurations to demonstrate their difference in performance. RF_1 includes one decision tree and RF_{100} constructs 100 trees.

The structure of FFNNs and 1D-CNNs is neither defined by rules nor by the given input data. Only the expected class number and the used NN layout constrain the structure. We set up various networks based on the classification task itself, different parameterizations, and the input data. Detecting materials based on capacitance spectra is a comparably simple classification task. Therefore, we use few hidden layers to mirror the complexity. We aim to construct networks that are able to perform as well as traditional k-NN approaches while being more robust to noisy input data. Moreover, NN can be adjusted by the used training optimizer and the defined loss function. We use a state-of-the-art optimizer called Adam [20], and the loss function is set to categorical cross-entropy, which is typical for multiclass classification. Additionally, we define the last layer of our networks to be consistent. Therefore, we set the number of neurons to the number of classes that we extract, and we use the softmax function, which performs well with the given categorical cross-entropy. We measure the capacitance spectra for 980 frequencies leading to $2 \times 980 = 1960$ features. The first network has three hidden layers (see Fig. 4.8). We use a linearly decreasing number of neurons per layer which is common practice for network structuring. Therefore, the first network $FFNN_{1s}$ has 1536, 1024, and 512 neurons. $FFNN_{1s}$ uses the sigmoid function as the activation function. We add the network $FFNN_{1r}$ with the same structure as $FFNN_{1s}$ except for the activation function to demonstrate the performance differences. $FFNN_{1r}$ uses rectified linear unit function (ReLU) as activation function. We use the same principle of

$FFNN_{1s}$ and $FFNN_{1r}$ for building two networks with six hidden layers, which leads to $FFNN_{2s}$ and $FFNN_{2r}$ with 1536, 1280, 1024, 768, 512, and 256 neurons. All the mentioned networks are built upon fully connected layers as they represent the standard network layers.

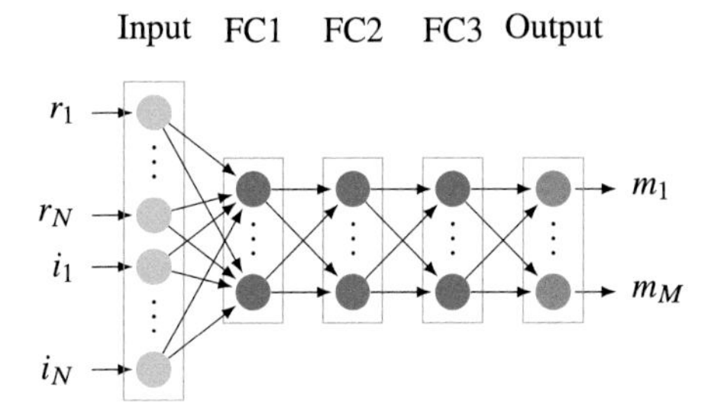

FIGURE 4.8 *FFNN* with $2N$ inputs for real and imaginary values of the spectrum and $M = 11$ outputs.

Furthermore, we demonstrate using just a few neurons in $FFNN_{3s}$ (256, 128, 64) with the same structure as $FFNN_{1s}$. $FFNN_{4s}$ uses the same structure but includes more neurons (2560, 1280, 640) than $FFNN_{3s}$. The structure of *1DCNN* differs slightly from *FFNN* as the input data consist of two stacked feature vectors (see Fig. 4.9). The 1D classifier has the advantage of reading a data matrix instead of a single feature vector. We halve the input vector from the input dataset, e.g., $Set_{Re:Im}$, into two single rows: one for real and one for imaginary values. Afterwards, we stack both rows to one input matrix, thus being able to learn input features in parallel network streams.[2] We put three groups of 1D layers into the network. The first and last groups have 100 neurons each, and the second group has 256 neurons for each layer. We put a max-pooling layer after the first group to prevent overfitting and an average-pooling after the last group to combine the two feature layers.

Section 4.3.1.2 describes the dataset adjustments to generate training datasets for state-of-the-art image classifiers. However, these classifiers cannot be used out of the box for our classification task. Each neural network classifier includes a fixed structure of different types of layers and predefined layer sizes. The differences arise partly from the different concepts behind the overall structures, as well as the different classification tasks for which the classifiers were developed. We have already limited the resolution of the input images. Another step is to adjust the size of the output vector of the neural networks. Therefore, we exchange every output vector of the classifier respectively the last layer in the CNN with a dense layer. The size of the new dense layer equals the number of different classes. Thus, our image classifiers output a vector of predictions where each value corresponds to one class in the dataset.

[2] It has to be mentioned that the stacking structure is adjusted to the $Set_{Re:Im}$ dataset. There are multiple configurations possible to do similar adjustments to other datasets. However, they are out of the scope of this work.

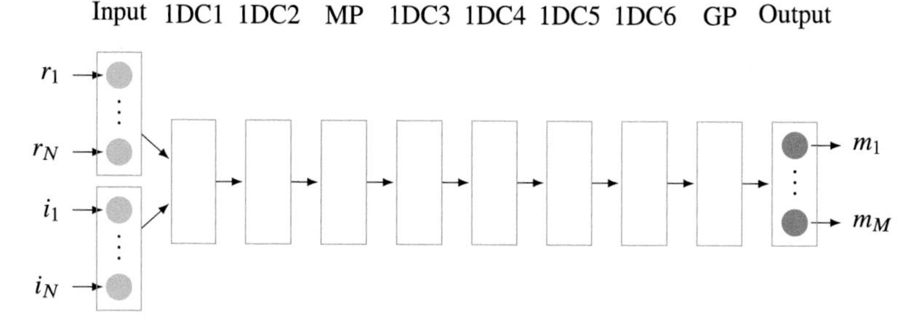

FIGURE 4.9 *1DCNN* with 1D convolutional layers (1DC), a max-pooling layer (MP), and a global average-pooling (GP) layer.

We use the stochastic gradient descent (SGD) implemented in Keras [21]. SGD generates fast parameter updates with high variances to increase the chance of reaching the global minimum. Furthermore, SGD is computationally fast and does not consume much memory at one optimization step, which is advantageous when large and deep neural networks are used, e.g., *InceptionV3*.

4.4 Experiments

Table 4.1 lists all the materials we are testing in this work. Our previous work showed successful identification of coarse material groups. The focus lies now on two material groups, metals and plastics, in order to achieve more nuanced identification of the materials within these groups.

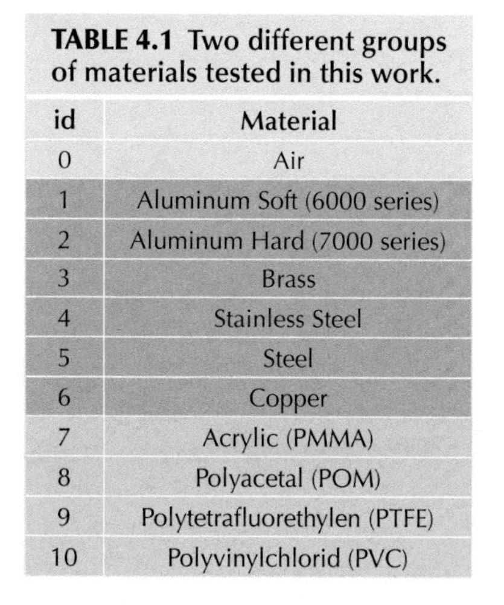

TABLE 4.1 Two different groups of materials tested in this work.

id	Material
0	Air
1	Aluminum Soft (6000 series)
2	Aluminum Hard (7000 series)
3	Brass
4	Stainless Steel
5	Steel
6	Copper
7	Acrylic (PMMA)
8	Polyacetal (POM)
9	Polytetrafluorethylen (PTFE)
10	Polyvinylchlorid (PVC)

We examine two datasets, one with ideal measurements and one with noise introduced by sensor motion. The ideal measurements represent measurements where the sensors are placed completely flat on the objects and are at rest, while the second dataset contains mainly measurements where the sensor is in motion during spectrum acquisition or is not properly aligned on the surface with slight deviations. As Fig. 4.10 (A) illustrates, a spectrum without motion during the measurements is only affected by offsets in the curve, and the material-specific information is still preserved. The ideal dataset can achieve very reliable results with most classification algorithms (Table 4.2, Set_{Still} and $Set_{Partial}$). However, motions introduced during measurements change the curves drastically (Fig. 4.10 (B)). Since the spectrum is scanned over time, offsets due to changing distances from vibrations alter the spectrum and make classification challenging. The second dataset aims to test for the robustness of the classification algorithms in harsh real-world scenarios.

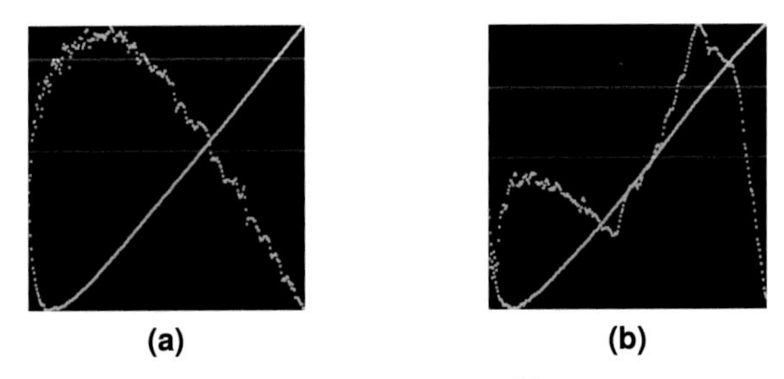

FIGURE 4.10 (a) Ideal capacitance spectrum of copper with measurement at rest. (b) The same measurement with noisy spectrum due to motion during measurement.

We prepared datasets for the evaluation process. Each dataset contains capacitance spectra as feature vectors or as feature images and the corresponding materials. The training dataset is used to train the classifiers, and a subset of the complete dataset is used to evaluate the performances. We use the mAP as the performance metric. The mAP combines the relation of predicted true positives and false positives into one value and is therefore useful for comparing different classifiers. A mAP of one corresponds to an ideal classifier, whereas a mAP of zero corresponds to the worst performance.

Table 4.2 shows the mAP results of the used datasets and the classifiers. The performance scores of conventional, artificial neural networks and image-based classifiers are promising, showing that the classifiers can identify materials by their impedance spectrum. The RF_{100} variation performs best in the conventional group. However, RF_1 performs worst, showing the advantage of the adjustable *RF* tree size. The *SVM* classifier SVM_L based on a linear kernel performs worse than SVM_{RBF} based on a radial classifier, showing that the input features are not easily separable with hyperplanes. $KNN2_5$ performs better than $KNN1_5$, showing that the frequency-wise classification is more precise and robust. The different NN structures have the same performances. Thus, no definite statement can be made about the structure or the general design of layer sizes and types. However, the *1DCNN* performs worst, either due to poor network structure or because the data cannot be well classified by these network types.

This shows that the structure of NNs cannot be defined either by the feature values or the feature size. This can cause problems when new or different features vectors are classified, leading to further experiments in adjusting network configurations. The group of image-based classifiers performs similarly, showing their robustness. *ResNet18* performs best in this group.

Overall the RF_{100} classifier predicts the materials with the highest precision for each dataset. However, the performance of *ResNet18* is almost as good as that of RF_{100}, which shows that the converting feature vectors to image data works and does not reduce classification results. However, it should be mentioned that conversion to image data has a higher flexibility because images can contain more additional feature types and are not restricted to numerical features, e.g., blocks can be constructed of different data types.

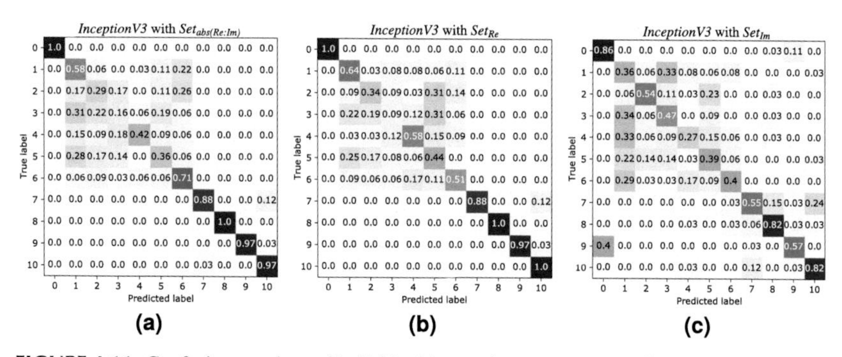

FIGURE 4.11 Confusion matrices of individual image datasets corresponding to the real, imaginary, and absolute impedance values classified by the *InceptionV3* image classifier.

Now we discuss the performance of the classification algorithms. Classifications rely on different sets of features. However, not every feature affects the classification performance in the same way and contributes to the distinctness of classes. Therefore, an evaluation of each feature is useful for further investigation and reduces the overall information size. Low-dimensional information reduces the computational effort. Fig. 4.11 visualizes the confusion matrices resulting from the *InceptionV3* image classifier using different datasets. The confusion matrices visualize the classification performance per class which enables a more detailed evaluation of failing predictions. An ideal classification results in a unit matrix. The comparison of Fig. 4.11 and the results from the complete dataset in Fig. 4.12 (B) show that using all measured features results in the best performance. Table 4.2 demonstrates that the use of imaginary values leads to low classification performances for every classifier, leading to the assumption that imaginary values are less distinguishable than other features. By contrast, real values provide outstanding classification results, showing the importance of these values for the distinctiveness of material classes. $Set_{abs(Re:Im)}$ combines the real and imaginary capacitance spectra into one feature and reduces the dimension and data size. However, the results are slightly worse compared to using real values. Therefore, dimension reduction can decrease the amount of information but can also mix essential features with insignificant features, causing degraded performances. $Set_{Re:Im}$ incorporates essential and insignificant features in parallel and outperforms the other datasets. It shows that the best classification performances can be achieved using the full feature set in the final application.

TABLE 4.2 mAP of different classification algorithms on different composed datasets. Yellow: conventional classification; green: artificial neural networks; blue: image-based convolutional neural networks.

Algorithm	$Set_{Re:Im}$	Set_{Re}	Set_{Im}	$Set_{abs(Re:Im)}$	Set_{Still}	$Set_{Partial}$
$KNN1_5$	0.79	0.78	0.55	0.7	0.98	0.98
$KNN2_5$	0.84	0.84	0.65	0.81	1	1
SVM_L	0.76	0.77	0.37	0.74	0.9	0.9
SVM_{RBF}	0.9	0.82	0.75	0.81	1	1
RF_1	0.67	0.54	0.46	0.48	1	0.99
RF_{100}	**0.95**	**0.87**	**0.9**	**0.86**	1	1
$FFNN_{1s}$	0.83	0.74	0.62	0.75	0.99	0.99
$FFNN_{1r}$	0.89	0.74	0.61	0.76	1	1
$FFNN_{2s}$	0.77	0.66	0.6	0.69	0.96	0.92
$FFNN_{2r}$	0.88	0.77	0.63	0.79	1	1
$FFNN_{3s}$	0.7	0.66	0.51	0.7	0.97	0.97
$FFNN_{4s}$	0.84	0.74	0.63	0.76	1	1
1DCNN	0.73	0.68	0.41	0.69	0.97	0.97
InceptionV3	0.89	0.78	0.61	0.77	0.99	0.99
ResNet18	0.92	0.84	0.74	0.85	0.99	0.99
VGG16	0.89	0.79	0.55	0.76	0.99	0.99

Next, we examine the overall material classification results from the confusion matrices in Fig. 4.12. The classification methods detect plastics with high precision and reliability, although they are difficult to measure due to their low permittivity and capacitances. Metals, on the other hand, are easier to detect by the sensor but difficult to classify. The classification relies mainly on the metals' electrical conductivity rather than the varying permittivity. Some metals differ only in their alloys, making them very similar in their characteristics, e.g., aluminum. Nevertheless, some classification algorithms are able to identify the metallic materials. The conclusion is that the varying frequency-dependent permittivity of plastics contains sufficient information for a proper identification with the capacitance spectrum. However, with ideal, undistorted data, most classification algorithms identify even metals with high precision, which shows that minimal material differences are visible with the ideal dataset. When robustness is required, *CNN*- and *RF*-based classification methods provide adequate results.

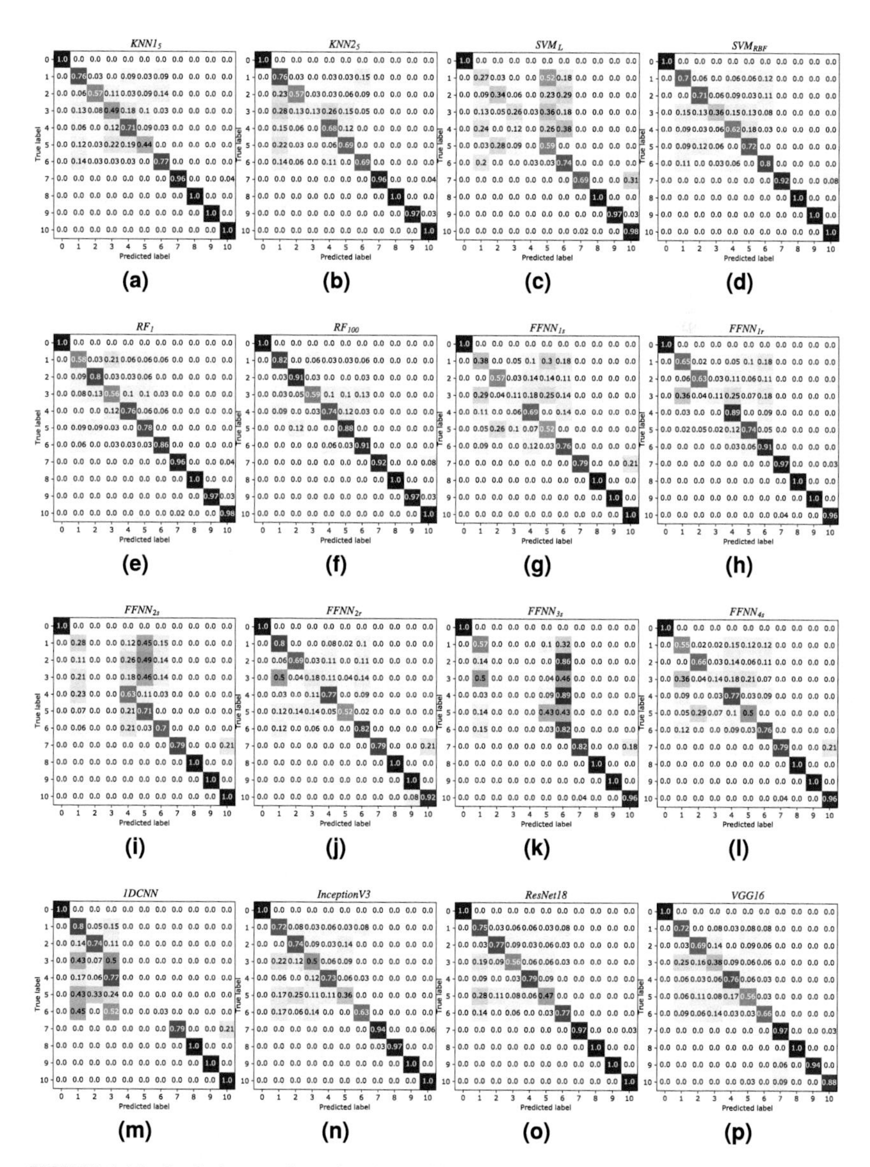

FIGURE 4.12 Confusion matrices of each classifier using the $Set_{Re:Im}$ dataset.

4.5 Conclusion

Our proposed pipeline is particularly accurate in identifying materials based on their capacitance spectra, especially when the signal contains little noise. In our previous work, each class contained several different variations of materials within a single class. The more granular segmentation of materials in this work reduces each class's variations, contributing to better results. We can distinguish coarse classes of materials, e.g., metals and plastics, as well as more nuanced identification of materials, e.g., aluminum with various alloys. Our sensor measures the capacitance spectra of materials. The differences in the capacitance spectra result, among other things, from the permittivity. Metals have an almost infinite permittivity and a high electric conductivity. Thus the effect on the measured capacitance spectra is low, making classification difficult. However, small variations in the imaginary values (electrical resistance) can still be classified by our method. This results in an overall accurate and robust material classifier. We demonstrate the usage of different state-of-the-art as well as conventional classifiers using feature vectors as input data. Furthermore, we explored the conversion from feature vectors to images which image classifiers can use. The conversion of the impedance spectra into images is flexible and can be easily extended with other data, leading to a better performance of the image-based classification methods. However, RF_{100} shows slightly better performance than other methods. However, possible sensor changes or the use of new materials may change the overall results. The latter is also a drawback of our method as an extended dataset leads to a renewed classifier training. We hope that fellow researchers can implement their systems based on the findings of this broad investigation of different algorithms.

References

[1] Y. Ding, H. Kisner, T. Kong, U. Thomas, Using machine learning for material detection with capacitive proximity sensors, in: 2020 IEEE/RSJ International Conference on Intelligent Robots and Systems (IROS), 2020, pp. 10424–10429.

[2] Y. Ding, H. Zhang, U. Thomas, Capacitive proximity sensor skin for contactless material detection, in: 2018 IEEE/RSJ International Conference on Intelligent Robots and Systems (IROS), 2018, pp. 7179–7184.

[3] J.L. Novak, J.J. Wiczer, A high-resolution capacitative imaging sensor for manufacturing applications, in: 1991 IEEE International Conference on Robotics and Automation, vol. 3, 1991, pp. 2071–2078.

[4] N. Kirchner, D. Hordern, D. Liu, G. Dissanayake, Capacitive sensor for object ranging and material type identification, Sensors and Actuators A: Physical 148 (1) (2008) 96–104, https://doi.org/10.1016/j.sna.2008.07.027.

[5] R. Jenkins, J. De Vries, Physics of X-Rays, Springer US, New York, NY, 1969.

[6] H. Alagi, A. Heilig, S.E. Navarro, T. Kroeger, B. Hein, Material recognition using a capacitive proximity sensor with flexible spatial resolution, in: 2018 IEEE/RSJ International Conference on Intelligent Robots and Systems (IROS), 2018, pp. 6284–6290.

[7] L. Chin, J. Lipton, M.C. Yuen, R. Kramer-Bottiglio, D. Rus, Automated recycling separation enabled by soft robotic material classification, in: 2019 2nd IEEE International Conference on Soft Robotics (RoboSoft), 2019, pp. 102–107.

[8] Y. Xie, C. Chen, D. Wu, W. Xi, H. Liu, Human-touch-inspired material recognition for robotic tactile sensing, Applied Sciences 9 (12) (2019), https://doi.org/10.3390/app9122537.

[9] M. Kaboli, D. Feng, G. Cheng, Active tactile transfer learning for object discrimination in an unstructured environment using multimodal robotic skin, International Journal of Humanoid Robotics 15 (01) (2018) 1850001, https://doi.org/10.1142/S0219843618500019.

[10] F. Pastor, J.M. Gandarias, A.J. García-Cerezo, J.M. Gómez-de Gabriel, Using 3d convolutional neural networks for tactile object recognition with robotic palpation, Sensors 19 (24) (2019), https://doi.org/10.3390/s19245356.
[11] J.M. Gandarias, A.J. García-Cerezo, J.M. Gómez-de-Gabriel, Cnn-based methods for object recognition with high-resolution tactile sensors, IEEE Sensors Journal 19 (16) (2019) 6872–6882, https://doi.org/10.1109/JSEN.2019.2912968.
[12] M. Alameh, Y. Abbass, A. Ibrahim, G. Moser, M. Valle, Touch modality classification using recurrent neural networks, IEEE Sensors Journal 21 (8) (2021) 9983–9993, https://doi.org/10.1109/JSEN.2021.3055565.
[13] A. Helwan, J.B. Idoko, R.H. Abiyev, Machine learning techniques for classification of breast tissue, Procedia Computer Science 120 (2017) 402–410, https://doi.org/10.1016/j.procs.2017.11.256, 9th International Conference on Theory and Application of Soft Computing, Computing with Words and Perception.
[14] S. Kiranyaz, T. Ince, M. Gabbouj, Real-time patient-specific ecg classification by 1-d convolutional neural networks, IEEE Transactions on Biomedical Engineering 63 (3) (2016) 664–675, https://doi.org/10.1109/TBME.2015.2468589.
[15] K. He, X. Zhang, S. Ren, J. Sun, Deep residual learning for image recognition, in: 2016 IEEE Conference on Computer Vision and Pattern Recognition (CVPR), 2016, pp. 770–778.
[16] C. Szegedy, V. Vanhoucke, S. Ioffe, J. Shlens, Z. Wojna, Rethinking the inception architecture for computer vision, CoRR, arXiv:1512.00567 [abs], 2015.
[17] K. Simonyan, A. Zisserman, Very deep convolutional networks for large-scale image recognition, arXiv preprint, arXiv:1409.1556, 2014.
[18] N. Hatami, Y. Gavet, J. Debayle, Classification of time-series images using deep convolutional neural networks, CoRR, arXiv:1710.00886 [abs], 2017.
[19] F. Pedregosa, G. Varoquaux, A. Gramfort, V. Michel, B. Thirion, O. Grisel, M. Blondel, P. Prettenhofer, R. Weiss, V. Dubourg, J. Vanderplas, A. Passos, D. Cournapeau, M. Brucher, M. Perrot, E. Duchesnay, Scikit-learn: machine learning in Python, Journal of Machine Learning Research 12 (2011) 2825–2830.
[20] D.P. Kingma, J. Ba, Adam: a method for stochastic optimization, arXiv:1412.6980, 2014.
[21] F. Chollet, et al., Keras, https://keras.io, 2015.

Part II

Skill representation and learning

Chapter 5

Admittance control: learning from humans through collaborating with humans

Ning Wang and Chenguang Yang
Bristol Robotics Laboratory, University of the West of England, Bristol, United Kingdom

5.1 Introduction

Nowadays, an industrial robot is most likely to be programmed to perform tasks in structured environments. To some extent, robots liberate humans from the mind-numbingly repetitive work routines. The robot is controlled under a fixed position control mode without much flexibility and adaptability. In some cases, tasks are either too complex to automate or too heavy to manipulate manually. It is difficult to address this problem by humans working alone or by automated robots. This kind of robotic manufacturing systems cannot gradually meet the increasing requirements of *high-mix*, *low-volume*, and *short-cycle* production in the market [1]. One promising solution to this problem is to integrate human factors into the robotic manufacturing systems in order to construct human-in-the-loop human-cyber-robot-systems (HCRSs) [2]. By taking the advantages of both robots (e.g., good repeatability) and humans (e.g., flexibility and adaptability), it has a great potential to improve the state-of-the-art robotic production and to remove the barriers toward the new generation of intelligent manufacturing.

A number of approaches have been recently developed for the enhancement of robot learning in order to improve the robotic manipulation abilities (e.g., [3,4]). Specifically, learning from human LfD with human–robot collaboration has been considered as an effective and efficient way to bring together humans and robots advantages [5,6]. It allows to conveniently transfer human skills to a robot without the need of an expert's specific knowledge.

Most of the previous studies in LfD have only concentrated on the learning of motion movements. These approaches can be utilized to address the encoding of motion profiles in a specific task. However, for force-dominant tasks these approaches may be insufficient. Even in a simple robotic pick-and-place task, for instance, when it comes to the consideration of the task dynamics compliant manipulation not only the motion planning should be addressed. Very recently, some researchers in the society of robotics have developed force/impedance-

Tactile Sensing, Skill Learning, and Robotic Dexterous Manipulation
https://doi.org/10.1016/B978-0-32-390445-2.00013-1

based approaches to enable the learning of compliant behaviors from humans [7,8]. The core idea of impedance control is to control the dynamic interaction between motion and contact force as needed, rather than controlling these variables separately. Impedance control can be used for all control stages, including free motion, constrained motion, and transient processes between them, without switching between different control modes. Impedance control offers the possibility of controlling both motion and contact forces by designing appropriate interactions between the manipulator and its environment. What should be emphasized is that the variable impedance control strategy has nearly become a common view that could help to achieve this point [9,10]. However, it is not easy and continent to obtain variable impedance profiles, and a time-consuming complex process is often required.

The ideal collaboration between humans and robots is that there is no separation and no guardrail. Therefore, a safer and more effective control strategy is needed for optimal HRC. Compared with hybrid force/position control and iterative learning control, impedance control is adaptable to the transition between free motion and constrained motion, and it shows satisfied tracking capability when the external constraints are known. There are two possible forms of impedance control; one is impedance control based on robot end point position control, i.e., admittance control, and the other is impedance control based on torque control in joint space [11]. Admittance control is one form of impedance control, when external forces exerted by the operator are measured as input and positions are taken as feedback to the operator.

Until now, there are four main ways in the literature to obtain proper stiffness trajectories for robotic variable admittance control, i.e., the electromyography (EMG)-based, optimization-based, force-based, and biomimetic control approaches, which are separately introduced below.

(i) *EMG-based*: The EMG signals detected from human arms can be utilized to extract human limb stiffness features. Therefore, the human arm stiffness profile can be estimated based on EMG during the interactions with robots. A number of studies have reported their results on this point. Typically, [9] proposed an EMG-based tele-impedance concept which could enable to transfer the human arm stiffness to a teleoperated robot. [12] and [13] proposed an EMG-based human–robot stiffness transfer interface that could allow robots to imitate both motion and impedance behaviors from humans.

Most of the studies utilized the EMG signals to estimate the diagonal elements in the human arm end point stiffness matrix. In [14], a model-based estimator was developed to extract human arm complete joint stiffness. However, EMG-based approaches need a complex process to estimate the parameters of the EMG impedance mapping model, which may sometimes be time consuming. The parameters vary from one human user to another due to the different arm characteristics, and it is quite difficult to learn a general model for multiple different human demonstrators. Besides, the human arm configuration would have a large effect on the estimation results.

(ii) *Optimization-based*: The optimization-based approaches prefer to learn a proper stiffness profile for variable impedance control by using optimization techniques such as reinforcement learning [15], black-box evolution [16], and adaptive control [17]. A constant reference stiffness trajectory is used for the initialization of models, and then a number of trials are often required to learn a decent stiffness profile. The disadvantage of this method is that it is sometimes not easy to define a good reward/cost function, especially for a complex task, resulting in the need of a large process of trial and error which could be harmful to the robotic platforms.

(iii) *Force-based*: Force-based approaches which require offline experiments refer to the use of a force sensor mounted onto the robotic end point to measure the interaction force, based on which the stiffness is estimated. Typically, [18] used a Gaussian mixture model to encode the joint dataset (position and force) and then used Gaussian mixture regression to get the stiffness profile based on the learned model. [19] extended the work to use a hidden semi-Markov model to model the correction between the position/rotation and the force/torque. The advantage is that it can be used over and over again after offline testing, but it needs to be remeasured for different operators.

(iv) *Biomimetic control*: Biomimetic control approaches are inspired by the human motor learning, which is employed to obtain the desired robotic compliant behaviors by online adapting the impedance profiles and the feedforward torques simultaneously. The robot can easily acquire motion skills from a human tutor by kinematics demonstration. It argues that the impedance and feedforward torque/force should be concurrently adapted in order to deal with stable and unstable situations in unknown environments [20,21]. In [22], a biomimetic controller was proposed based on this argument and implemented on a robot with one degree of freedom (DoF).

5.2 Learning from human based on admittance control

In this section, the biomimetic control-based LfD will be introduced. The admittance control model is presented and the biomimetic controller is given for the adaptation of the impedance profiles, as well as the feedforward torque.

5.2.1 Learning a task using dynamic movement primitives

Dynamic movement primitives (DMPs) is a well-known model which is able to efficiently represent a skill/task and has been widely used in a large number of articles. It can model and generate human-like movements. For the sake of completeness, here we give a brief introduction to DMPs. For more details, please refer to [23–25]. Basically, the DMPs model can be separated into the following two parts.

5.2.1.1 Constructing a second-order nonlinear system

First, a second-order nonlinear system is constructed to model a specific motion trajectory. Based on the motion types, i.e., the *Discrete* movements and the *Rhythmic* movements, different nonlinear systems are needed for different kinds of tasks. For a 1-DoF discrete movement trajectory, the system is defined by the following equations [24]:

$$\tau \dot{y} = k(g - x) - d\dot{x} - k(g - x_0)s + kf(s), \tag{5.1}$$

$$\tau \dot{x} = y, \tag{5.2}$$

$$\tau \dot{s} = -\alpha_s s, \tag{5.3}$$

$$f(s) = \frac{\sum_{i=1}^{N} \gamma_i \phi_i(s) s}{\sum_{i=1}^{N} \phi_i(s)}, \tag{5.4}$$

$$\phi_i(s) = \exp(-h_i(s - c_i)^2), \tag{5.5}$$

where x and y represent the angle in joint space and the corresponding velocity of the 1-DoF movement trajectory, and x_0 and g represent the initial value and the goal of the angle trajectory, respectively. Eq. (5.1) can also be considered as a spring–damper system with spring parameter k and damping parameter d, which are often properly chosen in advance as $d = 2\sqrt{k}$. The temporal constant τ is used to control the evolution duration of the system. The whole system is driven by the phase variable s generated from Eq. (5.3) instead of directly using time such that the evolution of the system can be efficiently edited. Phase $s \in (0, 1]$ starts from 1 and monotonically converges to 0 along with the duration of the motion trajectory, granting that the motion finally converges to a goal point. The constant α_s has to be the predefined coefficient.

The nonlinear force term $f(s)$ in Eq. (5.1) is determined by Eq. (5.4). Gaussian basis functions $\phi_i(s)$ has been widely used as the basis with a width $h_i > 0$ and center c_i which is evenly distributed along with the phase variable s. The total number N of the Gaussian basis needs to be set in advance. Parameters of the DMPs model denoted by γ_i can be utilized to regulate the shape of the force term and thus to regulate the shape of the motion trajectory. It can be seen that the specific task/skill can be parametrized by a set of parameters associated with corresponding motion variables.

Note that Eqs. (5.1), (5.2), and (5.4) are used for each separate DoF, where Eq. (5.3) is shared across all the DoFs. For example, for the encoding of a 7-DoF robot arm movements, all seven movement trajectories (represented by Eq. (5.1)) are driven by the same phase variable such that the duration synchronization of the whole system can be strictly guaranteed. Furthermore, Eqs. (5.1)–(5.3) can be coupled with additional spatial and temporal terms for specific usages [26–28].

5.2.1.2 Learning the DMPs model

The learning of the DMPs model here refers to the learning of the parameters γ_i as described above. Given one demonstration data consisting of a movement angle and a velocity trajectory $\{x_i, \dot{x}_i, \ddot{x}_i\}_{i=1}^T$, the following three steps are performed accordingly to adapt the parameters of the DMPs model. If the joint velocity and acceleration are not available, we can directly derive the joint angle x at each time step.

(i) *First step*: $s(t)$ is computed by integrating the canonical system as shown in Eq. (5.3).

(ii) *Second step*: we construct a target function f_{target} based on Eq. (5.1).

(iii) *Third step*: locally weighted linear regression is utilized to solve the following equation, and thus to obtain the model parameter γ_i:

$$\min(\sum (f_{target}(s) - f(s))^2). \tag{5.6}$$

We choose the DMPs as the task representation model because of a number of advantages. The first one of these lies in that it can be efficiently learned and generalized to other similar task situations. The second one is that it can represent any shape of trajectories theoretically. Furthermore, the optimization of the parameters can be easily formed as a reinforcement learning problem [15,29,30], which, however, will not be considered in this work.

5.2.2 Admittance control model

Considering a robotic arm with n DoFs. Its dynamics can often be expressed in joint space as follows [31]:

$$M(x)\ddot{x} + C(x, \dot{x})\dot{x} + G(x) = \tau_c + J^T F, \tag{5.7}$$

where x, $\dot{x}$, and $\ddot{x}$ represent the joint angle, velocity, and acceleration, respectively. $M(x)$ represents the inertia matrix. $C(x, \dot{x})$ denotes the Coriolis and centrifugal forces, and $G(x)$ is the gravity force. F represents the force applied by the environment in a specific interaction. The robotic arm dynamics $\tau_{dyn} = M(x)\ddot{x} + C(x, \dot{x})\dot{x} + G(x)$ are assumed known; they are provided by the robot manufacturer, or they are identified based on nonlinear adaptive control techniques (see, e.g., [32]). J represents the robotic arm Jacobian matrix. τ_c represents the input control torque which will be detailed in the following section.

5.2.3 Learning of compliant movement profiles based on biomimetic control

5.2.3.1 Robotic compliant movement representation

Given the above robotic arm dynamics, we separate the control input τ_c into two parts. Inspired by the human arm motor learning regulations, the control command can be represented by the sum of a feedforward command and a feedback command [33,34]:

$$\tau_c = u + \upsilon, \tag{5.8}$$

where u represents the feedforward torque vector and υ represents the impedance (i.e., the feedback command vector), which is defined as a PD form in this work:

$$\upsilon = Ke + D\dot{e}, \tag{5.9}$$

with the angle error and the velocity error

$$e = x_r - x, \tag{5.10}$$

$$\dot{e} = \dot{x}_r - \dot{x}, \tag{5.11}$$

where x_r and $\dot{x}_r$ represent the reference joint angle and the reference joint velocity, respectively, which are the outputs of the DMPs model as explained in Section 5.2.1. K and D represent the stiffness matrix and the damping matrix, respectively. The stiffness is a diagonal matrix, i.e.,

$$K = diag\{k_1, k_2, \cdots, k_n\}, \tag{5.12}$$

where each of the elements corresponds to each joint stiffness of the robotic arm, and will be adapted according to the task requirements. The damping matrix is also a diagonal matrix determined by

$$D = diag\{d_1, d_2, \cdots, d_n\}. \tag{5.13}$$

Until now, the compliant movements include the movement trajectories, the stiffness profiles, and the feedforward torque profiles. We conclude the compliant movements as follows:

$$\Omega = \{x_i, \dot{x}_i, K_i, \upsilon_i\}_{i=1}^{T}. \tag{5.14}$$

5.2.3.2 Adaptation law

The adaptation strategy of the variable impedance control is shown in Fig. 5.1. It shows that the feedforward torque and the impedance need to be updated at the same time within one control loop.

In the human motor learning, the goal is to minimize the movement error and the effort. Accordingly, we consider the following cost function [22]:

$$J_{cost} = \frac{\alpha}{2}\upsilon^T \upsilon + \gamma \sum_{i=1}^{N} u_i, \tag{5.15}$$

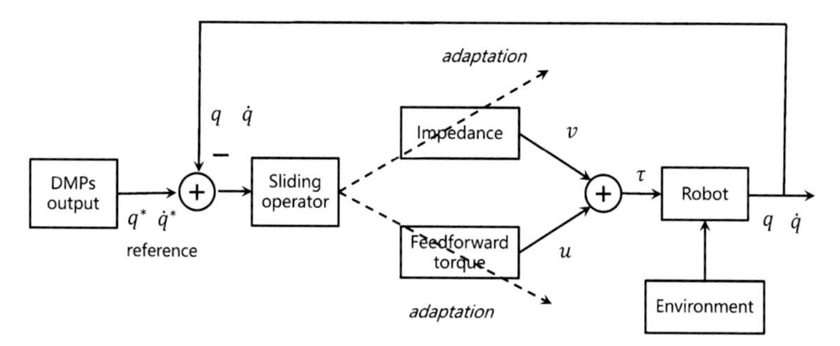

FIGURE 5.1 The control diagram for the learning of impedance and feedforward force; they are simultaneously learned based on the errors between the reference and the current robotic motion states. This figure is adapted from [20].

where the forward term is the cost for the movement feedback and the last term is the cost for the feedforward. α and γ are positive constant coefficients which will be later extended as vectors for our usage.

With [22], each element in the feedback vector is assumed as a linear function increasing in both directions:

$$v_i = \varepsilon_{i,+} + \zeta \varepsilon_{i,-}, \quad \zeta \in (0, 1), \tag{5.16}$$

where $\varepsilon_{i,+}$ and $\varepsilon_{i,-}$ represent the positive part and the negative part, respectively.

The sliding error is defined as

$$\varepsilon_i = \pi(e_i + \delta \dot{e}_i) \tag{5.17}$$

with the positive constant coefficients π and δ.

The learning problem can be solved by the gradient descent law

$$\triangle u^t = \alpha v^t - \gamma \begin{bmatrix} 1 \\ \vdots \\ 1 \end{bmatrix}_N . \tag{5.18}$$

Then, based on the assumption (5.16), the above equation can be split into three parts, i.e.,

$$\triangle u^t = \frac{\alpha}{2}(1-\zeta)\varepsilon^t + \frac{\alpha}{2}(1+\zeta) \mid \varepsilon^t \mid - \gamma \begin{bmatrix} 1 \\ \vdots \\ 1 \end{bmatrix}_N , \tag{5.19}$$

with

$$\mid \varepsilon \mid = (\mid \varepsilon_1 \mid, \mid \varepsilon_2 \mid, \cdots, \mid \varepsilon_N \mid), \tag{5.20}$$

finally yielding the following update law [22]:

$$\triangle K^t = \beta \mid \varepsilon^t \mid - \gamma, \tag{5.21}$$

$$\triangle u^t = \alpha \varepsilon^t - (1 - \mu) u^t, \tag{5.22}$$

where β is a positive constant gain coefficient and $\mu \in (0, 1)$ is a relaxation factor. The stiffness K_i may become negative, therefore; K_i are limited to a proper range $[K_{i,min}, K_{i,max}]_{i=1}^N$.

In this work, the following three aspects are modified for our usage:

(i) For the convenient control of a robotic manipulator with multiple DoFs, we first extend the constant coefficients to vectors. α, β, and γ are shared for all the muscles in the motor learning. However, the joints are separated and not coupled together for the robot arm. Accordingly, the objective function is thus adapted to [35]

$$J_{obj} = \min(\frac{\alpha}{2} v^T v + \sum_{i=1}^{N} \gamma_i u_i) \tag{5.23}$$

with N dimension vectors α and γ.

(ii) The last term of Eq. (5.21) is adjusted based on the sliding error instead of constant values by

$$\gamma_i = \frac{a}{1 + b \mid \varepsilon_i \mid}, \tag{5.24}$$

where a and b are predefined positive constant coefficients. With this formulation, γ_i can regulate the increment impedance of the corresponding joint.

(iii) The relaxation factor is also not fixed but adapted based on the error. Eq. (5.22) is accordingly modified as

$$\triangle u^t = \alpha \varepsilon^t - \frac{1}{\exp(\mid \varepsilon \mid)} u^t. \tag{5.25}$$

The stiffness and feedforward torque are updated by using Eqs. (5.21), (5.24), and (5.25) at each time step along the movement trajectories.

5.3 Experimental validation

In order to verify the effectiveness of the proposed approach, the following three experiments have been performed. For all experiments, the robotic arm is controlled in joint space under the torque control model.

5.3.1 Simulation task

The first experiment is a simulation task performed based on a simulated Baxter robot in the Gazebo environment.[1] The Baxter robots have two arms, and each of them has seven joints, i.e., a 2-DoF shoulder joint (S0, S1), a 2-DoF elbow

---

[1] http://sdk.rethinkrobotics.com/wiki/Baxter_Simulator.

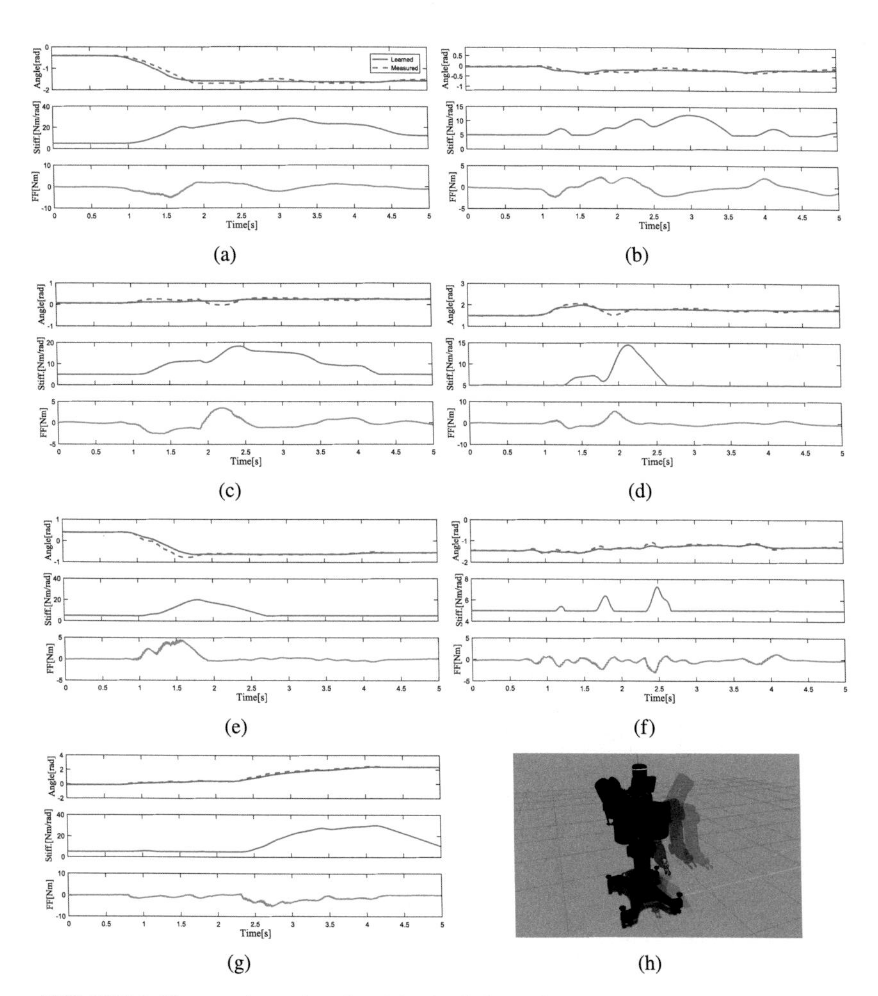

FIGURE 5.2 The experimental results of the simulation task.

joint (E0, E1), and a 3-DoF wrist joint (W0, W1, W2). The task is a simulation of a "water-pouring" movement in which all the joints of the robot are involved. In the simulation, the robotic arm is controlled under a free motion manner (see Fig. 5.2 (H)), i.e., no external force is applied onto the manipulator.

The parameter settings for the DMPs model are: $\tau = 1$, $\alpha_s = 1$, $k = 100$. The parameter settings for this simulation task are: $\pi = 1.2$, $\delta = 0.008$, $\beta = [0.3, 0.3, 0.3, 0.3, 0.3, 0.3, 0.5]^T$, $\alpha = [5.0, 5.0, 5.0, 5.0, 5.0, 5.0, 8.0]^T$, $a = 0.05$, $b = 10$, and $[K_{i,min}, K_{i,max}] = [2, 200]_{i=1}^{7}$.

The simulation results are shown in Fig. 5.2 (A–G). It shows the movement trajectories, the stiffness, and the feedforward torque profiles of the seven joints. The straight dark lines are the trajectories learned by the DMPs model, and the dash ones are the measured angle trajectories during the reproduction of the task. It also shows the adaptation of both stiffness and feedforward during the evolution of the movement trajectories. The movement, impedance, and force/torque of all the joints are adapted. Almost all the stiffness profiles follow the same pattern: increasing from a small value and then decreasing to a certain value, which is basically consistent with the human experience. Besides, the adaptation in time coordinate is also demonstrated as expected. Taking the last joint (W2) as an example, the stiffness and feedforward keep constant during the reaching phase and thereafter they adapt to complete the "pouring" step.

5.3.2 Handover task

The second task is implemented on a real-word Baxter robot which has the same structure as the simulated Baxter robot in the Gazebo. First, a human demonstrator teaches the robot how to hand over an object to another human partner, during which the robot arm states are recoded. The recorded data are then modeled by DMPs with the same parameters used in the first task. Subsequently, the robot plays back the handover movement of the handover task without the human guidance again (see Fig. 5.3 (H)), during which the stiffness and the feedforward are learned at each time step.

The parameter settings for this task are as follows: $\pi = 1.3$, $\delta = 0.008$, $\beta = [4.0, 2.5, 1.0, 3.5, 0.3, 0.3, 0.3]^T$, $\alpha = [5.0, 5.0, 5.0, 5.0, 5.0, 5.0, 5.0]^T$, $a = 0.8$, $b = 10$, and $[K_{i,min}, K_{i,max}] = [15, 200]_{i=1}^{7}$.

The experimental results of this task are shown in Fig. 5.3 (A–G). Again, it shows the robot is able to complete the task while keeping as compliant as possible: increasing stiffness if needed to compensate for the movement error and keeping it low if not necessary. Unlike in the first task, not all the joints are needed to adjust their impedance and feedforward values. If one joint is not particularly involved, its impedance keeps at the smallest value.

5.3.3 Sawing task

The third task is the HRC sawing task. The setup for this task is shown in Fig. 5.4 (E). A saw is connected to one of the robotic end points through a specifically designed module. The robot and the human partner collaborate to saw a piece of wood which is mounted onto the table. In this task, the reference angles remain unchanged and the reference velocities remain zero.

The settings for the sawing task are given as follows: $\pi = 1.3$, $\delta = 0.01$, $\beta = [5.0, 2, 0.4, 0.75, 0.4, 0.6, 0.75]^T$, $\alpha = [5.0, 5.0, 5.0, 5.0, 5.0, 5.0, 5.0]^T$, $a = 0.6$, $b = 12$, and $[K_{i,min}, K_{i,max}] = [5, 200]_{i=1}^{7}$.

The experimental results of this task are shown in Fig. 5.4 (A–D). It shows the measured angles, the joint torques, and the stiffness of the seven joints. Three joints (i.e., S1, E1, and W1) are mainly involved during the task execution, while the others (i.e., S0, E0, W0, and W2) almost keep constant (see Fig. 5.4 (D))

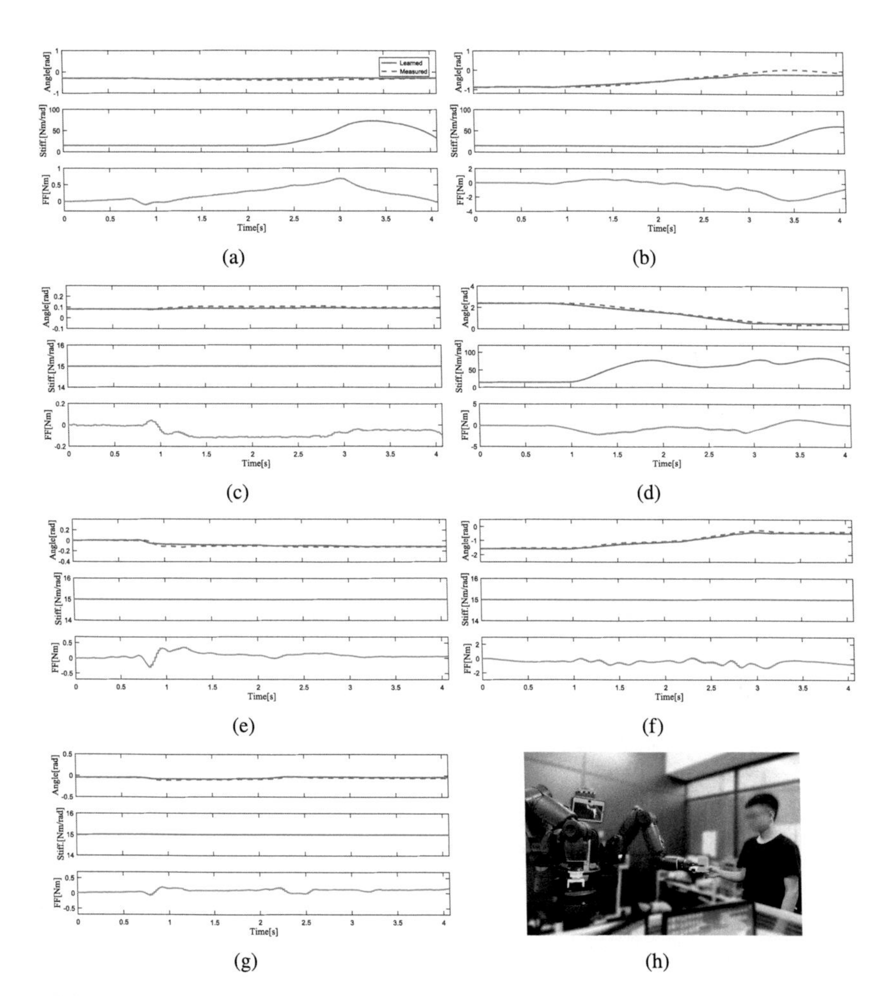

FIGURE 5.3 The experimental results and setup of the handover task.

since these joint angles do not change much during the sawing periods. It can be seen that the stiffness profiles of the three joints could be automatically adapted to the human partner during the sawing process. When the human partner increases his/her strength to pull the saw, the robot arm impedance increases gradually. When the robot arm impedance becomes large to some extent, the robot would start to pull it back while the human partner reduces his/her arm strength. This period then repeats over and over until the task is finished finally.

5.4 Human robot collaboration based on admittance control

This section will introduce a flexible HRC framework based on variable admittance control.

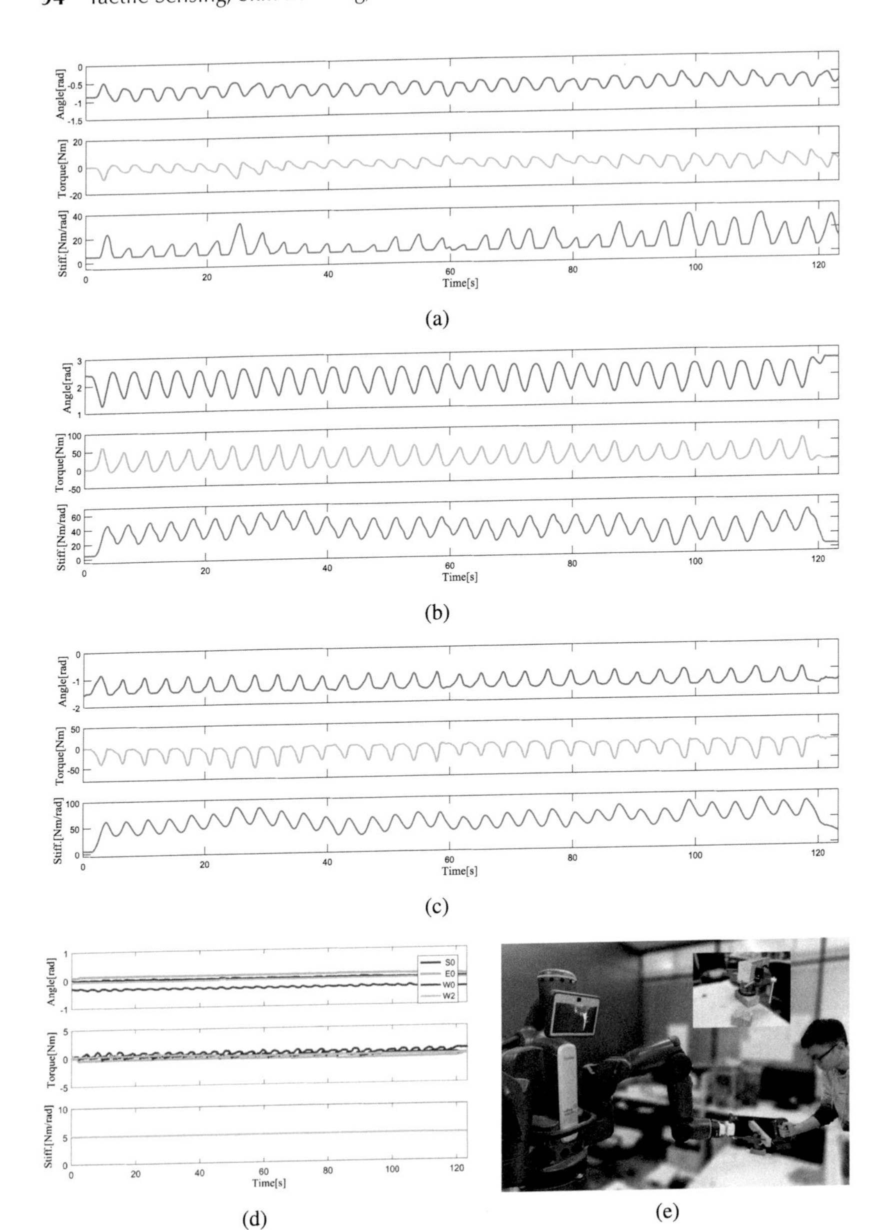

FIGURE 5.4 The experimental results and the setup of the collaborative sawing task.

5.4.1 Principle of human arm impedance model

Without loss of generality, the dynamic model of the human arm impedance can be described as follows:

$$M_H \ddot{x}_h + D_H \dot{x}_h + K_H x_h = F, \tag{5.26}$$

where x_h, $\dot{x}_h$, and $\ddot{x}_h$ denote the human arm end point position, velocity, and acceleration, and M_H, D_H, and K_H represent the human arm mass, damping, and stiffness terms. F denotes the interaction force.

The mass term M_H of the human arm is regarded as a constant in the following description by ignoring the negligible effect of the muscle mass distribution on M_H in a certain range of a predetermined arm configuration [36]. A dimensionality-reduction estimation method of the human arm stiffness is utilized to estimate K_H in (5.26) [37]:

$$K_H(p,q) = J_H^{+T}(q)[v(p)\overline{K_J} - G(q)]J_H^{+}(q), \tag{5.27}$$

where p is the coactivation index of upper arm muscles, q is the joint angle of the human arm, $J_H(q)$ is the human arm Jacobian matrix, $v(p)$ is a muscle-activation-dependent index, and $\overline{K_J}$ is regarded as a constant matrix representing the human arm minimal joint stiffness. $G(q) = \frac{\partial J_H^T(q)F}{\partial q}$ denotes the influence of the interaction forces on the stiffness transformation. In this representation, it is clear that the stiffness term would vary with different muscle activation levels and human arm configurations. In addition, the processing of damping matrix D_H is introduced in the next subsection.

5.4.2 Estimation of stiffness matrix

According to [38], the human arm configurations can be determined by three parameters of a triangle model, i.e., the direction of the human arm plane, the direction of the upper arm, and the angle between forearm and upper arm.

As shown in Fig. 5.5, a representative human arm Denavit–Hartenberg (D-H) model [39] is modified to denote the simplified human arm kinematics. The base coordinate is located at the shoulder. Directions x_0 and z_0 represent the axis of the frame, horizontal right and horizontal upwards. L_{ua} and L_{fa} represent the lengths of the upper arm and forearm. The angles of the first four joints, i.e., three shoulder joints and a single elbow joint, can be calculated through the IK algorithm [38].

Generally, we need to track the joint angles of the human arm to calculate the Jacobian matrix J_H. The real-time tracking of human arm joint angles can be achieved by transforming quaternions obtained by gyroscopes of Myo armbands to joint angles in a triangle model mention above. Therefore, the Jacobian matrix J_H of the human arm can be obtained according to the human arm D-H model.

In order to identify the minimal joint stiffness matrix $\overline{K_J}$, multiple identification experiments are conducted with the classic perturbation method under different arm configurations and muscle activation levels (see [40] for details). The restoring force is measured by the force sensor and the dynamic relationship between the end point displacement of the human arm and the recorded

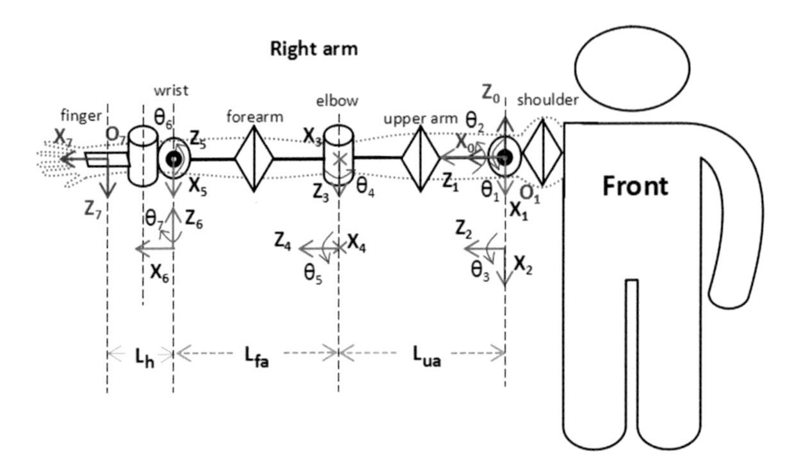

FIGURE 5.5 The human arm D-H model (modified from [38]).

force is described as follows [37]:

$$F = \begin{bmatrix} F_{x_c} \\ F_{y_c} \\ F_{z_c} \end{bmatrix} = \begin{bmatrix} L_{xx} & L_{xy} & L_{xz} \\ L_{yx} & L_{yy} & L_{yz} \\ L_{zx} & L_{zy} & L_{zz} \end{bmatrix} \begin{bmatrix} \triangle X_c \\ \triangle Y_c \\ \triangle Z_c \end{bmatrix}, \tag{5.28}$$

where F_{x_c}, F_{y_c}, and F_{z_c} denote the human arm end point interaction force and $\triangle X_c$, $\triangle Y_c$, and $\triangle Z_c$ denote the displacement of the end point. We employed a second-order linear model to identify the transfer function L_{ij},

$$L_{ij} = M_{Hij}s^2 + D_{Hij}s + K_{Hij}, \quad s = 2\pi f\sqrt{-1}. \tag{5.29}$$

The parameters M_H, D_H, and K_H of the transformation function L are identified using the least squares method. Consequently, all the stiffness matrices K_H identified by (5.29) using the minimum coactivation level experimental results are utilized to obtain $\overline{K_J}$ by minimizing the Frobenius norm:

$$\| \overline{K_J} - J_H^T(q)K_H(p,q)J_H(q) - G(q) \| . \tag{5.30}$$

In addition, the coactivation index p of upper arm muscles is obtained through the following method. As shown in Fig. 5.6, the envelope is extracted from the EMG signals using a moving average process and a low pass filter. We use only two channels close to the triceps and biceps for convenience. Therefore, the following equation is used to indicate the coactivation level p of the upper arm muscles:

$$p(s_p) = \frac{1}{W_s}(\sum_{s_p=1}^{W_s-1} A_B(t-s_p) + \sum_{s_p=1}^{W_s-1} A_T(t-s_p)), \tag{5.31}$$

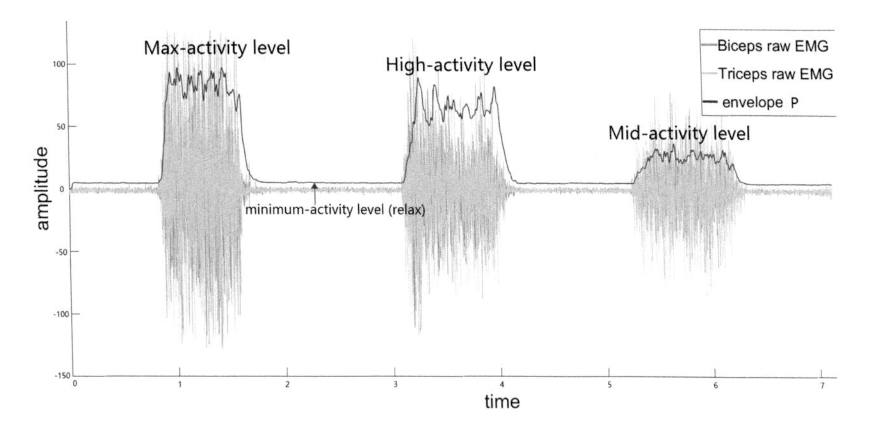

FIGURE 5.6 The envelope extracted from EMG signals using the proposed method.

where W_s is a predetermined window size, A_B and A_T are the amplitude of p, and s_p and t represent the current sample point and sampling time.

Furthermore, the restoring force and the human arm end point displacement recorded from experiments with medium and high coactivation levels are employed to calculate the constant parameters of $v(p)$ in (5.33), i.e., χ_1 and χ_2, by minimizing the Frobenius norm in (5.32):

$$\| v(p)\overline{K_J} - J_H^T(q)K_H(p,q)J_H(q) - G(q) \|, \tag{5.32}$$

$$v(p) = -\frac{\chi_1[e^{-\chi_2 p} - 1]}{e^{-\chi_2 p} + 1} + 1. \tag{5.33}$$

Therefore, the human arm end point stiffness matrix K_H is calculated online based on (5.27) using the human arm Jacobian matrix J_H, the muscle-activation-dependent index $v(p)$, and the minimal joint stiffness matrix K_J.

Utilizing (5.28) and (5.29), all the damping matrices D_H of the minimum-activity, mid-activity, and high-activity level trails can be obtained. Based on the analysis of the offline experimental results of the damping matrices, it can be observed that the variation of the damping value is not obvious within a certain range of the muscle activation level centered at each level. Corresponding to a different muscle-activation-dependent index $v(p)$ in each arm configuration, we consider that the continuous numerical curve of the damping term can be discretized into variable constant matrices based on analysis and observation of the offline experimental results. Therefore, a look-up table among the damping matrices, arm configurations, and muscle activation levels in a certain range according to the offline experimental results can be established. In the human–robot sawing scenario, the human arm configuration can be kept to only change in a small range while the muscle activation level is required to change in a wide range. Therefore, the look-up table can only consider the corresponding relations between the muscle activation levels and the obtained damping matrices.

5.4.3 Stiffness mapping between human and robot arm

Generally, the interaction force F transforms to torque τ_c for control in joint space according to $\tau = J_r^T F$. Therefore, the stiffness in joint space of the cooperative robot arm can be derived from the human arm end point stiffness in Cartesian space according to the following equation [41]:

$$K_r^q \approx J_r^T K_H J_r, \tag{5.34}$$

where K_r^q denotes the mapping joint stiffness of the robot arm and J_r is the robot arm Jacobian matrix. The "approximately equal" symbol means that the joint stiffness of the robot arm is mapping from the end point stiffness of the human arm, not exactly mapping from the end point stiffness of the robot arm.

For the purpose of realizing an HRC with master–slave conversion, stiffness mapping between the human arm and the robot arm should be adjusted. The stiffness mapping strategy is adjusted as follows:

$$K_r^q \approx K_r^a - J_r^T K_H J_r, \tag{5.35}$$

where K_r^a is a constant stiffness matrix for adjustment. With the help of the time-varying human arm end point stiffness and the adjustment factor, the mapping robot arm joint stiffness can be obtained. In addition, for the stability of the robot arm, limitation ranges of the robot arm joint stiffness are presented as $K_{min} \leq K_r^q \leq K_{max}$. K_{max} and K_{min} are the maximum stiffness and the minimum stiffness of the given range of the robot arm.

We employ a PD controller with variable gains to drive the robot arm cooperating with the human arm under a good tracking performance,

$$\tau_r = D(\dot{q}_d - \dot{q}) + K_r^q (q_d - q), \tag{5.36}$$

where τ_r is the control input of the robot arm and q_d and q represent the desired robot arm joint position and the actual joint position. $D = k_s K_r^q$, with a properly chosen scale factor k_s, $k_s = [k_{s_1}, k_{s_2}, k_{s_3}, ..., k_{s_n}]$, and n represents the robot arm DoFs.

5.5 Variable admittance control model

Generally, a typical dynamic equation of the robot end-effector which is realized by admittance control is described as follows:

$$F = I_r \ddot{x} + D_r \dot{x} + K_r (x - x_0), \tag{5.37}$$

where F denotes the interaction force, x, $\dot{x}$, and $\ddot{x}$ denote the robot arm position, velocity, and acceleration, x_0 is the initial position of the robot arm end-effector, and I_r, D_r, and K_r are desired virtual inertia, damping, and stiffness matrices, respectively. In practical implementations, (5.37) is modified as a more simplified model retaining stiffness and damping:

$$F = D_r \dot{x} + K_r (x - x_0). \tag{5.38}$$

Consider the following general time-invariant linear system:

$$\dot{\zeta} = X\zeta(t) + Yu(t), \tag{5.39}$$

where ζ represents the robot arm state, $u(t)$ is the system input at t time, and X and Y are defined as known matrices. In addition, the robot arm state ζ is defined as

$$\zeta = \begin{bmatrix} \dot{x}^T & x^T & \delta^T \end{bmatrix}^T, \tag{5.40}$$

where δ denotes the state of a linear system to generate the reference task goal, x_f, which provides the feasibility to implement the optimized trajectory tracking. In particular, this linear system is described as follows:

$$\begin{cases} \dot{\delta} = U - \delta, \\ x_f = V\delta, \end{cases} \tag{5.41}$$

where U and V are known matrices needed to be determined. Considering the human arm state (5.26), the matrices X and Y are defined as follows in order to match the corresponding impedance parameters of the human arm:

$$X = \begin{bmatrix} -M_H^{-1} D_H & -M_H^{-1} K_H & 0 \\ I_n & 0 & 0 \\ 0 & 0 & U \end{bmatrix}, \qquad Y = \begin{bmatrix} -M_H^{-1} \\ 0 \\ 0 \end{bmatrix}. \tag{5.42}$$

The mass, damping, and stiffness of the human arm are all included in matrices X and Y, which are then used to solve the desired admittance model. Furthermore, an LQR is employed to minimize the following cost function [42], [43]:

$$C = \int_0^\infty [\dot{x}^T Q_1 \dot{x} + (x - x_f)^T Q_2 (x - x_f) + u^T R u] dt, \tag{5.43}$$

where R, Q_2, and Q_1 represent the input of the system, the trajectory tracking error, and the velocity weighting matrices.

Based on Eqs. (5.40), (5.41), and (5.43), the cost function (5.43) can be rewritten as

$$C = \int_0^\infty [\dot{\zeta}^T Q \dot{\zeta} + u^T R u] dt, \tag{5.44}$$

where $Q = \begin{bmatrix} Q_1 & 0 & 0 \\ 0 & Q_2 & -Q_2 V \\ 0 & -V^T Q_2 & V^T Q_2 V \end{bmatrix}$.

The essence of the LQR is to obtain an optimal feedback of the system to minimize the cost function (5.44). According to this, the system input of the robot arm is defined as follows:

$$u = -K_f \zeta, \tag{5.45}$$

where K_f is a state feedback gain matrix.

We use the interaction force F as the input u defined in (5.45). To understand (5.45) in the sense of admittance control, we assume that the optimal control has been achieved, and the desired admittance model is described as

$$\begin{aligned} F &= -K_f \zeta = -R^{-1} B^T P \zeta \\ &= -R^{-1} P_{11} \dot{x} - R^{-1} P_{12} x - R^{-1} P_{13} (V^T V)^{-1} x_f, \end{aligned} \tag{5.46}$$

where P_{11}, P_{12}, and P_{13} denote the three submatrices in the first row of the matrix P, which is the solution of the following equation:

$$PX + X^T P - PYR^{-1} Y^T P + Q = 0. \tag{5.47}$$

According to the measured interaction force and a predetermined reference task goal, the desired trajectory and velocity of the robot arm end-effector can be calculated. We can utilize the robotic IK algorithm to obtain the desired robot arm joint trajectory q_d.

5.6 Experiments

5.6.1 Test of variable admittance control

The test of fixed admittance control and the test of variable admittance control are conducted in this subsection under the left arm of the Baxter robot.

In the first test, the performance of the robot arm end-effector with different fixed admittance parameters (high/low stiffness and damping) in the y-axis is under consideration. The stiffness and damping terms of the admittance model are set as $K_r = 200$ N/m, $D_r = 10$ N/m and then they turn to $K_r = 50$ N/m, $D_r = 5$ N/m at 5 s. Two groups of stiffness and damping terms mention above are set according to the stable range of the robot arm, reflecting the relatively high stiffness and relatively low stiffness, respectively. The robot arm trajectory is set as 0.7 m $\rightarrow$ 0.3 m in the x-axis. The external force $F \approx 13$ N is applied in the process mentioned above at about 1.5 s and 6.5 s in the y-axis and it is kept for about a half second. The experimental results are shown in Fig. 5.7.

According to the experimental results, the influences of admittance parameters on responses of the robot arm to disturbances are reflected intuitively. At the

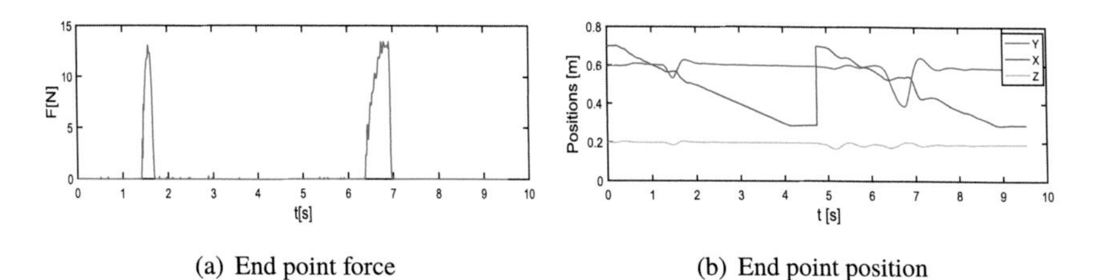

(a) End point force (b) End point position

FIGURE 5.7 The external force applied in the y-axis and the position of the robot arm end-effector.

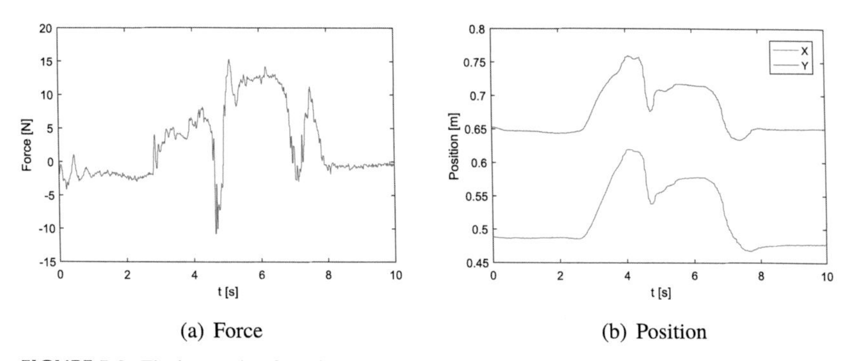

(a) Force (b) Position

FIGURE 5.8 The interaction force in the x-axis and the position of the end-effector with fixed high admittance parameters.

first stage, with the high stiffness and damping, the robot arm end point position is dropped to 0.53 m when the external force is applied. At the second stage, with the low stiffness and damping, the robot arm end point position is dropped to 0.398 m when the external force is applied. The robot arm with high stiffness and damping is less affected, and we can also see that the trajectory of the y-axis suffered less interference. In this case, giving the end-effector of the robot arm high stiffness and damping in directions without movement in an HRC task can strengthen the stability and enhance the performance.

In the second test, the performance of the robot arm end-effector with variable admittance parameters in the $x'(y')$-axis is under consideration. A comparative test which simulates the pulling back process is conducted to verify the validity of the variable admittance control method. The scenario of this experiment is that the operator brakes suddenly with high end point stiffness and damping in the process of the saw pulling back. We track the interaction force and position to verify the proposed method. The robot arm trajectory targets are set as 0.76 m to 0.65 m in the x-axis and as 0.62 m to 0.48 m in the y-axis. Thus the initial position x_0' is set as 0.98 m and the trajectory task goal is 0.98 m to 0.81 m. The robot arm reference trajectory is determined by (5.41) with $U = 1$ and $V = 0.81$. Therefore, the reference end point trajectory is $x_f = 0.81 + 0.18e^{-t}$ and its goal position is $x_f = 0.81$ m ($t \to \infty$).

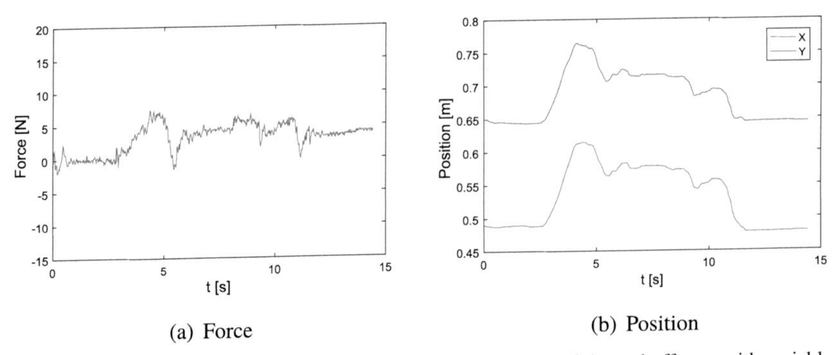

(a) Force (b) Position

FIGURE 5.9 The interaction force in the x-axis and the position of the end-effector with variable admittance parameters.

The experimental results are shown in Fig. 5.8 and Fig. 5.9. Fig. 5.8 shows the interaction force and end point trajectory of the robot arm with fixed high admittance parameters. We can see that the operator brakes suddenly at about 5 s and the interaction force changes from 5 N to -10 N and it bounces back to about 15 N. It is obvious that the change range of the interaction force is too large and it would cause unnecessary vibrations. Fig. 5.9 shows the interaction force and end point trajectory of the robot arm with variable admittance parameters. With the help of variable admittance control, the robot arm admittance model changes corresponding to the impedance of the human arm. Once the operator's arm suddenly becomes rigid, the admittance parameters of the robot arm shrink in order to cater to the operator's changes. As a result, we can see that the operator brakes suddenly at about 5 s and 9 s and the interaction force changes from about 5 N to 0 N and it bounces back to about 4 N. Fig. 5.8 (B) shows that when the operator braked at about 5 s, the position dropped from 0.75 m to 0.66 m in the y-axis and from 0.62 m to 0.53 m in the x-axis, which corresponds to the high interaction force between the robot arm and the human arm at about 5 s in Fig. 5.8 (A). In addition, Fig. 5.9 (B) shows the smooth braking process with variable admittance control. This test shows that variable admittance control can result in a satisfied balance between the interaction force and the robot arm end-effector position.

5.6.2 Human–robot collaborative sawing task

Based on general experience, the performance in a two-person sawing task under a master–slave structure could be better. In this case, the HRC sawing task can be split into two stages. In the first stage, the operator plays the role of the master to pull the saw along the blade and the robot arm is compliant to the master in the motion axis in order not to oppose the operator's effort. In the second stage, the robot arm is changed to be the master to pull the saw along the blade and the operator is compliant to the robot arm in turn.

In this subsection, a set of cooperative experiments is conducted to verify the proposed control method. The extracted human arm configurations and the muscle activation level data are used to regulate the human arm impedance parameters and the end point stiffness of the human arm is utilized to map the joint stiffness of the robot arm. The force feedback is employed as the input of the robot arm end point admittance control model.

For convenience, the robot end point task goal in the $x'(y')$-axis is predetermined at a point $x = 0.75$ m and $y = 0.65$ m. Therefore, the robot arm pulls the saw to the task goal when it is the robot arm's turn to be master. Conversely, when the human arm pulls back the saw, the robot arm is compliant and allows the human arm to pull the saw to anywhere within the control range. The rotational stiffness of the human arm end point is set at a high value, $K_{H_{rot}} = 1000$ N/m, in order to maintain the orientation of the robot arm end-effector with a high mapping rotational joint stiffness. According to the fixed admittance control test results, the parameters of the admittance model along the direction perpendicular to the direction of motion $x'(y')$ is set in high stiffness and damping in order to avoid interference with the sawing performance.

The first experiment is conducted utilizing a fixed high–low stiffness switching method. In this case, the stiffness of each joint can be set to an arbitrary value under the premise of stability. According to the actual situation, the stiffness of joint $S1$ is fixed as $K_{max} = 200$ Nm/rad and $K_{min} = 50$ Nm/rad, the stiffness of joint $E1$ is fixed as $K_{max} = 150$ Nm/rad and $K_{min} = 40$ Nm/rad, and the stiffness of joint $W1$ is fixed as $K_{max} = 35$ Nm/rad and $K_{min} = 10$ Nm/rad. The trigger value of the stiffness switching is $K_x = 1$ kN/m, which means that the robot joint stiffness switches to the maximum value when the human arm stiffness $K_x \geq 1$ and it switches back to minimum value when $K_x \leq 1$.

The results of the first experiment are shown in Fig. 5.10. The first figure shows the estimated human arm end point stiffness, while the second figure shows the stiffness of three key joints $S1$, $E1$, and $W1$ used in the sawing task. The stiffness of the robot joints $S1$, $E1$, and $W1$ switches to 200 Nm/rad, 150 Nm/rad, and 35 Nm/rad when the human arm stiffness $K_x \geq 1$ and they switch back to 50 Nm/rad, 40 Nm/rad, and 10 Nm/rad when $K_x \leq 1$. These two figures reflect the master–slave structure of the HRC sawing task. The third figure shows the interaction force between the human arm end point and the robot arm end-effector. We can see the interaction force at the switching point, for example at about 11 s and 12 s, it bounces to 67 N and drops down to −46 N. The interaction force is very high at the switching moment, and it would cause the feeling of discomfort to the operator. In addition, the sawing performance will also be affected. The fourth figure shows the position of the robot arm end-effector in Cartesian space. According to Fig. 5.10, we can see that the sawing performance is not smooth enough and the blade is stuck at about 55 s in the sawing process.

The second experiment is conducted using the proposed method. The robot arm joint stiffness changes corresponding to the human arm end point stiff-

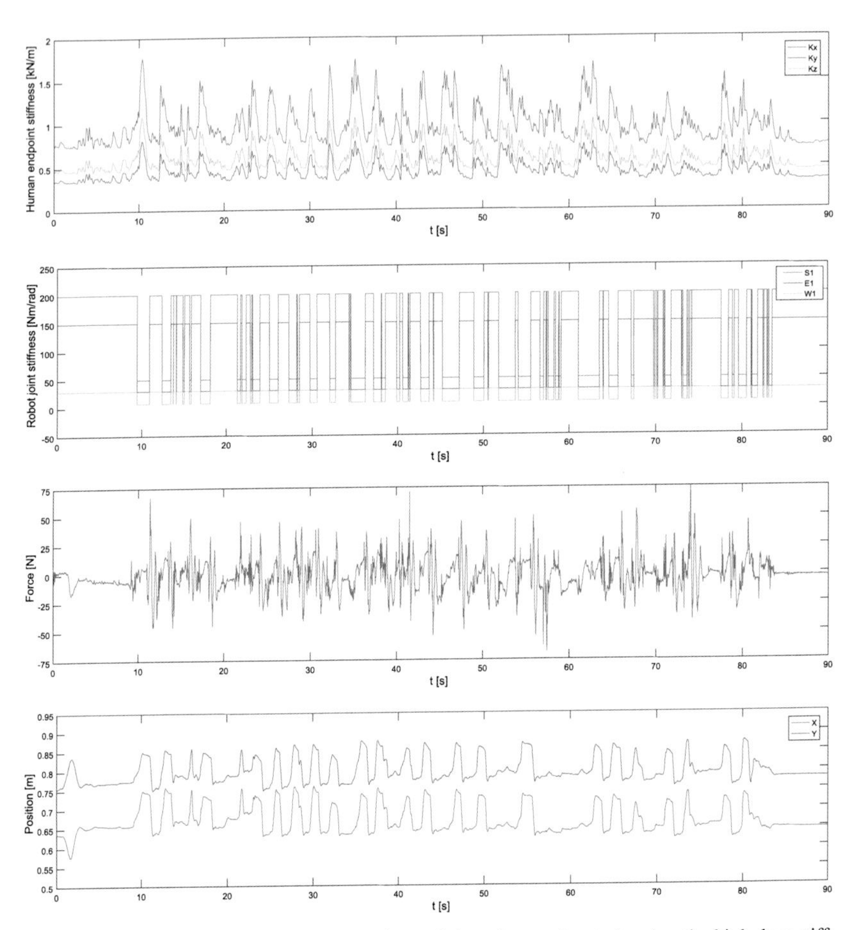

FIGURE 5.10 The results of the human–robot collaborative sawing task using the high–low stiffness switching method. The first figure shows the estimation of human arm end point stiffness. The second figure shows the stiffness of three key joints used in the sawing task. The third figure shows the interaction force between the human arm end point and the robot arm end-effector. The fourth figure shows the position of the robot arm end-effector in Cartesian space.

ness in real-time. According to the variable admittance control method test, this method is employed in the process of the robot arm pulling back.

The experimental results are shown in Fig. 5.11. The first and second figures show the estimated human arm end point stiffness and the mapping stiffness of the three key joints based on (5.35) and the constant matrix K_r^a is set as $diag\{230, 150, 50\}$. The third figure shows the interaction force during the sawing process. We can see that the interaction force is maintained between 25 N and −25 N during the master–slave transition moments. The fourth and fifth figures show the position of the robot arm end-effector. Since the direction of the sawing task is along the $x'(y')$-axis, we present the position of the x-axis and

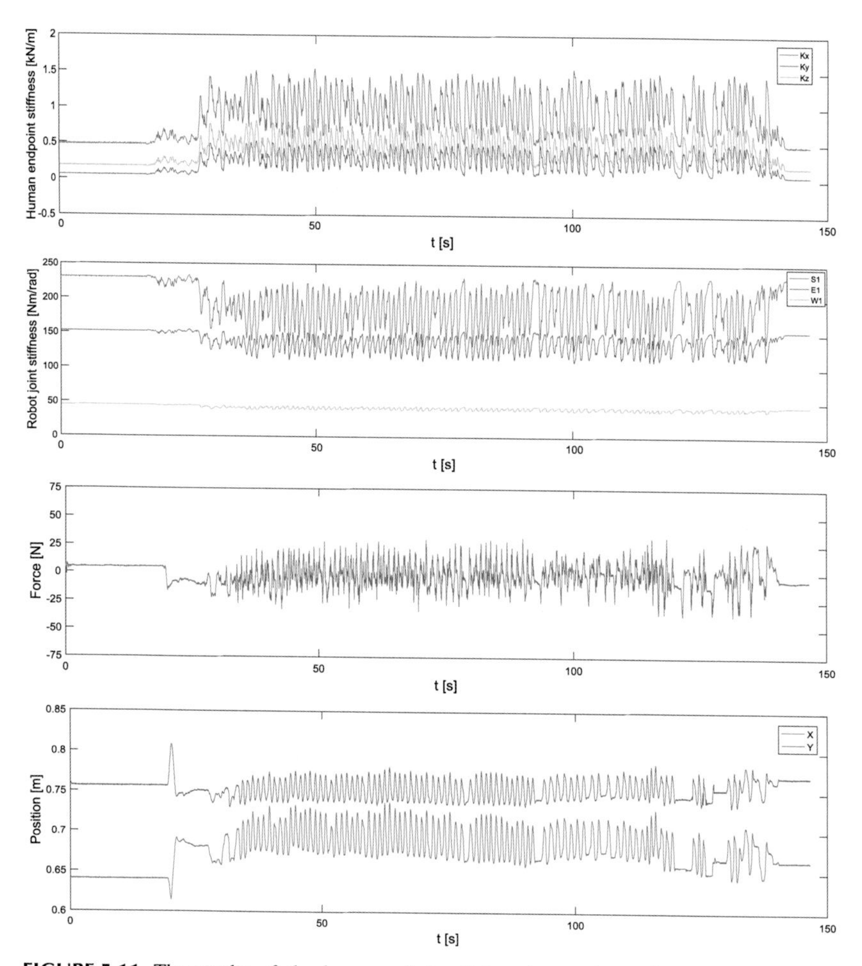

FIGURE 5.11 The results of the human–robot collaborative sawing task using the proposed method. The first figure shows the estimation of human arm end point stiffness. The second figure shows the mapping stiffness of three key joints used in the sawing task. The third figure shows the interaction force between the human arm end point and the robot arm end-effector. The fourth figure shows the position of the robot arm end-effector in Cartesian space.

the y-axis as an alternative. According to Fig. 5.11, it is obvious that the sawing performance of the proposed method is much smoother than the performance of the high–low stiffness switching method; the interaction force does not bounce to a very high level or drop down with the control law of the admittance model. We can observe from the third and fourth figures at about 75 s to 110 s that the operator is allowed to increase or decrease the sawing frequency, or even stop and resume the sawing task at any time (the smooth curve at about 125 s and 135 s in the fourth figure). And the interaction force would not vary greatly during the stop and resume process.

5.7 Conclusion

In this chapter, we introduce the application of admittance control in LfD and HRC. The experimental results show that the effect of admittance control in the two application scenarios is considerable, which provides a reference work scheme for the implementation of admittance control in the actual application scenarios.

References

[1] F. Chen, K. Sekiyama, F. Cannella, T. Fukuda, Optimal subtask allocation for human and robot collaboration within hybrid assembly system, IEEE Transactions on Automation Science and Engineering 11 (4) (2014) 1065–1075.
[2] Z. Ji, P. Li, Y. Zhou, B. Wang, J. Zang, M. Liu, Toward new-generation intelligent manufacturing, Engineering 4 (1) (2018) 11–20.
[3] H. Liu, F. Sun, X. Zhang, Robotic material perception using active multi-modal fusion, IEEE Transactions on Industrial Electronics (2018).
[4] H. Liu, F. Wang, F. Sun, B. Fang, Surface material retrieval using weakly paired cross-modal learning, IEEE Transactions on Automation Science and Engineering 16 (2) (2018) 781–791.
[5] A. Billard, S. Calinon, R. Dillmann, S. Schaal, Robot programming by demonstration, in: Springer Handbook of Robotics, Springer, 2008, pp. 1371–1394.
[6] B.D. Argall, S. Chernova, M. Veloso, B. Browning, A survey of robot learning from demonstration, Robotics and Autonomous Systems 57 (5) (2009) 469–483.
[7] S. Calinon, P. Kormushev, D.G. Caldwell, Compliant skills acquisition and multi-optima policy search with em-based reinforcement learning, Robotics and Autonomous Systems 61 (4) (2013) 369–379.
[8] M. Deniša, A. Gams, A. Ude, T. Petrič, Learning compliant movement primitives through demonstration and statistical generalization, IEEE/ASME Transactions on Mechatronics 21 (5) (2016) 2581–2594.
[9] A. Ajoudani, N. Tsagarakis, A. Bicchi, Tele-impedance: teleoperation with impedance regulation using a body–machine interface, The International Journal of Robotics Research 31 (13) (2012) 1642–1656.
[10] F. Ficuciello, L. Villani, B. Siciliano, Variable impedance control of redundant manipulators for intuitive human–robot physical interaction, IEEE Transactions on Robotics 31 (4) (2015) 850–863.
[11] K. Wen, D. Necsulescu, G. Basic, Development system for a haptic interface based on impedance/admittance control, in: Proceedings. Second International Conference on Creating, Connecting and Collaborating Through Computing, 2004, pp. 147–151.
[12] C. Yang, C. Zeng, P. Liang, Z. Li, R. Li, C.-Y. Su, Interface design of a physical human–robot interaction system for human impedance adaptive skill transfer, IEEE Transactions on Automation Science and Engineering 15 (1) (2018) 329–340.
[13] C. Yang, C. Zeng, C. Fang, W. He, Z. Li, A DMPs-based framework for robot learning and generalization of humanlike variable impedance skills, IEEE/ASME Transactions on Mechatronics 23 (3) (2018) 1193–1203.
[14] C. Fang, A. Ajoudani, A. Bicchi, N.G. Tsagarakis, Online model based estimation of complete joint stiffness of human arm, IEEE Robotics and Automation Letters 3 (1) (2018) 84–91.
[15] J. Buchli, F. Stulp, E. Theodorou, S. Schaal, Learning variable impedance control, The International Journal of Robotics Research 30 (7) (2011) 820–833.
[16] F. Stulp, O. Sigaud, Robot skill learning: from reinforcement learning to evolution strategies, Paladyn, Journal of Behavioral Robotics 4 (1) (2013) 49–61.
[17] W. He, Y. Dong, Adaptive fuzzy neural network control for a constrained robot using impedance learning, IEEE Transactions on Neural Networks and Learning Systems 29 (4) (2018) 1174–1186.

[18] L.D. Rozo, S. Calinon, D. Caldwell, P. Jiménez, C. Torras, Learning collaborative impedance-based robot behaviors, in: AAAI Conference on Artificial Intelligence, 2013.
[19] M. Racca, J. Pajarinen, A. Montebelli, V. Kyrki, Learning in-contact control strategies from demonstration, in: 2016 IEEE/RSJ International Conference on Intelligent Robots and Systems (IROS), IEEE, 2016, pp. 688–695.
[20] C. Yang, G. Ganesh, S. Haddadin, S. Parusel, A. Albu-Schaeffer, E. Burdet, Human-like adaptation of force and impedance in stable and unstable interactions, IEEE Transactions on Robotics 27 (5) (2011) 918–930.
[21] Y. Li, G. Ganesh, N. Jarrassé, S. Haddadin, A. Albu-Schaeffer, E. Burdet, Force, impedance, and trajectory learning for contact tooling and haptic identification, IEEE Transactions on Robotics (99) (2018) 1–13.
[22] G. Ganesh, A. Albu-Schäffer, M. Haruno, M. Kawato, E. Burdet, Biomimetic motor behavior for simultaneous adaptation of force, impedance and trajectory in interaction tasks, in: 2010 IEEE International Conference on Robotics and Automation, IEEE, 2010, pp. 2705–2711.
[23] A.J. Ijspeert, J. Nakanishi, S. Schaal, Movement imitation with nonlinear dynamical systems in humanoid robots, in: IEEE International Conference on Robotics and Automation, 2002. Proceedings, ICRA, 2002, pp. 1398–1403.
[24] P. Pastor, H. Hoffmann, T. Asfour, S. Schaal, Learning and generalization of motor skills by learning from demonstration, in: 2009 IEEE International Conference on Robotics and Automation, IEEE, 2009, pp. 763–768.
[25] A.J. Ijspeert, J. Nakanishi, H. Hoffmann, P. Pastor, S. Schaal, Dynamical movement primitives: learning attractor models for motor behaviors, Neural Computation 25 (2) (2013) 328–373, https://doi.org/10.1162/NECO_a_00393.
[26] J. Kober, B. Mohler, J. Peters, Learning perceptual coupling for motor primitives, in: Intelligent Robots and Systems, 2008. IROS 2008. IEEE/RSJ International Conference on, IEEE, 2008, pp. 834–839.
[27] A. Gams, B. Nemec, L. Zlajpah, M. Wächter, A. Ijspeert, T. Asfour, A. Ude, Modulation of motor primitives using force feedback: interaction with the environment and bimanual tasks, in: 2013 IEEE/RSJ International Conference on Intelligent Robots and Systems, IEEE, 2013, pp. 5629–5635.
[28] A. Gams, B. Nemec, A.J. Ijspeert, A. Ude, Coupling movement primitives: interaction with the environment and bimanual tasks, IEEE Transactions on Robotics 30 (4) (2014) 816–830.
[29] F. Stulp, J. Buchli, A. Ellmer, M. Mistry, E.A. Theodorou, S. Schaal, Model-free reinforcement learning of impedance control in stochastic environments, IEEE Transactions on Autonomous Mental Development 4 (4) (2012) 330–341.
[30] Z. Li, T. Zhao, F. Chen, Y. Hu, C.-Y. Su, T. Fukuda, Reinforcement learning of manipulation and grasping using dynamical movement primitives for a humanoidlike mobile manipulator, IEEE/ASME Transactions on Mechatronics 23 (1) (2018) 121–131.
[31] L. Sciavicco, B. Siciliano, Modelling and Control of Robot Manipulators, Springer Science & Business Media, 2012.
[32] E. Burdet, A. Codourey, L. Rey, Experimental evaluation of nonlinear adaptive controllers, IEEE Control Systems Magazine 18 (2) (1998) 39–47.
[33] E. Burdet, R. Osu, D.W. Franklin, T.E. Milner, M. Kawato, The central nervous system stabilizes unstable dynamics by learning optimal impedance, Nature 414 (6862) (2001) 446–449.
[34] E. Burdet, G. Ganesh, C. Yang, A. Albu-Schäffer, Interaction force, impedance and trajectory adaptation: by humans, for robots, in: Experimental Robotics, Springer Berlin Heidelberg, 2014, pp. 331–345.
[35] Z. Chao, Y. Chenguang, C. Zhaopeng, Bio-inspired robotic impedance adaptation for human-robot collaborative tasks, Science China Information Sciences 63 (7) (2020) 170201.
[36] A. Ajoudani, Teleimpedance: Teleoperation with Impedance Regulation Using a Body-Machine Interface, Sage Publications, Inc., 2012.
[37] A. Ajoudani, C. Fang, N.G. Tsagarakis, A. Bicchi, A reduced-complexity description of arm endpoint stiffness with applications to teleimpedance control, in: 2015 IEEE/RSJ International Conference on Intelligent Robots and Systems (IROS), 2015, pp. 1017–1023.

[38] C. Fang, X. Ding, A set of basic movement primitives for anthropomorphic arms, in: 2013 IEEE International Conference on Mechatronics and Automation, 2013, pp. 639–644.

[39] X. Ding, C. Fang, A novel method of motion planning for an anthropomorphic arm based on movement primitives, IEEE/ASME Transactions on Mechatronics 18 (2) (2013) 624–636, https://doi.org/10.1109/TMECH.2012.2197405.

[40] X. Chen, C. Yang, C. Fang, Z. Li, Impedance matching strategy for physical human robot interaction control, in: 2017 13th IEEE Conference on Automation Science and Engineering, CASE, 2017, pp. 138–144.

[41] C. Yang, C. Zeng, P. Liang, Z. Li, R. Li, C. Su, Interface design of a physical human-robot interaction system for human impedance adaptive skill transfer, IEEE Transactions on Automation Science and Engineering 15 (1) (2018) 329–340, https://doi.org/10.1109/TASE.2017.2743000.

[42] R. Johansson, M.W. Spong, Quadratic optimization of impedance control, in: Proceedings of the 1994 IEEE International Conference on Robotics and Automation, vol. 1, 1994, pp. 616–621.

[43] M. Matinfar, K. Hashtrudi-Zaad, Optimization-Based Robot Compliance Control: Geometric and Linear Quadratic Approaches, Sage Publications, Inc., 2005.

Chapter 6

Sensorimotor control for dexterous grasping – inspiration from human hand

Ke Li

Institute of Intelligent Medicine Research Center, Department of Biomedical Engineering, Shandong University, Jinan, Shandong, China

The human hand is a versatile tool to explore and modify the external environment. It represents both the cognitive organ of the sense of touch and the most important end-effector in object manipulation and grasping. Our brain can cope efficiently with the high degree of complexity of the hand, which arises from the huge number of actuators and sensors. This allows us to perform a large number of daily life tasks, from the simple ones, such as determining the ripeness of a fruit or drive a car, to the more complex ones, as for example performing surgical procedures, playing an instrument, or painting. Not surprisingly, much research effort has been devoted to understand the neurophysiological mechanisms underpinning the sensorimotor control of human hands and to attempt to reproduce such mechanisms in artificial robotic systems. This chapter describes the sensorimotor control for dexterous grasping by the human hand. First, an overview of sensorimotor control for dexterous grasping is presented. Next, the sensorimotor control for grasping kinematics is presented. Finally, the sensorimotor control for grasping kinetics is demonstrated. The descriptions of sensorimotor control for dexterous grasping by the human hand should provide novel insights into manipulation, prosthetics, and rehabilitation.

6.1 Introduction of sensorimotor control for dexterous grasping

The human hand is a miraculous instrument with a sophisticated biomechanical structure and intricate sensorimotor function. There are 27 bones in the hand, with 8 carpal bones constituting the wrist, 5 metacarpal bones in the palm, and 14 phalangeal bones that make up the digits. There are 38 muscles responsible for hand movement control, although some of these muscles are divided into distinct parts with separate tendons, such as the flexor digitorum profundus, which

Tactile Sensing, Skill Learning, and Robotic Dexterous Manipulation
https://doi.org/10.1016/B978-0-32-390445-2.00014-3

sends tendons to the distal phalanx of all four fingers. There are 22 joints and totally 25 degrees of freedom of movement in the hand. Controlling the large number of elements of the hand is a challenging task for the central nervous system (CNS). A critical issue of hand control is how the CNS effectively organizes the multiple elements, changes the interaction among the elements, and makes them work together towards different motion goals.

Object manipulation using the hand is subject to sensorimotor control mechanisms. Multiple sensory modalities, such as visual and tactile sensation, play a role in spatial and temporal regulation for object manipulation. For example, visual information about the position and characteristics of the object may facilitate the formation of an appropriate sequence of motor commands for specific manipulation goals [1–3]. Tactile sensors innervating fingertips can detect the physical properties of the object, including the curves and friction of the contact area, and can encode the information about the weight, moving speed, and center of mass (COM) of the object [4–7]. Studies on the sensorimotor control of grasping are mostly confined to the prototyped grasping postures (e.g., a precision grip opposing the thumb and index finger) or nonfunctional testing (e.g., holding the object stably or moving the object between two points) with an instrumented apparatus (e.g., a specially designed object for the experiment) [8–12]. It remains an intriguing issue how functional sensorimotor neural control strategies affect a full hand grasping to perform a daily task.

The characteristics of object manipulation can be examined in kinetics and kinematics. The kinetic parameters, such as the amount and direction of fingertip forces and moments, are adaptable to the object's weight, COM, and movement status during lifting, holding, or releasing the object [13–15]. By examining the digit forces for stably holding a glass of water, Sun et al. found that the indices of digit force coordination were highly sensitive to the changes of the water volume [16]. Kinematic parameters, such as the attitude and joint angles of the grasping hands and the moving speed and trajectory of the target object, have been mostly analyzed during reach-to-grasp tasks [17–19]. The initiation of grasp closure has been found to be correlated with the onset of the deceleration phase of transport [20]; the contact points can be modulated according to the object's COM under visual or tactile feedbacks [21]; faster movements and decreased grip aperture were achieved with simultaneous availability of vision and tactile sensation [22].

The hand plays a critical role in activities of daily life. The study of the grasping kinematics and kinetics for object manipulation may provide new approaches for detection or evaluation of neuromuscular disorders after neurological or traumatic injuries and inspire the design of novel instruments for hand functional rehabilitation. In the remainder of the present chapter, examples of our studies of grasping kinematics and kinetics will be presented, showing how we performed the experiments and analyses of the grasping function of the human hand.

6.2 Sensorimotor control for grasping kinematics

Grasping a glass and pouring water is one of the most frequently performed manual tasks in our daily life. This apparently effortless task involves multiple subgoals which include lifting the glass, pouring the water out of the glass, and placing the glass back on the table. To seamlessly integrate and successfully achieve these subgoals, sensorimotor mechanisms are indispensable [23–25]. For example, to move the glass smoothly, to lift the glass up to an appropriate altitude, or to pour the water out of the glass scrupulously requires visual information about the position and attitude of the glass and tactile information about the glass's weight for the closed-loop control. Investigating the performance of a grasp-to-pour task of a glass of water could thus shed light on the sensorimotor control mechanisms underlying daily manual tasks and may provide an avenue for evaluating hand function through behavioral observation. Despite a number of studies on the grasping kinematics, relatively little is known about how the different sensory modalities like visual or tactile sensations influence the kinematic performance for a daily manual task that usually involves multiple consecutive subgoals.

With this aim, this study investigated the effects of sensorimotor control on the kinematic performance during the grasp-to-pour task of a glass of water. The transparence of the glass and the volume of water in the glass were altered to examine the effects of visual and tactile sensation. Kinematic parameters including the attitude, speed, and position of the glass were examined for both the dominant and the nondominant hand. It was hypothesized that both the visual and tactile feedback could affect the kinematic parameters during the grasp-to-pour task of a glass of water. It was also hypothesized that the dominant and nondominant hands could be equally influenced by the sensory feedback for achieving the subgoals for the grasp-to-pour task.

Thirteen healthy right-handed subjects (six females and seven males, age 22.5 ± 1.6 years) participated in the experiment. All participants had normal or correct-to-normal vision and were naïve to the purpose of the study. The exclusive criteria were as follows: (1) any history of musculoskeletal or neurological disorders; (2) severe cervical spondylosis; (3) vestibular system disease; (4) severe depression, anxiety, or cognitive difficulties. This study was carried out in accordance with the recommendations of local ethics guidelines. All subjects gave written informed consent in accordance with the Declaration of Helsinki. All protocols were approved by the Institutional Review Board of Shandong University.

A 3D motion capture system (Opti TrackTM, USA) was used to record the kinematic signals of the digits. The motion capture system was calibrated before data collection. The position between the camera and the ground, roll angle of the camera, exposure time, frames per second, threshold, and brightness were adjusted in order to improve the accuracy and reliability of tracking the retro-reflective markers. The retro-reflective markers were glued to the center of the

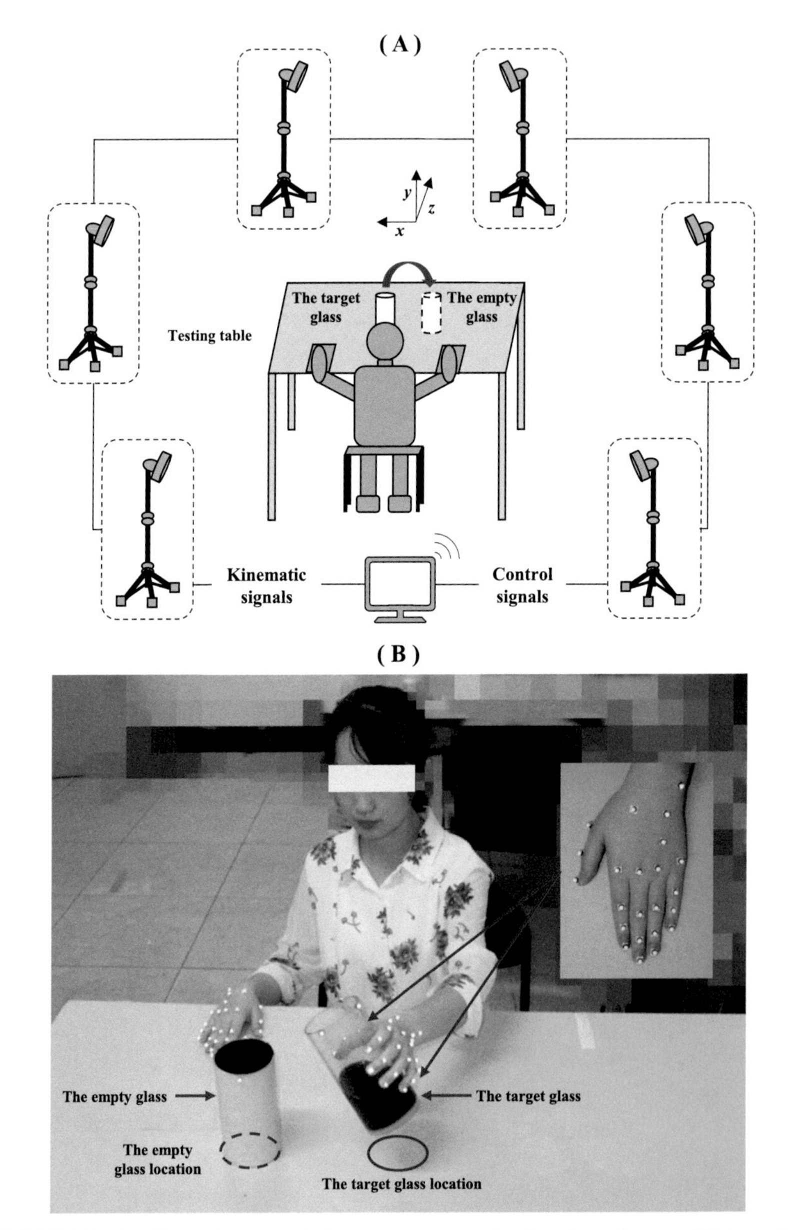

FIGURE 6.1 Experimental setup and the grasp-to-pour task of a class of water. (A) A motion capture system consisting of six cameras to record kinematic signals. (B) the posture for the grasp-to-pour task and the markers utilized for computing digit kinematics.

nails of all the digits to compute and analyze their real-time 3D positions. The experimental setup is shown in Fig. 6.1 (A).

Subjects were seated on an adjustable chair at a testing table (120 cm × 60 cm, Fig. 6.1 (B)). A styrofoam glass (15 cm in height, 8 cm in diameter) was placed 36 cm in front of the subject. Another empty glass of the same size was positioned 30 cm from the target glass and 36 cm to the initial position of the grasping hand. Both the left and right hands were tested in the experiment. Two factors – vision and weight – were taken into account in the experiment. Regarding vision, the target glass was either transparent so that the water inside was visible (with visual feedback [VF]) or opaque (no visual feedback [NVF]); regarding weight, the liquid inside the target glass was either 200 ml (lighter weight [LW]) or 400 ml (heavy weight [HW]). The liquid inside the target glass was dyed black to avoid unintended reflection disturbing motion capture.

Subjects were instructed to grasp and lift a glass of water, pour the water out of the glass, and place the empty glass back in the initial position. To successfully perform the task, subjects were instructed to grasp the target glass with the full hand as they usually do in daily life, lift the glass at normal speed, and pour the water carefully without spilling. Both the left and right hands were tested under VF and NVF conditions and under LW and HW conditions. The testing sequences for the conditions were randomized. For each condition six trials were performed by each subject.

The kinematic data from all the retro-reflective markers were rectified and then run through a low pass Butterworth filter (cut-off frequency, 6 Hz). To evaluate the behavior during the grasp-to-pour task, the action was divided into a few subactions by the following crucial turning points:

(1) Lifting onset (LO). The LO was the onset of lifting the target glass. The LO was identified as the moment when the velocity of the target glass first exceeded 5 mm/s for at least 200 ms during upward motion.

(2) Pouring onset (PO). The PO was the onset of pouring the water into the empty glass. The PO was identified as the moment when the derivative of the curve in the y-axis was equal to zero and the rim of the target glass was greater than 0.22 cm in the vertical direction (the y-axis) for the first time.

(3) Pouring done (PD). The PD was defined as the moment when the rim of the target glass (recorded by the reflective markers) was at the lowest vertical position (in the y-axis) after PO.

(4) Replacing done (RD). The RD was defined as the moment when the tangential velocity of the rim of the target glass was less than a threshold speed (5 mm/s) for more than 200 ms.

The entire process was divided into three phases. Phase I, lifting, was defined as the session between the LO and PO. Phase II, pouring, was defined as the session between the PO and PD. Phase III, replacing, which was defined as the session between the PD and RD.

In the experiment, we extracted the duration of each phase and the three crucial positions. In each phase the curve of the deflection angles of the glass was fitted to a linear function. The slope of each fitting function was figured out for each phase. In this way, totally nine kinematic parameters were derived from the nine phases, including the slopes of the deflection angle curves during the lifting ($\alpha 1$), pouring ($\alpha 2$), and replacing ($\alpha 3$) phases; the duration of lifting (T1), pouring (T2), and replacing (T3) phases; and the maximal vertical positions in the three phases (P1, P2, and P3) (Fig. 6.2).

The position of the glass was determined by the x-, y-, and z-coordinates, which was along the sagittal axis, the coronal axis, and the vertical axis, respectively (Fig. 6.1 (A)). The glass did not show any significant changes in the z-axis. For simplicity, only the x- and y-coordinates of the glass were examined in the following analyses.

Statistical analyses were performed using SPSS 20.0 (SPSS Inc., Chicago, IL). The Kolmogorov–Smirnov test was used to test the normality of data. Three-way repeated-measures ANOVAs (weight × vision × hands) were performed to evaluate the effects of weight (LW vs. HW), vision (VF vs. NVF), and hand (dominant vs. nondominant) on the nine kinematic parameters. The Huynh–Feldt correction was used when the assumption of sphericity was violated. A paired t-test was used to examine the difference between the HW and LW conditions, between the VF and NVF conditions, and between the dominant and nondominant hands. A P-value of less than 0.05 was considered statistically significant.

Results of the lifting phase. There were significant main effects of the weight ($F_{1,77} = 86.93$, $P < 0.001$), vision ($F_{1,77} = 54.02$, $P < 0.001$), and hand ($F_{1,77} = 15.87$, $P < 0.05$) on $\alpha 1$ (Fig. 6.3 (A)). Significant interactions were found between weight and vision ($F_{1,77} = 54.68$, $P < 0.001$), weight and hand ($F_{1,77} = 55.75$, $P < 0.001$), and vision and hand ($F_{1,77} = 80.09$, $P < 0.001$). A significantly lower $\alpha 1$ was found for the heavy glass than for the light glass, regardless of vision or hand conditions ($P < 0.001$, Fig. 6.3 (A)). The dominant hand showed significantly higher $\alpha 1$ than the nondominant hand when grasping the heavy glass under both VF ($t = -2.54$, $P < 0.05$) and NVF ($t = -9.338$, $P < 0.001$) conditions. However, no significant difference was found between the dominant and nondominant hands when grasping the light glass either with ($P = 0.124$) or without visual feedback ($P = 0.122$). The effects of vision could be only observable in the dominant hand when grasping the heavy glass. Specifically, $\alpha 1$ under VF conditions was significantly lower than that under NVF conditions ($t = 14.89$, $P < 0.001$).

There were main effects of weight ($F_{1,77} = 19.41$, $P < 0.001$) and hand ($F_{1,77} = 5.56$, $P < 0.05$) on T1. No significant difference was observed in T1 between VF and NVF ($P = 0.202$). No interaction was found between any of the factors. The nondominant hand showed significantly higher T1 for HW than for LW, regardless of vision conditions ($P < 0.001$, Fig. 6.3 (B)). For the dominant hand, no significant difference was found in T1 between the HW and LW conditions. Compared with the dominant hand, the nondominant hand showed significantly increased T1 as grasping the heavy glass with visual feedback ($t = 3.441$, $P < 0.005$, Fig. 6.3 (B)).

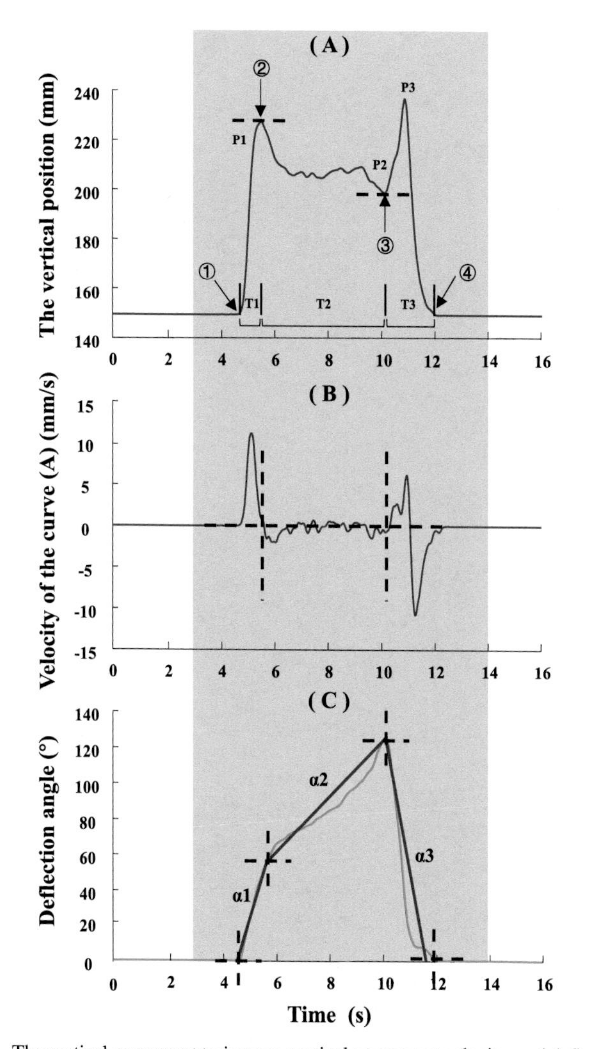

FIGURE 6.2 The vertical movement trajectory, vertical movement velocity, and deflection angle of the glass over the grasp-to-pour cycle from a representative subject. (A) The trajectory of the vertical movement of the glass. ①, ②, ③, and ④ represent the lifting onset, pouring onset, pouring done, and replacing done, respectively. P1 indicates the maximal vertical position of the glass during lifting; P2 indicates the minimal vertical position of the glass during pouring of the water; and P3 indicates the maximal vertical position of the glass during the replacing phase. T1, T2, and T3 are the duration of the lifting, pouring, and replacing phases, respectively. (B) The velocity of the vertical movement of the glass. (C) The deflection angles of the glass. $\alpha 1$, $\alpha 2$, and $\alpha 3$ are the slopes of deflection angles fitted by linear functions during the lifting, pouring, and replacing phases, respectively.

There were main effects of weight for P1 ($F_{1,77} = 16.81$, $P < 0.001$, Fig. 6.3 (C)). Compared with the LW, the dominant hand showed significantly lower P1 for the HW with visual feedback ($t = 2.956$, $P < 0.05$), and the nondominant

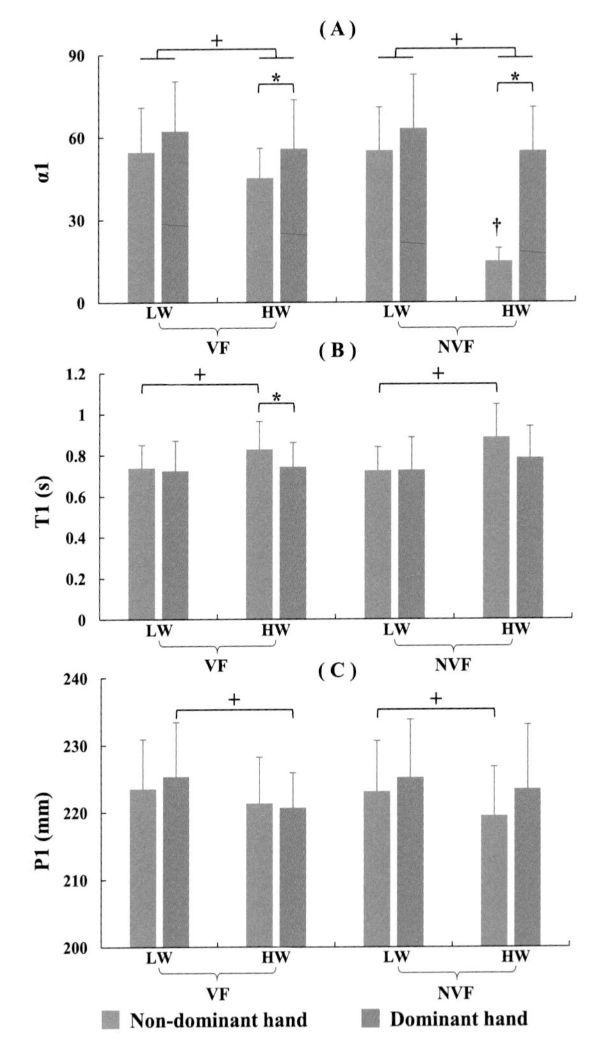

FIGURE 6.3 Statistical analysis of the kinematic parameters during the lifting phase. (A) Slopes of deflection angles ($\alpha 1$). (B) The duration (T1). (C) The maximal vertical position of the glass (P1). * Significant differences between the dominant and nondominant hand; + significant differences between the light weight (LW) and heavy weight (HW) conditions; † significant differences between the two vision conditions – with visual feedback (VF) vs. without visual feedback (NVF).

hand showed significantly lower P1 without visual feedback ($t = 2.255$, $P < 0.05$). No effects of vision ($P = 0.928$) and hands ($P = 0.217$) were observed for P1.

Results of the pouring phase. There were significant main effects of weight ($F_{1,77} = 72.77$, $P < 0.001$) and hand ($F_{1,77} = 5.34$, $P < 0.05$) on $\alpha 2$. No significant difference was observed in $\alpha 2$ between the two vision conditions

($P = 0.372$). Significant interactions were found between the weight, vision, and hands ($F_{1,77} = 6.27$, $P < 0.05$). There were significantly lower $\alpha 2$ values for the HW than for the LW, regardless of vision or hand conditions ($P < 0.001$, Fig. 6.4 (A)). When grasping the heavy glass, the nondominant hand showed lower $\alpha 2$ values than the dominant hand, under both the VF ($t = -2.311$, $P < 0.05$) and NVF ($t = -2.844$, $P < 0.05$) conditions. No significant difference was found between the two hands when grasping the light glass ($P \geqslant 0.067$).

A significant main effect of weight was found on T2 ($F_{1,77} = 56.98$, $P < 0.001$, Fig. 6.4 (B)). No significant difference of T2 was observed either between the VF and NVF conditions ($P = 0.823$) or between the dominant and nondominant hands ($P = 0.792$). An interaction of the weight and hands was found on T2 ($F_{1,77} = 8.032$, $P < 0.05$). The higher T2 was associated with the heavier glass, regardless of vision or hand conditions ($P < 0.05$). No effects of weight ($P = 0.06$), vision ($P = 0.084$), or hand ($P = 0.062$) were found on P2 (Fig. 6.4 (C)).

Results of the replacing phase. The results of the kinematic parameters during the replacing phase are shown in Table 6.1. There was no effect of weight, vision, or hand on either $\alpha 3$, T3, or P3 during the replacing phase (Table 6.1).

TABLE 6.1 The kinematic parameters during replacing the glass.

Vision		Nondominant hand*			Dominant hand		
	Weight	$\alpha 3$	T3 (s)	P3 (mm)	$\alpha 3$	T3 (s)	P3 (mm)
VF	LW	78.3±17.1	1.41±0.18	245.6±12.4	80.5±14.2	1.57±0.29	242.0±23.7
	HW	77.7±16.8	1.58±0.26	264.6±13.0	82.1±16.4	1.53±0.34	241.1±17.4
NVF	LW	76.4±17.7	1.6±0.28	244.9±13.9	77.5±18.0	1.54±0.27	240.7±13.8
	HW	76.6±15.0	1.58±0.24	245.1±13.1	82.6±12.6	1.46±0.19	239.8±15.9

* *Results are presented as mean ± standard deviation.*

This study explored the potential effects of sensorimotor control on kinematic performance during a grasp-to-pour task of a glass of water – a frequently performed manual task involving a series of subgoals. The subgoals include grasping and lifting the glass, pouring the water out of the glass, and placing the glass back on the table. To achieve these subgoals, the kinematic parameters were supposed to be precisely controlled in accordance with the movement status of the glass. Vision and tactile sensations are two sensory modalities that potentially affect the kinematic performance. In this study, the effects of vision and tactile sensation were examined by alteration of the transparence and the weight of the glass. The slopes of the deflection angle, the duration, and the vertical position were kinematic parameters utilized for statistical analyses.

It has been extensively reported that the fingertip tactile sensation could precisely obtain the information of an object's weight for grasping, lifting, or holding it [26–28]. The CNS could integrate the sensory information and deliver motor commands to produce suitable kinetic parameters, such as forces,

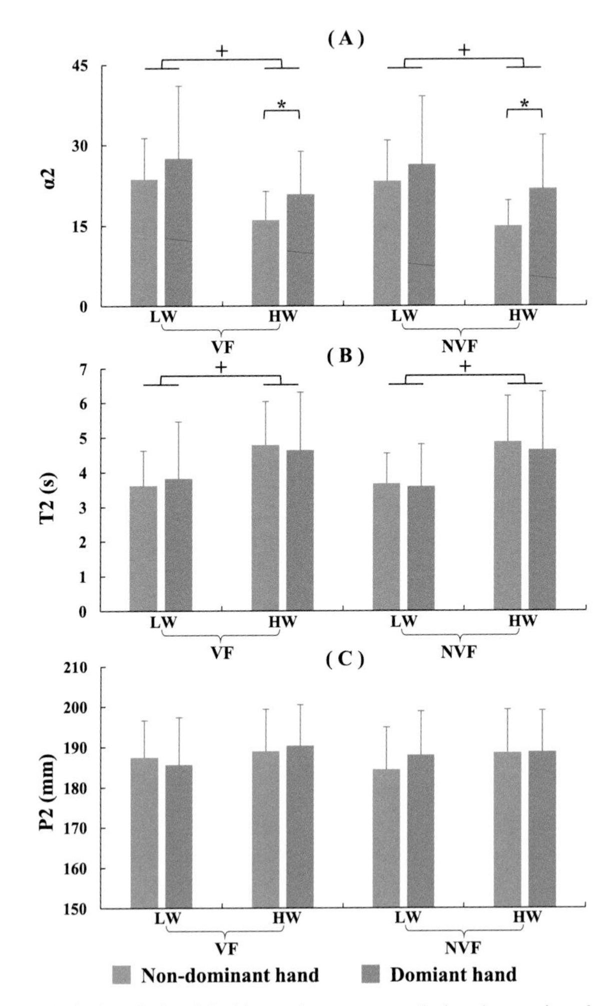

FIGURE 6.4 Statistical analysis of the kinematic parameters during the pouring phase. (A) Slopes of deflection angles ($\alpha 2$). (B) The duration (T2). (C) The maximal vertical position of the glass (P2). + Significant difference between the light weight (LW) and heavy weight (HW) conditions.

moments, or grip-to-load force ratios, in accordance with the object's weight [29–33]. By changing the weight of the glass, this study found significant main effects of weight on $\alpha 1$, $\alpha 2$, T1, T2, and P1 during the lifting and pouring phases (Figs. 6.3 and 6.4), which confirms our hypothesis that the tactile information about the object's weight may remarkably influence the kinematic parameters for the grasp-to-pour task of a glass of water. The lower slopes of the deflection angles $\alpha 1$ and $\alpha 2$ for the heavy glass suggest less tilt during the lifting and pouring processes, reflecting a higher protection against the potential spillage of water. Meanwhile, grasping the heavier glass required a longer time in the lifting (T1) and pouring (T2) phases, suggesting a lower speed that results in higher security in manipulation [34,35]. Results further showed that the object's weight could affect the terminal point of the vertical movement in the lifting phase. For the heavier glass it required a higher vertical position. These findings provided evidence that the kinematic parameters, including the attitudes, movement speed, and positions during the grasp-to-pour task of a glass of water, are highly reliant on the touch perception of the weight.

Results further showed an effect of vision on $\alpha 1$ during lifting of the heavier glass, which, however, was highly dependent on the hand used. Without visual feedback, the nondominant hand showed much lower $\alpha 1$ values than with visual feedback (Fig. 6.3 (A)). But the dominant hand did not exhibit any significant difference in $\alpha 1$ between the two visual conditions (Fig. 6.3 (A)). A previous study has found that in the absence of visual feedback, significant differences between the dominant and nondominant hands were observable in the amount and structural variability of digit forces and the interdigit force coordination during a sustained precision pinch [9,36,37]. The current study further indicated that the dominant and nondominant hands may have differential reliance on the visual feedback. The nondominant hand is usually not as dexterous as the dominant hand for manipulation, thereby requiring more information about the moving status of the object through visual feedback [38,39]. It is noteworthy that the effect of visual feedback was only observed in $\alpha 1$ during lifting the glass, rather than in the other kinematic parameters at the following stages (Figs. 6.3 and 6.4). Previous studies from kinetic analysis showed that the grip force can be characterized by more feedback-driven corrections when grasping at self-selected contact points at the early stage of manipulation [11,40]. Consistent with these findings, the current study further revealed that the visual feedback may also play a role in control of kinematics at the early stage of manipulation. It could also suggest that the deflection angle would be one of the most important kinematic parameters that the nervous system needs to control for the grasp-to-pour task of a glass of water.

Furthermore, the dominant hand gave significantly higher $\alpha 1$ but lower T1 values, indicating more tilt and higher speed than for the nondominant hand during lifting of the heavier glass. When pouring the water out of the glass, the nondominant hand also showed significantly lower $\alpha 1$ values than the dominant hand. These results substantiate the differences between the dominant and nondominant hands in control of kinematics for the grasp-to-pour task of a glass of water and are in line with the previous findings that the dominant hand was more skilled in grasping and manipulating an object [39,41]. Facing a higher risk of tilt, the nondominant hand with lower dexterity could have a lower deflection angle ($\alpha 1$, $\alpha 2$) and longer time (T1) than the dominant hand to guarantee grasping safety. The kinematic differences between the dominant and nondominant hands were only observable in grasping the heavy rather than the light glass (Figs. 6.3 and 6.4), implying a disparity between hands for kinematic control that could be more evidently exhibited in more challenging tasks.

6.3 Sensorimotor control for grasping kinetics

It often takes more than one finger to make a precise movement in many activities of daily life. However, many factors, such as aging, carpal tunnel syndrome, and stroke with hemiplegia, will seriously affect the performance of human hands [42]. The evaluation of the digit force coordination ability in manual tasks is of great significance, which can improve our understanding of neuromuscular function and provide some help for hand rehabilitation [43,44]. Grasping, divided into hook grasping, spherical grasping, and column grasping, plays a vital role in various hand movements [45]. Good fingertip force coordination is the basis for effective manipulation of objects [8,46].

The influence of a variable COM on sensorimotor learning and force control of digits during grasping has been investigated. Extensive researches have focused on the effects of different COMs on pinch grip when the COM of an apparatus was unpredicted [47]. Some properties of an object, such as its COM or weight, can be predicted before somatosensory feedback is processed [48]. Nevertheless, when the COM of an apparatus is unpredicted before grasping, a default distribution of fingertip force is exerted, and the force control mechanisms are modulated during manipulation of the object [13]. Some researchers also explored the coordination between fingertip force and the COM of an object during multidigit grasping [14,49]. Some studies have researched the coordination of digits during pinch grasping using the center of pressure (COP) of fingertips [50–52]. However, there are few researches on torque perception-based digit force coordination control in multidigit precision grasping with variable height motion, which may provide new ideas for object manipulation and precise grasping.

Mechanically, the applied force by digits on a hand-held object can be described by two sets of forces, i.e., the manipulation forces and the internal forces [53]. The manipulation forces may cause movement of the hand-held object, and the internal forces cannot affect the object because they cancel each other out. The net force is equal to zero in a static case of multidigit grasping [54]. That means the thumb opposes other fingers while holding a water glass using multiple digits. The normal force of the thumb and other fingers must cancel each other out. Arbib et al. put forward the concept of virtual finger (VF), an imagined finger with the mechanical action equal to that of other fingers combined which are opposite the thumb [55,56]. Hence the net force of the thumb and the VF equals zero in a static case. However, for multidigit grasping during object manipulation, the net force of the thumb and VF is not equal to zero, which means that the normal force of the thumb is not equal to that of the VF anymore, and the manipulation force (the net force) is equal to the difference in normal force between the thumb and VF [57].

The purpose of this study is to investigate the coordination control of the digit normal force based on different COMs during multidigit precision grip. A grasping apparatus was designed with SolidWorks to simulate a water cup as well as to measure and record each digit force during multidigit grasping with three kinds of digit combinations and five kinds of different COM. The mean (MN), standard deviation (SD), and COP area were applied to evaluate the control of the digit forces. We hypothesized that the different number of participating digits during multidigit grasping may have different robustness dealing with different COMs of the grasping apparatus. And different digits may have different contributions during multidigit precision grasping.

Thirty right-handed healthy subjects (15 female and 15 male, age 23.29 ± 1.60 years, height 168.55 ± 8.95 cm, weight 62.53 ± 7.61 kg) participated in the experiment. The Edinburgh Handedness Inventory scores were 89.59 ± 10.96, which suggested all the subjects are strong right-handed. All subjects had normal or correct-to-normal vision. Subjects were excluded if they had: (1) neurological or musculoskeletal injury on an upper limb; (2) cardiovascular or cerebrovascular diseases; (3) severe cervical or lumbar spondylosis; (4) cognitive difficulty or anxiety disorders. All participants signed informed consent, and the study was approved by the Institutional Review Board of Shandong University.

A novel apparatus was designed to record the applied forces of each engaged digit during grasping. The apparatus consisted of a base, a cylindrical surface, four six-axis force/torque transducers (Nano17, ATI Industrial Automation, Apex, NC, USA), a cylindrical container, and a top cover (Fig. 6.1). An extra load (weight 50 g) was placed at one of the following locations of the base: middle (M), proximal (P), distal (D), ulnar (U), and radial (R) positions (Fig. 6.5). A three-dimensional motion capture system (Fastrak, Polhemus, Colchester, VT, USA) was used to record the real-time motion of the apparatus. The transducers were mounted on the cylindrical surface by precise positioning so that the x-axis and the y-axis were along the vertical and horizontal directions in the contact surface of each transducer and the z-axis was in the perpendicular direction to the contact surface (Fig. 6.5).

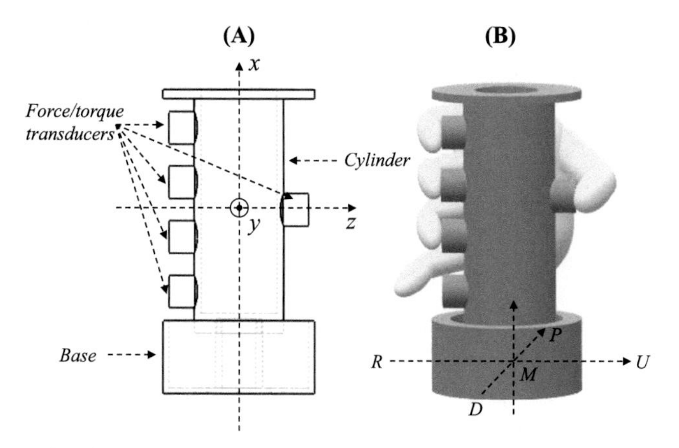

FIGURE 6.5 Grasping apparatus. The cylindrical grasping apparatus designed with SolidWorks. (A) The grasping apparatus model. (B) The posture for grasping the apparatus.

Subjects sat at a testing table with their left hand resting on the table and their right hand grasping the apparatus. There were three digit combinations: (1) two-digit grasping with the thumb and index finger; (2) three-digit grasping with the thumb, index, and middle fingers; and (3) four-digit grasping with the thumb, index, middle, and ring fingers.

Each subject was required to reach and grasp the apparatus and complete a lift-hold-descend task. Specifically, the apparatus was left from the table and the height was changed between a low and a high reference position (about 10 cm and 30 cm above the table, respectively). When the apparatus reached the low or high reference position, it was held stably until the subject heard a cue for the next motion. The apparatus was lowered at a self-selected comfortable speed. There were five consecutive cycles for each trial of the lift-hold-descend task. After one trial of the lift-hold-descend task, the extra load position was randomly changed among the M, P, D, U, and R positions without the subject's knowledge. For each COM condition, the three-digit combinations were used in random order for grasping. The grasping with each digit combination was repeated five times. Each subject was given a 1-min rest period between digit combinations and a 3-min rest between COM conditions. Subjects were noticed to try their best to maintain a stable grasp without obvious tile or vibration, to keep the uninvolved digits away from the grasping apparatus, and to return the apparatus to the original position after each trial. Subjects were not informed of the COM conditions of the apparatus before each trial.

All force data were filtered by a low pass fourth-order Butterworth filter with a cutoff frequency of 5 Hz. The altitude of the apparatus and the digit forces during grasping from one representative subject are depicted in Fig. 6.6. To simplify the analysis, a VF whose force and torque were equivalent to those of the grasping fingers was calculated [56]. The MN and SD of the forces of the thumb and VF were calculated. In addition, the COP signals of each finger and the VF were calculated based on the force and torque components from the transducer [52]:

$$COP_x = -\frac{T_y}{F_z}, \tag{6.1}$$

$$COP_y = \frac{T_x}{F_z}, \tag{6.2}$$

where T_x and T_y are the torque of each digit in the x- and y-directions, respectively, and F_z is the normal force of each digit. The area of the COP was estimated using principal component analysis [57,58]. The normal force of the VF was defined as

$$F_{VF}^n = \sum_{j=1}^{3} F_j^n,\ j = [\mathrm{I, M, R}], \tag{6.3}$$

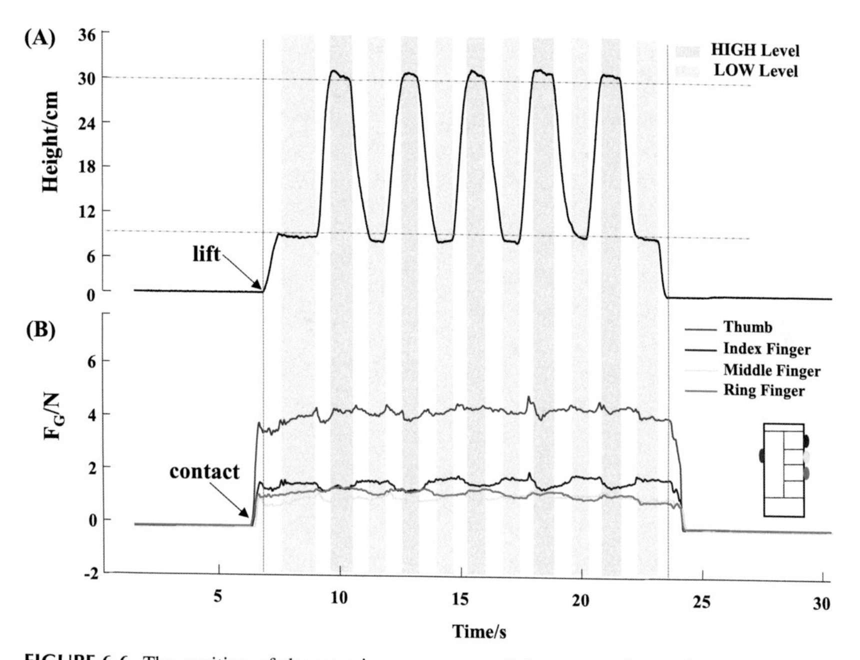

FIGURE 6.6 The position of the grasping apparatus and the normal force of four finger digits during grasping from one representative. (A) The height of the grasping apparatus changing between 10 cm and 30 cm. (B) The normal force of four digits: the thumb (red line), index finger (blue line), middle finger (yellow line), and ring finger (green line).

where I is the index finger, M is the middle finger, and R is the ring finger; F_{VF}^{n} is the normal force of the VF, and F_{j}^{n} denotes the normal forces of the individual fingers.

Statistical analyses were performed using MATLAB® (MathWorks, Natick, MA, USA) and SPSS 23.0 (SPSS Inc., Chicago, IL). A two-way repeated-measures ANOVA was performed to examine the differences in MN and SD of digit forces across the digit combinations (two, three, and four digits) and COM conditions (M, P, D, U, and R). A two-way repeated-measures ANOVA was also employed to examine the differences in the COP of grasping digits among the digits (thumb, index, middle, and ring fingers) and COM conditions (M, P, D, U, and R). The Huynh–Feldt correction was used when the assumption of sphericity was violated. Post hoc pairwise comparisons were performed for the digit combinations and COM conditions using a Holm–Sidak test followed by a pairwise t-test. A P-value of less than 0.05 was considered statistically significant.

The ANOVA test showed significant main effects of the COM conditions on both the MN and SD of digit forces ($P < 0.01$). The post hoc pairwise comparisons showed significantly higher MN when the load was at the D position (the thumb: $t = 2.432$, $P = 0.021$; VF: $t = 2.159$, $P = 0.039$) but lower

MN when the load was at the P position than the other COM conditions (the thumb: $t = -2.325$, $P = 0.027$; VF: $t = -2.731$, $P = 0.011$). No significant difference was found among the other COM conditions or the digit combinations ($P > 0.05$). Similarly, post hoc pairwise comparisons showed significantly higher SD when the load was at the D position than in other directions during two-digit grasping (the thumb: $t = 2.635$, $P = 0.013$; VF: $t = 2.099$, $P = 0.045$). No significant differences were found in the SD of the forces on the other COM conditions or during three-digit and four-digit grasping ($P > 0.05$). Comparison between the thumb and VF showed that the thumb had relatively higher MN and SD than the VF.

Results of COP areas of each digit during three-digit combinations with the five COM conditions are shown in Fig. 6.7 ($F_{2,58} = 24.067$, $P < 0.01$). The COP areas of the thumb were significantly lower than those of the other digits at the five COM conditions during two- and three-digit grasping (take the COM in the M as an example; two digits: $t = -3.398$, $P = 0.002$; three digits: $t = -2.282$, $P = 0.030$; $t = -2.547$, $P = 0.016$). During four-digit grasping, the COP areas of the middle and ring fingers were significantly higher than that of the thumb (Fig. 6.7 (C); take the COM in the M as an example; $t = -2.173$, $P = 0.038$, $t = -3.554$, $P = 0.001$); no significant difference was found in the COP area between the thumb and the index finger, or between the middle and ring digits (Fig. 6.7 (C); $P > 0.05$). Furthermore, no significant difference was observable in the COP areas between the thumb and VF during the three- and four-digit grasping (Fig. 6.7 (B and C); $P > 0.05$).

The magnitude and stability of digit forces are related to the COM of the grasping apparatus and the number of digits involved. To learn more about the digit force control during multidigit grasping, this study investigated digit force control during a precision grasping task with different numbers of digits in different COMs. The MN, SD, and COP area of the thumb and VF force as well as COP area of each digit were analyzed. The results showed that (1) with the increase of the number of digits involved in grasping, the grasping stability was enhanced to a certain extent; (2) stable moment deflection can lead to the change of grasping stability by shifting the COM of the object, and the proximal–distal (P-D) direction is more sensitive to the change of deflection moment.

Precision grasping requires the close cooperation and coordination of multiple digits, and the number of digits involved in grasping significantly affects the robustness of the dynamic motor system. As the number of digits increased, the MN and SD of the thumb and VF normal forces as well as the COP area of the VF decreased. This result is consistent with the previous finding that the motion strategy actively utilizes additional degrees of freedoms (DoFs) to ensure the performance stability, which has to do with improving accuracy and precision of the task [54]. Kim et al. designed the simulation archery movement according to the DoF problem on multidigit synergies, adjusted the DoF by changing the number of digits, and quantified the resulting digit forces through the stability index. Their results showed that the accuracy and precision of the task increase with the increase of freedom. An increase in the number of digits seems to contribute to the stability and robustness of dexterous manipulation. Previous studies showed that digit interaction may show effects of the history of digit involvement in a task [59], and the ring and little fingers are relatively weak in generating force and have poor independence in joint movement [60]. The experimental evidence is consistent with the expectation that with the ring finger involved in grasping, the variance of the VF increased. It was indicated that the independence limit of the ring finger broke the original grasping structure and led to decreased stability.

Previous studies have pointed towards a possibility that symmetrical shape cues may influence the digits forces, as opposed to asymmetrical density cues [61]. Before lifting the object, participants used visual cues to generate fingertip forces and torques in an expected manner. During the initial lifting, the participants adjusted the fingertip forces to produce a compensating torque to avoid tilting when the density of the object was uneven. When the participant has a priori knowledge about the COM position of the grasping apparatus, the coordination of digit forces can be controlled by the feedforward mechanism, less dependent on the feedback processing after contacting with the apparatus [45]. It was speculated that the CNS gives the motor commands for the generation of digit forces, which requires successful sensorimotor transformations [50]. The symmetrical shape cues, as opposed to the asymmetrical density cues, force the participants to use the default force to grasp at the beginning of touching the apparatus. And to resist the tilt caused by the different COMs, they must generate the opposite torque to maintain the stability, even though the torque may not be perceived. The MN and SD of the thumb and VF normal force with the COM in the distal (D) position during two-digit grasping were larger than those of other conditions, and the MNs of thumb and VF normal forces with the COM in the proximal (P) position during two-digit grasping were smaller than those of other conditions, which may indicate that the digit force control was easily disturbed with the COM in the P-D direction during grasping with only two digits. The differences of the force in ulnar–radial (U-R) and P-D directions was related to the wrist joint movement strategy. Compared to the P-D direction, the anterior–posterior direction dominated the motor direction of the wrist joint [62]. Besides, the P direction is the principal direction of inertia of the wrist joint, where the inertia force must be overcome when the COM is in the D direction. The disturbance torque and flexor muscle are in the same direction when the COM is in the distal position. The inertia force can counteract a certain force produced by the load when the COM is in the P direction. So, the extensor torque offsets part of the disturbance torque when the COM is in the P direction, and vice versa. Therefore, greater magnitude and variation of the digit force with the COM in the distal position indicate that the control of digit force against the disturbance torque of the U-R direction is more robust than that of the P-D direction.

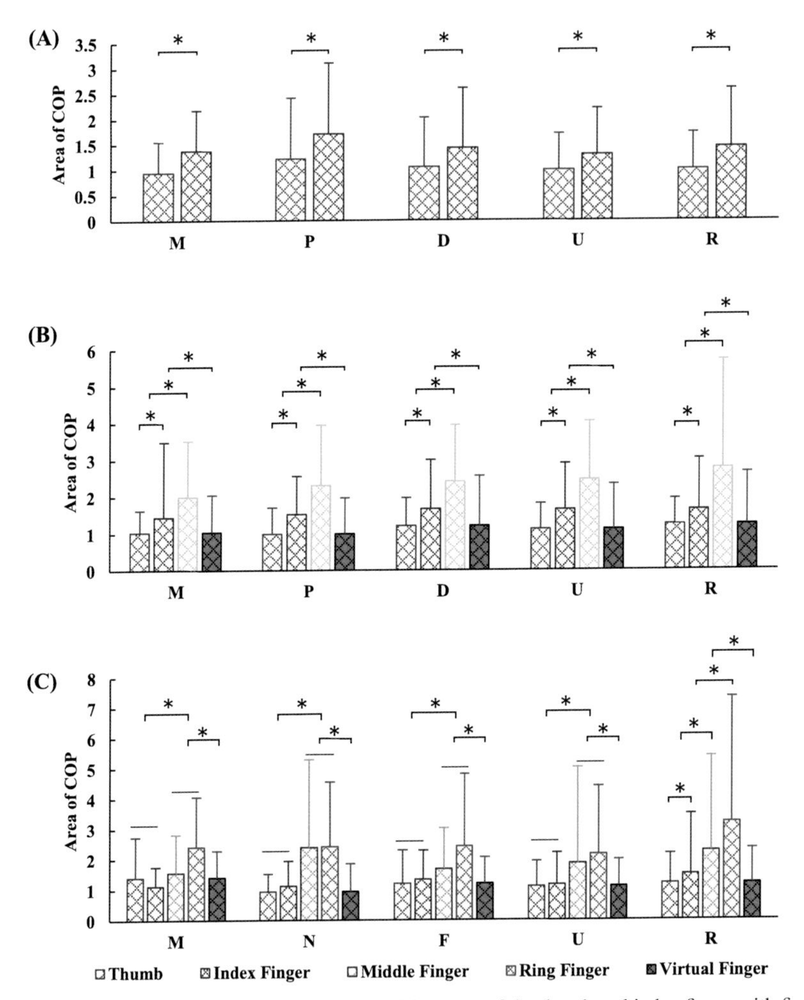

FIGURE 6.7 The COP areas of each digit. (A) COP areas of the thumb and index finger with five kinds of different centers of mass during the two-digit grasping gesture. (B) COP areas of the thumb, index, middle, and virtual fingers with five kinds of different centers of mass during the three-digit grasping gesture. (C) COP areas of the thumb, index, middle, ring, and virtual fingers with five kinds of different centers of mass during the four-digit grasping gesture. * a significant difference between the thumb and another finger, $P < 0.05$.

During multidigit precision grasping, the contribution of the digits to the grasping stability is different. The results showed that the MN and SD of the thumb normal force were respectively larger than those of the VF normal force under all conditions. The larger normal force and smaller COP area of the thumb may indicate that the thumb could be act as a "pivot" while other fingers are more flexible during grasping [45]. That means the thumb is more stable in controlling the object as it moves and the other fingers are more flexible to change the position and angles of the object. Johanson et al. found no statistical difference in the strength of the force of the thumb between stable or unstable tasks [63]. The abductor pollicis brevis and extensor longus were significantly more active in the unstable task, indicating their importance in controlling the force of the thumb tip, which may represent a strategy to increase lateral stability in tasks. Rachaveti et al. have shown that the relative contribution of the thumb to produce smoother motion is greater, and the smoothness of the thumb does not vary with the target, which confirmed that the thumb contributes more to the grasping stability [64].

6.4 Conclusions

In this chapter we presented our recent studies of sensorimotor control in grasping kinematics and kinetics for daily manipulative tasks. The methodologies developed in these studies may facilitate the examination of the hand sensorimotor function through behavioral observation. These studies may also play a role in the biologically inspired hand robotics. One example is the prosthetic hand for the amputees. Loss of a hand can lead to physical incapacity for maintaining daily life. There are 2.2 million amputees in China caused by a traumatic event or disease. Prosthetic hands can restore independence for these individuals. Developing the prostheses involves approaches to predict the motor intention from the signals acquired from the user and the execution of these motor intentions with the help of the prosthetic device. As we mentioned above, the human hand can integrate the motor commands with the sensory feedback in a goal-redirected motion; the performance of a unidirectional efferent prosthetic hand, however, is limited by the inadequate sensory afferents to the CNS. The mainly used noninvasive sensory feedback methods include mechanotactile feedback, vibration feedback, and electrotactile feedback; the invasive sensory feedback approaches include stimulation of the peripheral nerve or the somatosensory area of the brain with electrical pulses that transfer information from the grasped object to the body. However, even today to design a prosthetic hand integrating sensory feedback with motor commands and manipulate an object under bidirectional closed-loop control is still a challenging issue. Investigation of the hand sensorimotor function to better understand the control strategies for grasping kinetics and kinematics would inspire the design of advanced bionic prosthetic hands with a large range of motion, intuitive control, natural feeling, and dexterous manipulation.

Acknowledgments

The authors would thank the participations of all the individuals in the experiment. This study was supported by the National Natural Science Foundation of China (62073195), the National Key Research and Development Program (2020YFC2007904), and the Key Research & Development Programs of Guangdong Province (2020B0909020004) and Shandong Province (2019GSF108164, 2019GSF108127, 2019JZZY021010).

References

[1] J.R. Flanagan, G. Rotman, A.F. Reichelt, R.S. Johansson, The role of observers' gaze behaviour when watching object manipulation tasks: predicting and evaluating the consequences of action, Philos. Trans. R. Soc. Lond. B, Biol. Sci. 368 (1628) (Oct 19 2013) 20130063, https://doi.org/10.1098/rstb.2013.0063.

[2] Q. Fu, M. Santello, Context-dependent learning interferes with visuomotor transformations for manipulation planning, J. Neurosci. 32 (43) (Oct 24 2012) 15086–15092, https://doi.org/10.1523/JNEUROSCI.2468-12.2012.

[3] A. Battaglia-Mayer, R. Caminiti, Corticocortical systems underlying high-order motor control, J. Neurosci. 39 (23) (Jun 5 2019) 4404–4421, https://doi.org/10.1523/JNEUROSCI.2094-18.2019.

[4] R.S. Johansson, I. Birznieks, First spikes in ensembles of human tactile afferents code complex spatial fingertip events, Nat. Neurosci. 7 (2) (Feb 2004) 170–177, https://doi.org/10.1038/nn1177.

[5] G. Westling, R.S. Johansson, Factors influencing the force control during precision grip, Exp. Brain Res. 53 (2) (1984) 277–284 [online], available: https://www.ncbi.nlm.nih.gov/pubmed/6705863.

[6] M.P. Rearick, A. Casares, M. Santello, Task-dependent modulation of multi-digit force coordination patterns, J. Neurophysiol. 89 (3) (Mar 2003) 1317–1326, https://doi.org/10.1152/jn.00581.2002.

[7] H.Y. Chiu, et al., How the impact of median neuropathy on sensorimotor control capability of hands for diabetes: an achievable assessment from functional perspectives, PLoS ONE 9 (4) (2014) e94452, https://doi.org/10.1371/journal.pone.0094452.

[8] K. Li, N. Wei, M. Cheng, X. Hou, J. Song, Dynamical coordination of hand intrinsic muscles for precision grip in diabetes mellitus, Sci. Rep. 8 (1) (Mar 12 2018) 4365, https://doi.org/10.1038/s41598-018-22588-z.

[9] K. Li, N. Wei, S. Yue, Effects of tactile sensitivity on structural variability of digit forces during stable precision grip, BioMed Res. Int. 2016 (2016) 8314561, https://doi.org/10.1155/2016/8314561.

[10] J.S. Diamond, J.Y. Nashed, R.S. Johansson, D.M. Wolpert, J.R. Flanagan, Rapid visuomotor corrective responses during transport of hand-held objects incorporate novel object dynamics, J. Neurosci. 35 (29) (Jul 22 2015) 10572–10580, https://doi.org/10.1523/JNEUROSCI.1376-15.2015.

[11] K. Mojtahedi, Q. Fu, M. Santello, Extraction of time and frequency features from grip force rates during dexterous manipulation, IEEE Trans. Biomed. Eng. 62 (5) (May 2015) 1363–1375, https://doi.org/10.1109/TBME.2015.2388592.

[12] G.P. Slota, M.S. Suh, M.L. Latash, V.M. Zatsiorsky, Stability control of grasping objects with different locations of center of mass and rotational inertia, J. Mot. Behav. 44 (3) (2012) 169–178, https://doi.org/10.1080/00222895.2012.665101.

[13] W. Hu, N. Wei, Z.M. Li, K. Li, Effects of muscle fatigue on directional coordination of fingertip forces during precision grip, PLoS ONE 13 (12) (2018) e0208740, https://doi.org/10.1371/journal.pone.0208740.

[14] W. Zhang, et al., Effects of carpal tunnel syndrome on adaptation of multi-digit forces to object mass distribution for whole-hand manipulation, J. NeuroEng. Rehabil. 9 (Nov 21 2012) 83, https://doi.org/10.1186/1743-0003-9-83.

[15] K. Li, Z.M. Li, Cross recurrence quantification analysis of precision grip following peripheral median nerve block, J. NeuroEng. Rehabil. 10 (Mar 2 2013) 28, https://doi.org/10.1186/1743-0003-10-28.

[16] Y. Sun, V.M. Zatsiorsky, M.L. Latash, Prehension of half-full and half-empty glasses: time and history effects on multi-digit coordination, Exp. Brain Res. 209 (4) (Apr 2011) 571–585, https://doi.org/10.1007/s00221-011-2590-6.

[17] R. Nataraj, C. Pasluosta, Z.M. Li, Online kinematic regulation by visual feedback for grasp versus transport during reach-to-pinch, Hum. Mov. Sci. 36 (Aug 2014) 134–153, https://doi.org/10.1016/j.humov.2014.05.007.
[18] N. Hogan, D. Sternad, Sensitivity of smoothness measures to movement duration, amplitude, and arrests, J. Mot. Behav. 41 (6) (Nov 2009) 529–534, https://doi.org/10.3200/35-09-004-RC.
[19] Y. Hu, R. Osu, M. Okada, M.A. Goodale, M. Kawato, A model of the coupling between grip aperture and hand transport during human prehension, Exp. Brain Res. 167 (2) (Nov 2005) 301–304, https://doi.org/10.1007/s00221-005-0111-1.
[20] M. Jeannerod, The timing of natural prehension movements, J. Mot. Behav. 16 (3) (Sep 1984) 235–254, https://doi.org/10.1080/00222895.1984.10735319.
[21] J.R. Lukos, C. Ansuini, M. Santello, Anticipatory control of grasping: independence of sensorimotor memories for kinematics and kinetics, J. Neurosci. 28 (48) (Nov 26 2008) 12765–12774, https://doi.org/10.1523/JNEUROSCI.4335-08.2008.
[22] I. Camponogara, R. Volcic, Grasping movements toward seen and handheld objects, Sci. Rep. 9 (1) (Mar 6 2019) 3665, https://doi.org/10.1038/s41598-018-38277-w.
[23] D.A. Nowak, J. Hermsdorfer, C. Marquardt, H. Topka, Moving objects with clumsy fingers: how predictive is grip force control in patients with impaired manual sensibility?, Clin. Neurophysiol. 114 (3) (Mar 2003) 472–487 [online], available: https://www.ncbi.nlm.nih.gov/pubmed/12705428.
[24] A.W. Goodwin, P. Jenmalm, R.S. Johansson, Control of grip force when tilting objects: effect of curvature of grasped surfaces and applied tangential torque, J. Neurosci. 18 (24) (Dec 15 1998) 10724–10734 [online], available: https://www.ncbi.nlm.nih.gov/pubmed/9852607.
[25] M.J. Dodson, A.W. Goodwin, A.S. Browning, H.M. Gehring, Peripheral neural mechanisms determining the orientation of cylinders grasped by the digits, J. Neurosci. 18 (1) (Jan 1 1998) 521–530 [online], available: https://www.ncbi.nlm.nih.gov/pubmed/9412528.
[26] J.Z. Wu, R.G. Dong, S. Rakheja, A.W. Schopper, W.P. Smutz, A structural fingertip model for simulating of the biomechanics of tactile sensation, Med. Eng. Phys. 26 (2) (Mar 2004) 165–175, https://doi.org/10.1016/j.medengphy.2003.09.004.
[27] Z. Su, J.A. Fishel, T. Yamamoto, G.E. Loeb, Use of tactile feedback to control exploratory movements to characterize object compliance, Front. Neurorobot. 6 (2012) 7, https://doi.org/10.3389/fnbot.2012.00007.
[28] P. Jenmalm, S. Dahlstedt, R.S. Johansson, Visual and tactile information about object-curvature control fingertip forces and grasp kinematics in human dexterous manipulation, J. Neurophysiol. 84 (6) (Dec 2000) 2984–2997, https://doi.org/10.1152/jn.2000.84.6.2984.
[29] K. Koh, et al., The role of tactile sensation in online and offline hierarchical control of multi-finger force synergy, Exp. Brain Res. 233 (9) (Sep 2015) 2539–2548, https://doi.org/10.1007/s00221-015-4325-6.
[30] J.K. Shim, et al., Tactile feedback plays a critical role in maximum finger force production, J. Biomech. 45 (3) (Feb 2 2012) 415–420, https://doi.org/10.1016/j.jbiomech.2011.12.001.
[31] L.A. Jones, E. Piateski, Contribution of tactile feedback from the hand to the perception of force, Exp. Brain Res. 168 (1–2) (Jan 2006) 298–302, https://doi.org/10.1007/s00221-005-0259-8.
[32] V. Patel, M. Burns, R. Vinjamuri, Effect of visual and tactile feedback on kinematic synergies in the grasping hand, Med. Biol. Eng. Comput. 54 (8) (Aug 2016) 1217–1227, https://doi.org/10.1007/s11517-015-1424-2.
[33] D.M. Wolpert, Z. Ghahramani, Computational principles of movement neuroscience, Nat. Neurosci. 3 (Suppl) (Nov 2000) 1212–1217, https://doi.org/10.1038/81497.
[34] A. Kritikos, M. Beresford, Tactile interference in visually guided reach-to-grasp movements, Exp. Brain Res. 144 (1) (May 2002) 1–7, https://doi.org/10.1007/s00221-002-1004-1.
[35] R. Volcic, F. Domini, The endless visuomotor calibration of reach-to-grasp actions, Sci. Rep. 8 (1) (Oct 4 2018) 14803, https://doi.org/10.1038/s41598-018-33009-6.
[36] K. Li, et al., Coordination of digit force variability during dominant and non-dominant sustained precision pinch, Exp. Brain Res. 233 (7) (Jul 2015) 2053–2060, https://doi.org/10.1007/s00221-015-4276-y.

[37] Q. Fu, W. Zhang, M. Santello, Anticipatory planning and control of grasp positions and forces for dexterous two-digit manipulation, J. Neurosci. 30 (27) (Jul 7 2010) 9117–9126, https://doi.org/10.1523/JNEUROSCI.4159-09.2010.

[38] R. Tang, R.L. Whitwell, M.A. Goodale, Explicit knowledge about the availability of visual feedback affects grasping with the left but not the right hand, Exp. Brain Res. 232 (1) (Jan 2014) 293–302, https://doi.org/10.1007/s00221-013-3740-9.

[39] P. Boulinguez, V. Nougier, J.L. Velay, Manual asymmetries in reaching movement control. I: study of right-handers, Cortex 37 (1) (Feb 2001) 101–122 [online], available: https://www.ncbi.nlm.nih.gov/pubmed/11292156.

[40] D. Prattichizzo, F. Chinello, C. Pacchierotti, M. Malvezzi, Towards wearability in fingertip haptics: a 3-DoF wearable device for cutaneous force feedback, IEEE Trans. Haptics 6 (4) (Oct–Dec 2013) 506–516, https://doi.org/10.1109/TOH.2013.53.

[41] A. Grosskopf, J.P. Kuhtz-Buschbeck, Grasping with the left and right hand: a kinematic study, Exp. Brain Res. 168 (1–2) (Jan 2006) 230–240, https://doi.org/10.1007/s00221-005-0083-1.

[42] C. Colomer, R. Llorens, E. Noe, M. Alcaniz, Effect of a mixed reality-based intervention on arm, hand, and finger function on chronic stroke, J. NeuroEng. Rehabil. 13 (1) (May 11 2016) 45, https://doi.org/10.1186/s12984-016-0153-6.

[43] K. Li, P.J. Evans, W.H. Seitz Jr., Z.M. Li, Carpal tunnel syndrome impairs sustained precision pinch performance, Clin. Neurophysiol. 126 (1) (Jan 2015) 194–201, https://doi.org/10.1016/j.clinph.2014.05.004.

[44] K. Li, R. Nataraj, T.L. Marquardt, Z.M. Li, Directional coordination of thumb and finger forces during precision pinch, PLoS ONE 8 (11) (2013) e79400, https://doi.org/10.1371/journal.pone.0079400.

[45] M. Davare, P.J. Parikh, M. Santello, Sensorimotor uncertainty modulates corticospinal excitability during skilled object manipulation (in English), J. Neurophysiol. 121 (4) (Apr 2019) 1162–1170, https://doi.org/10.1152/jn.00800.2018.

[46] L.C. Kuo, S.W. Chen, C.J. Lin, W.J. Lin, S.C. Lin, F.C. Su, The force synergy of human digits in static and dynamic cylindrical grasps, PLoS ONE 8 (3) (2013) e60509, https://doi.org/10.1371/journal.pone.0060509.

[47] J. Lukos, C. Ansuini, M. Santello, Choice of contact points during multidigit grasping: effect of predictability of object center of mass location, J. Neurosci. 27 (14) (Apr 4 2007) 3894–3903, https://doi.org/10.1523/JNEUROSCI.4693-06.2007.

[48] M.K. Budgeon, M.L. Latash, V.M. Zatsiorsky, Digit force adjustments during finger addition/removal in multi-digit prehension, Exp. Brain Res. 189 (3) (Aug 2008) 345–359, https://doi.org/10.1007/s00221-008-1430-9.

[49] W. Zhang, et al., Effects of carpal tunnel syndrome on adaptation of multi-digit forces to object weight for whole-hand manipulation, PLoS ONE 6 (11) (2011) e27715, https://doi.org/10.1371/journal.pone.0027715.

[50] D. Shibata, A.M. Kappers, M. Santello, Digit forces bias sensorimotor transformations underlying control of fingertip position, Front. Human Neurosci. 8 (2014) 564, https://doi.org/10.3389/fnhum.2014.00564.

[51] D. Shibata, J.Y. Choi, J.C. Laitano, M. Santello, Haptic-motor transformations for the control of finger position (in English), PLoS ONE 8 (6) (Jun 2013) e66140, https://doi.org/10.1371/journal.pone.0066140.

[52] S. Dun, R.A. Kaufmann, Z.M. Li, Lower median nerve block impairs precision grip, J. Electromyogr. Kinesiol. 17 (3) (Jun 2007) 348–354, https://doi.org/10.1016/j.jelekin.2006.02.002.

[53] T. Yoshikawa, K. Nagai, Manipulating and grasping forces of multi-fingered hands, Trans. Soc. Instrum. Control Eng. 23 (11) (2009) 1206–1213.

[54] K. Kim, D. Xu, J. Park, Effect of kinetic degrees of freedom on multi-finger synergies and task performance during force production and release tasks, Sci. Rep. 8 (1) (2018), https://doi.org/10.1038/s41598-018-31136-8.

[55] A.W. Goodwin, I. Darian-Smith, Motor control of the hand, Science 229 (229) (1985) 752–753.

[56] G.P. Slota, M.L. Latash, V.M. Zatsiorsky, Grip forces during object manipulation: experiment, mathematical model, and validation, Exp. Brain Res. 213 (1) (Aug 2011) 125–139, https://doi.org/10.1007/s00221-011-2784-y.

[57] M.L. Latash, J. Friedman, S.W. Kim, A.G. Feldman, V.M. Zatsiorsky, Prehension synergies and control with referent hand configurations, Exp. Brain Res. 202 (1) (2009) 213–229, https://doi.org/10.1007/s00221-009-2128-3.

[58] Z.M. Li, Inter-digit co-ordination and object-digit interaction when holding an object with five digits, Ergonomics 45 (6) (May 15 2002) 425–440, https://doi.org/10.1080/00140130210129673.

[59] S. Li, M.L. Latash, V.M. Zatsiorsky, Finger interaction during multi-finger tasks involving finger addition and removal, Exp. Brain Res. 150 (2) (May 2003) 230–236, https://doi.org/10.1007/s00221-003-1449-x.

[60] V.M. Zatsiorsky, Z.M. Li, M.L. Latash, Enslaving effects in multi-finger force production, Exp. Brain Res. 131 (2) (Mar 2000) 187–195, https://doi.org/10.1007/s002219900261.

[61] T. Lee-Miller, M. Marneweck, M. Santello, A.M. Gordon, Visual cues of object properties differentially affect anticipatory planning of digit forces and placement, PLoS ONE 11 (4) (2016) e0154033, https://doi.org/10.1371/journal.pone.0154033.

[62] R.S. Maeda, T. Cluff, P.L. Gribble, J.A. Pruszynski, Compensating for intersegmental dynamics across the shoulder, elbow, and wrist joints during feedforward and feedback control (in English), J. Neurophysiol. 118 (4) (Oct 2017) 1984–1997, https://doi.org/10.1152/jn.00178.2017.

[63] M.E. Johanson, F.J. Valero-Cuevas, V.R. Hentz, Activation patterns of the thumb muscles during stable and unstable pinch tasks, J. Hand Surg. Am. 26 (4) (Jul 2001) 698–705, https://doi.org/10.1053/jhsu.2001.26188.

[64] D. Rachaveti, N. Chakrabhavi, V. Shankar, V. Skm, Thumbs up: movements made by the thumb are smoother and larger than fingers in finger-thumb opposition tasks, PeerJ 6 (2018) e5763, https://doi.org/10.7717/peerj.5763.

Chapter 7

From human to robot grasping: force and kinematic synergies☆

Close comparison between human and robotic hands in both force and kinematic domain

Abdeldjallil Naceri[a], Nicolò Boccardo[c], Lorenzo Lombardi[c], Andrea Marinelli[c], Diego Hidalgo[a,b], Sami Haddadin[a,b], Matteo Laffranchi[c], and Lorenzo De Michieli[c]

[a]*Chair of Robotics and Systems Intelligence, Munich Institute of Robotics and Machine Intelligence (MIRMI), Technical University of Munich (TUM), Munich, Germany,* [b]*Centre for Tactile Internet with Human-in-the-Loop (CeTI), Dresden, Germany,* [c]*Rehab Technologies Lab, Istituto Italiano di Tecnologia, Genova, Italy*

> *"With all its technical sophistication, the photographic camera remains a coarse device compared to the human hand and brain."*
>
> Claude Levi-Strauss

7.1 Introduction

Handmade labels usually refer to superior quality products compared to those made by a tool or a machine. It is a paradoxical situation where we trust the human more than automated precise machines. The hand is considered to be one of the most complex and beautiful pieces of natural engineering in the human body. Indeed, it gives us a sense of touch, feeling, textures, and power and is capable of precision grasps of manipulated objects as well as many other manual tasks. This versatility is unique and sets human apart from every other creature on the planet [1].

The complexity, inherent to manipulation, makes it remain as one of the open topics in robotics. Several research endeavors have looked into not only improvements of robotic manipulation in general, but also a deeper understanding of human manipulation. Humans are naturally capable of using objects and tools in multiple interaction scenarios. We learn from an early age how to adapt

☆ An overview of human to robotic capabilities.

Tactile Sensing, Skill Learning, and Robotic Dexterous Manipulation
https://doi.org/10.1016/B978-0-32-390445-2.00015-5

our grasping and manipulation strategies depending on the goals in a given situation. Furthermore, human hands serve not only as our most used part to interact with the world, but also as a means of establishing social bonds with other people. There is little our hands cannot accomplish after enough training. It is, thus, crucial to understand how human hands work and how we use them, in order to significantly improve robotics manipulation.

The extensive number of degrees of freedom (DoFs) present on the human hand contributes to making it an apparatus which allows a tremendous versatility and high dexterity in almost any task. Such complexity, however, demands a convoluted control strategy from the central nervous system (CNS) [2], as well as a high mechanical complexity. This has triggered research endeavors to look into how humans really control their hands. The biomechanical properties and constraints of the human hand make it difficult for the brain to individually control each one of its muscles or joints. This has promoted the idea of synergies and synergy control, which entails that groups of elements in a system act as a single unit depending on different circumstances and can be controlled using few coordination patterns [2].

One of the first studies on the concept of synergies in human manipulation was that of the authors of [3], whose main finding was that a few linear combinations of DoFs might account for most of the variance of human hand postures. The importance of these findings was twofold. On the one hand, they unveiled the fact that even our deeply complex brains and bodies are not able to control each individual motion of human hands and that perhaps it might not be necessary. On the other hand, these findings endowed the research community with new tools for the design and control of robotic hands.

The robotics community has long focused its attention into resembling performance and morphological configuration of human hands. However, this is a challenging task, as the human hand is an extremely complex system. Consequently, robotic hands remain systems which are not able to match human-like performance, and in many instances, may not even be applicable as real-world solutions. In recent years, innovations in mechanical actuation principles and control strategies, some of which take advantage of synergies, have led to a new stream of research focus, which is the one of robust, easily configurable, and financially viable hands, which can reliably complete a subset of tasks of the broad spectrum which human hands can cover.

7.1.1 Human hand synergies

During manipulation, the human hand performs a number of grasps, whose purpose is to manipulate objects or tools in a given way for a period of time. A stable grasp is achieved when the overall forces the human hand exerts are capable of counteracting external forces or disturbances such as gravity. Research endeavors to try to understand how humans grasp objects have pointed to a hierarchical force control. This means that the CNS not necessarily controls all of the fingertip forces individually, but rather controls more global force patterns, which make individual fingertip forces dependent on each other [4,5]. These force patterns are usually learned through daily activities and hand usage. Hierarchical force control shows another area in which synergy control is thought to be employed by humans while using their hands.

Two main frameworks have been proposed in order to explain how the CNS solves the redundancy problem, namely motor synergies [2,3] and task-optimal control [6,7]. Moreover, it has been suggested that the CNS might be combining both [8] in solving the redundancy problem as they are not mutually exclusive. Synergy can be defined as a correlation between a large set of motor variables as found for instance in the force domain, "force synergies," [9] as well as in the kinematic domain, "postural (kinematics) synergies," [3]. The synergy approach has received broad interest in both the neuroscience [2] and robotic communities [10].

The authors of [11] provided a definition for the word synergy, which describes it well for the manipulation field. Their definition is as follows: "a collection of relatively independent degrees of freedom that behave as a single functional unit – meaning that the internal degrees of freedom take care of themselves, adjusting to their mutual fluctuations and to the fluctuations of the external force field, and do so in a way that preserves the function integrity of the collection." On a more practical level regarding human hands, a synergy is a collection of individual muscle and joint configurations, which depend on the biomechanical constraints and the goal of motion of a human hand. Synergies can account for most of the motions exerted by a human hand. Hence, it is important to understand them. Synergies can be used to effectively reduce the representation space of human hand configurations to a more tractable one, where less configuration information is needed to understand how a human hand behaves.

On a mathematical level, synergies can be represented by mathematical reduction techniques, such as principal component analysis (PCA) or singular value decomposition (SVD). On a kinematic level, this dimension reduction implies that only a small set of kinematic configurations can be enough to reconstruct the span of possible configurations and postures of the human hand. It has been shown that the first two synergies (i.e., principal components) accounted for more than 90% of the finger force variance [12,9,13] and 80% of kinematic (postural) synergies.

Research results have proven that synergies are likely to be the control method used by humans to control their hands. This has considerable implications for a wide variety of fields, specially in robotics, as it entails that individual joint and muscle control is not the optimal way to go when controlling such a complex system as the human hand. Synergies hint at the possibility of a more hierarchical approach which can be taken advantage of when designing artificial hands. This has the added benefit of reducing the complexity required to control every DoF, while retaining satisfactory performance on artificial robotic hands.

7.1.2 The impact of the synergies approach on robotic hands

In a general sense, dexterity in grasping and manipulation tasks is central to robotic interaction with the environment. The benefits of anthropomorphic robotic hands become self-evident when considering robotic manipulators that are used to interact with the *built* environment, i.e., structures and objects made for and by humans. The closer the hand design is to being anthropomorphic, the broader its capacity to grasp and manipulate made-for-human objects, e.g., mechanical tools, natural mechanical interfaces (door handles, steering wheels, valves), etc. Added to this, the concept of underactuation using postural synergies provides any robotic hand with desirable properties of reduced design and control complexity, sufficient DoFs, and adaptability to a variety of interaction tasks.

As previously described, human hands have an extensive range of applications. Due to the dexterity and maneuverability that hands can achieve, humans use their hands for almost every task. In robotics, dexterous manipulation is a crucial factor yet to be solved [10]. Dexterous manipulation is the missing gap between general purpose robotics and robots that can cooperate with humans in every aspect of their lives [10]. Humans can benefit greatly from robotic systems and their applications, for example in prosthetic development and geriatrics.

The logical step to solve the issue of dexterous manipulation is to develop robotic systems which are capable of resembling the capabilities of the human hand. This has pushed several research groups to strive to mimic the human hand. The authors of [14] provided a substantial overview of the technological developments that have been achieved during the last decades regarding artificial robotic hands. The extensive number of research works on the direction of artificial hands has endowed the research community with several possibilities for the design and development of such devices. For instance, there are artificial hands developed using rigid, flexible, dislocatable, soft joints. Several advances have also been achieved regarding the methods of actuation of artificial hands. There are artificial hands which are fully actuated, coupled, or underactuated.

Fully actuated artificial hands are those where each joint can be individually controlled independently from the others. Coupled actuation refers to systems where the movement of a joint depends on or is correlated to the movement of another one. This means that the number of joints is higher than the number of DoFs. Underactuated hands allow passive motions between the DoFs. This allows them to be adaptable to passive elements such as grasping objects, springs, among others. In general, underactuated systems are characterized by a smaller number of degrees of actuation compared to the number of DoFs.

Traditionally full actuation has been the predominant actuation method in artificial hands. However, during the last couple of decades, the number of underactuated hands has increased. This correlates with the hypothesis of synergies, as it might not be necessary to individually control each joint in order to achieve a good manipulation performance on an artificial hand. The use of underactuated systems has additional benefits, including the reduction of weight,

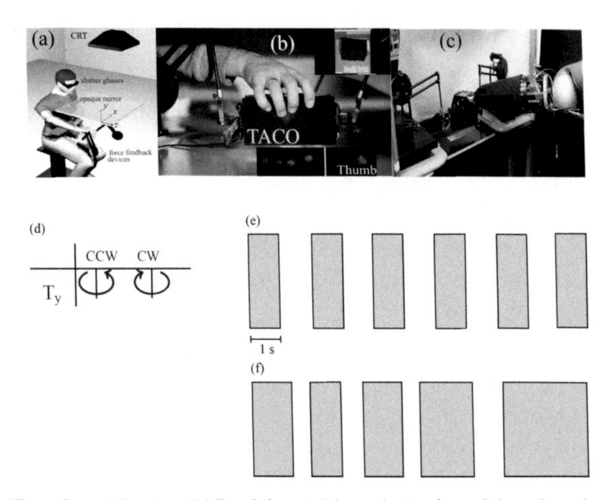

FIGURE 7.1 Experimental setup. (a) Participants binocularly viewed the mirror image of the visual scene. (b) The TACO attached to the PHANToM force feedback devices. Top right, 3D virtual scene, where the purple cube color indicates to participants the desired height (red color otherwise, as shown in (a)). Bottom right, TACO output image, where a red square represents digit pressure. (c) Perturbation torques T_y clockwise (CW) and counterclockwise (CCW) applied for both human and robot hands. (e) and (f) Examples of periodic (P) and aperiodic (Ap) trials, respectively. Gray-shaded boxes represent intervals when torque perturbations $T_y^{CW,CCW}$ are applied.

components, and complexity, among others. The downside of these hands is, however, that dexterity is fully dependent on the underactuation previously chosen, i.e., they are not capable of adapting to new grasping requirements.

In this book chapter we will revisit our previous contributions where we compared human hands versus robot hands in both the kinematics domain [15] and he force domain [16] in order to better understand the gap between human and actual robot hands. To the best of our knowledge, none of the previous contributions addressed such a comparison between robot and human hands in both the kinematics and the force domain.

7.2 Experimental studies

7.2.1 Study 1: force synergies comparison between human and robot hands

In the study conducted in [16], we compared human hands during unconstrained grasping with robotic hands (Shadow Dexterous Hand). In this experiment, human participants and a robot performed the same grasping task of a sensorized tactile object called TACO designed for unconstrained grasping research [17]. Briefly, both participants and the robotic hand grasped the TACO and controlled it to counteract the external perturbations (Fig. 7.1).

The human participants' (six right-handed males) view of their hands and the TACO was occluded by a mirrored screen. This was done to remove visual feedback from their grasps. The mirror faced a computer monitor with a resolution of 1280 × 1024 pixels and a refresh rate of 100 Hz. The display was used to show a visual stimulus consisting of a virtual rectangular cuboid of the same dimensions as the TACO. Participants wore liquid-crystal shutter glasses (CrystalEyes) providing binocular disparity. The TACO was attached to two PHANToM (SensAble Technologies) force-feedback devices to track its position and allow force and torque perturbations for the experiment. Due to the constraints of this attachment, the TACO had only five DoFs (translations on x, y, z and rotations about the y- and z-axes).

The advantage of using the aforementioned devices was that they allowed an unconstrained grasping environment for participants. It is crucial to understand how they truly grasp an object and the correlations between the forces they exert on it. During the experiment, participants needed to grasp, lift, and hold the TACO for a period of time. During the holding phase, external disturbances were applied, and participants were required to hold the height and configuration of the object. Disturbances were applied in clockwise (CW) and counterclockwise (CCW) torques around the y-axis ($T_y^{CW,CCW}$). Additionally, perturbations were applied with periodical and aperiodical frequencies. The purpose of this was to evaluate how humans adapt when they can and cannot predict the behavior of external disturbances.

On the robotic counterpart, the authors replicated the task using the Shadow hand, which is a robotic hand with a comparable number of DoFs and kinematic structure to the human hand. The Shadow hand was attached to a KUKA-LWR4+ robot as the end effector. Once the hand successfully grasped the TACO, the experimenter lifted the TACO using KUKA-LWR4+. When the arm reached the intended height, the KUKA-LWR4+ was grounded to avoid any oscillations. Afterwards, two types of perturbations were applied to the robotic hand, namely, CW and CCW torques on the y-axis (Fig. 7.1). They were applied with periodic and aperiodic frequencies, as in the humans experiment. The reason for the usage of only these two perturbations was due to the low friction forces between the Shadow hand and the TACO. It is important to remark that the robot hand was not commanded to actively control individual digit grasping forces. Instead, the hand control strategy was to fix the stiffness of the robotic hand as a whole during each grasping trial.

Our rationale behind such a comparison is to investigate whether the human hand control is full hand stiffness-based, reacting to the external perturbation, or it is based more on individual finger forces control, predicting the direction of the external perturbations. Since the passive compliance of the artificial actuation of the robotic hand resembles the mechanical properties of the muscles of the human hand and forearm, the Shadow hand is a plausible model to evaluate a grasping strategy based on the control of the global hand stiffness. A global hand stiffness, in turn, has a direct correlation with a synergistic control, which was the hypothesis behind this experiment.

To do so, we used the Shadow robot hand controller developed by [18] which is based on two control variables: joint position and joint stiffness. Specifically, each joint is controlled by a pair of air pressure-controlled artificial muscles. The difference in pressure correlates with the joint position, whereas the pressure sum correlates to the joint's stiffness. The passive compliance of the muscles physically realizes an impedance control whose stiffness is adjustable. In this experiment, we chose a fixed stiffness of the grasping posture that allowed successful grasping and lifting of the TACO.

7.2.2 Results of force synergies study

Similar force patterns were observed when comparing human hand forces control coordination with the robotic hand that was controlled via full hand stiffness. Fig. 7.2 shows the asymmetry in the normal force between fingers that we observed in human grasp can be qualitatively reproduced by the robotic hand, where the only controlled variable was the global hand stiffness. Fig. 7.3 shows the mean fingers' forces across trials and for both human participants and robot. It can be seen that the force trends recorded in human participants (black solid lines) were similar to the one recorded for the robot hand. Specifically, the index finger exerted a larger force relative to the other fingers in condition T_y^{CW} and the little and ring fingers exerted a larger force in T_y^{CCW}. Because of the larger torque, this difference was larger in the robot hand compared with that of the human hand. The hand-stiffening strategy successfully compensated the external torque regardless of the external perturbation frequency and type. The grasp controller employed in the robot hand did not actively control the grasping forces of the individual fingers to counteract disturbances. Therefore, the difference in force among fingers was due to their locations on the TACO. The larger the moment arm is, the higher is the reactive finger force opposing the external torque (Fig. 7.3).

7.2.3 Study 2: kinematic synergies in both human and robot hands

After comparing digit force synergies, we tackle now our recent study comparing human hand kinematics versus an anthropomorphic prosthetic hand named "Hannes" [15]. In this work we investigated the question of how much prosthetic hand kinematic synergies are comparable to the ones of the human hand. Such a comparison seems to be evident and trivial since the Hannes hand is underactuated using one single motor. The rationale behind this question is to investigate whether the prosthetic Hannes hand would replicate key biological properties of the human hand, among which are the synergies manifested in the human counterpart. In fact, this study perfectly compliments our force synergies comparison in both human and robot hands. We have to note that the Hannes hand was produced as a prosthetic hand satisfying the following criteria:

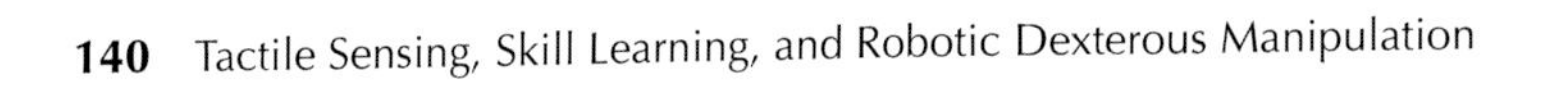

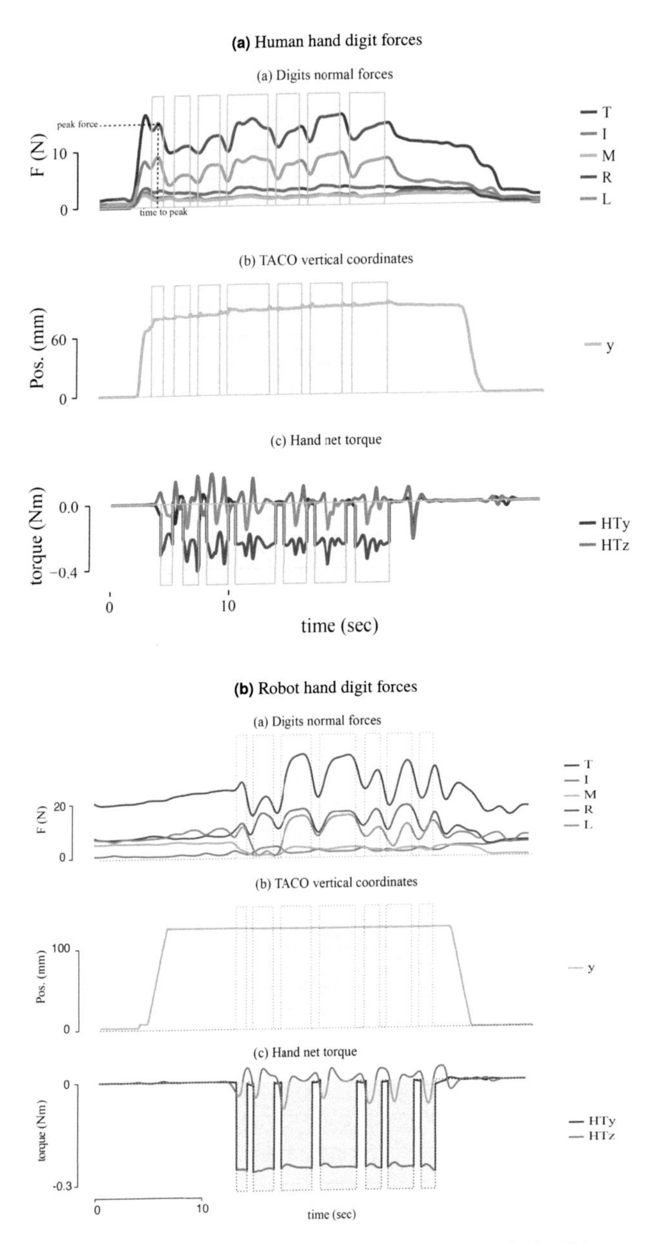

FIGURE 7.2 Digit force profiles of both human and robot hands in a single trial.

- has anthropomorphic properties, including kinematics, size, and weight, among others,
- has a human-like speed, force, and appearance, and

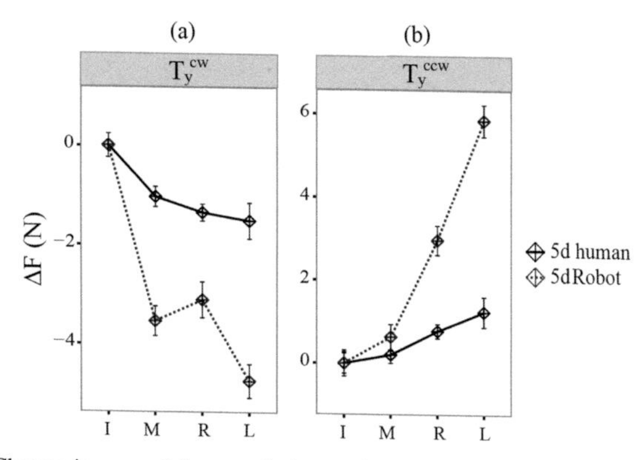

FIGURE 7.3 Change in normal forces relative to the normal force exerted by the index finger. Mean values are given for the different digits in the human hand and robot hand (Shadow Dexterous Hand). Error bars represent standard errors.

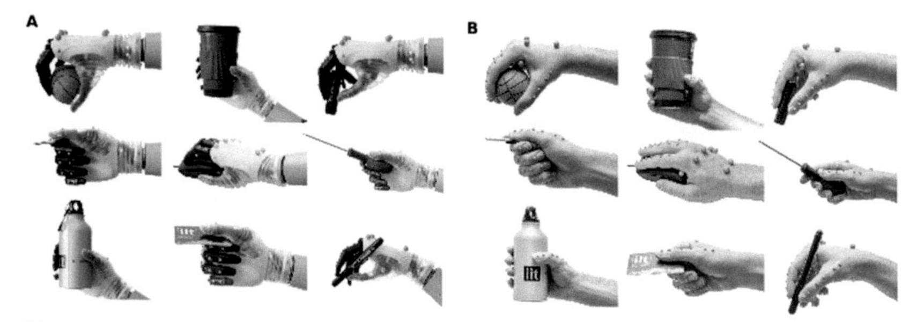

FIGURE 7.4 Human and Hannes hands grasping nine experimental objects. (A) Hannes robotic hand. (B) Human hand.

- allows robust and synergistic grasping capabilities.

In this experiment, the participants (three healthy right-handed participants and one healthy participant using Hannes) sat on a chair with their forearm lying on a table in a relaxed initial position before the experimental grasping tasks. Participants were asked to grasp nine objects as shown in Fig. 7.4. As soon as the participants felt ready to start a new trial, they were asked to grasp one of the nine objects (presented by the experimenter) as appropriate as possible, lift and hold it for 1 s, and then place it back at the starting position. Each object was grasped five times, giving a total of 45 trials completed by each participant and the Hannes patient (a healthy participant using Hannes). The order of trials was randomized for each participant. For the healthy participant using the Hannes right hand model, Hannes was placed in a fixed configuration by attaching the prosthesis to a stump locked to a fixed frame. Hannes was commanded to open/close using the EMG interface consisting of two EMG sensors mounted onto the participant's right arm with an elastic band. The participant conducted the same task as the healthy participants. The Hannes hand

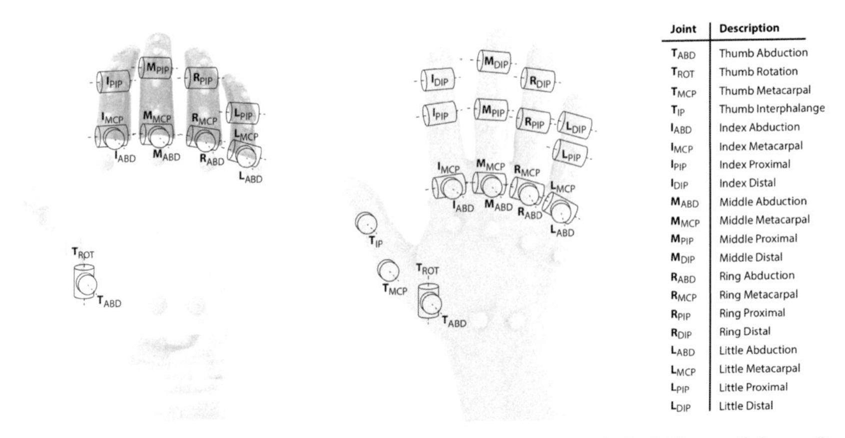

Joint	Description
T_{ABD}	Thumb Abduction
T_{ROT}	Thumb Rotation
T_{MCP}	Thumb Metacarpal
T_{IP}	Thumb Interphalange
I_{ABD}	Index Abduction
I_{MCP}	Index Metacarpal
I_{PIP}	Index Proximal
I_{DIP}	Index Distal
M_{ABD}	Middle Abduction
M_{MCP}	Middle Metacarpal
M_{PIP}	Middle Proximal
M_{DIP}	Middle Distal
R_{ABD}	Ring Abduction
R_{MCP}	Ring Metacarpal
R_{PIP}	Ring Proximal
R_{DIP}	Ring Distal
L_{ABD}	Little Abduction
L_{MCP}	Little Metacarpal
L_{PIP}	Little Proximal
L_{DIP}	Little Distal

FIGURE 7.5 Kinematic models showing different tracked joint angles in both Hannes (left panel) and the human hand (right panel). The table illustrates each joint abbreviation.

had 15 DoFs, whereas the human hand had 20 DoFs that we tracked and processed during the data processing and analysis phase (Fig. 7.5).

7.2.4 Results of kinematic synergies study

The linear regression analysis of the human postures, shown in Fig. 7.4, revealed correlations of $r > 0.8$, $P < 0.01$ between the MCP flexion adjacent fingers and the abduction of neighboring fingers, with the exception of the little finger. Similar correlations were observed on the Hannes hand. The Hannes hand showed large correlations even for the index–little fingers pair. This is consistent with the finger adjacency observed for the human hand. Additionally, the PIP and DIP joints' correlations within each finger and with neighboring fingers could be observed. The aforementioned correlations hinted at the possibility of dimensionality reduction for the postures of the Hannes hand. Hence, PCA was conducted. Specifically, the first synergy of the human hand posture accounts for about 39% of the total variance. The second and third principal components account for 23% and 13%, respectively. The first two principal components accounted for 62% of the variance, in cumulative terms. If the cumulative contribution is added up to the third and fifth components, the percentage rises to 76% and 87%, respectively. In the Hannes hand, on the other side, the first synergy is more predominant and accounts for about 75% of the variance. The second synergy only accounts for 12% of the variance. Similarly, in cumulative terms, the first two synergies account for 86% of the data variance. When the cumulative percentages are added up to the third and fifth components, the variance rises to 97% and 99%, respectively. This results is somehow expected due to the underactuated system with moderate intertrial variability for the Hannes hand compared to the human hand.

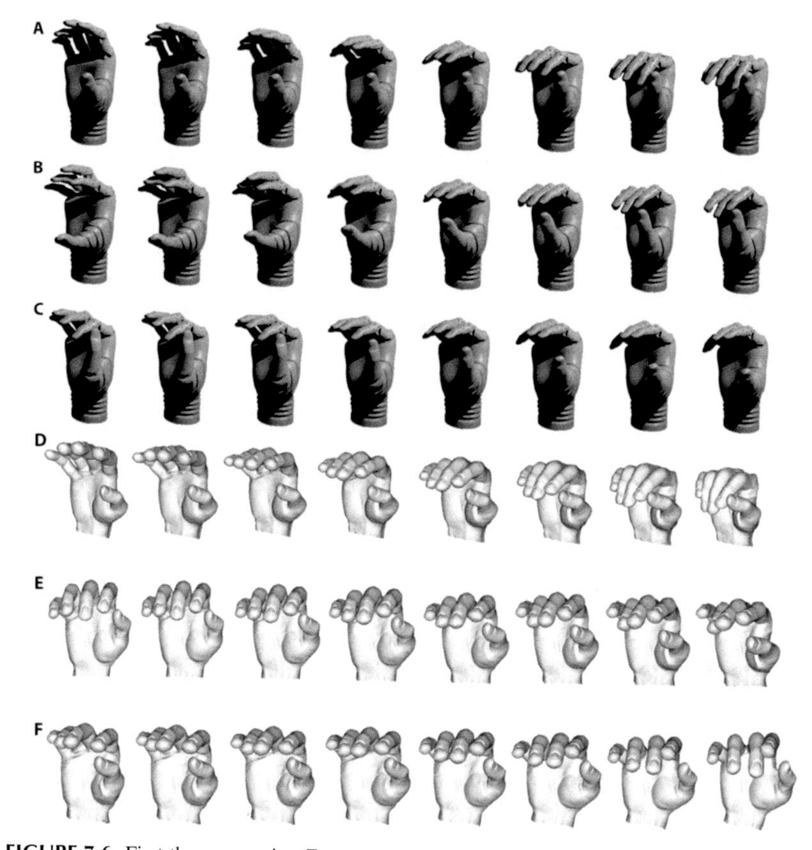

FIGURE 7.6 First three synergies. From top to bottom: Hannes (A) PC1, (B) PC2, and (C) PC3 and the human hand (D) PC1, (E) PC2, and (F) PC3 over the range of the minimum (far right) and maximum (far left) observed scores.

Fig. 7.6 shows the three principal components for both Hannes and human hand. The first synergy (principal component 1 [PC1]), the opening and closing of both human hands and the Hannes hand, was characterized by means of finger MCP flexion combined with a little thumb rotation and abduction. The second synergy (PC2) corresponds to the PIP flexion and extension of the fingers with thumb abduction also in both human and Hannes hands. The third synergy (PC3) corresponds to thumb rotation and thumb abduction for the Hannes hand whereas it corresponds to thumb rotation for the human hand. From the results of the analysis of the Hannes hand, it can be seen that PC1 present in the Hannes hand is consistent with the state of the art in human hand grasping [3,19]. The second and third synergies in the Hannes hand appear to contribute to the same DoFs involved in human synergies for grasping tasks, with the exception of the DIP joints and thumb movements. Overall, the latter result indicates that the observed synergies for the Hannes hand were similar to those of the human hand.

7.3 Discussion

In this work, we aimed to compare the human hand to a robotic hand in both the kinematic and the force domain. By doing so, we aimed to characterize the gap between human and state-of-the-art robotic hands. This study allowed us to interpret the functionality of human motor control and to which extent this is reproducible on robotic hands. The studies [16,15] endeavored to correlate a synergistic concept with experimental results from human participants. These two works showed how synergies could be the main concept that the CNS uses for controlling the human hand. From the force control standpoint, it could be stated that when a robotic system use a whole-hand force control approach to demonstrate the performance of human participants, similar force patterns as those adopted by the human hand appear, suggesting that humans could indeed use a synergistic force control strategy to achieve stable grasp. Regarding the kinematic behavior, the inclusion of a synergistic design approach concluded with the creation of the Hannes hand, which is not only anthropomorphic and biomimetic, but also replicates human-hand-like performance during the grasping of different objects. On the other hand, adopting the synergy concept allowed the resulting prosthetic device to have a reduced actuation complexity, which directly translates into a reduction of the number of components required for large objects variance.

7.3.1 Force synergies: human vs. robot

The experimental design of the work of the authors in [16] allowed participants to place their fingertips in an unconstrained manner, which had not been done in works preceding this experiment. The results of the study showed that the force responses in participants' fingertips had little time differences, suggesting a synchronous response, which is consistent with a synergistic control strategy of the human hand. Participants had a better performance when trying to stabilize the TACO object when periodic external perturbations were administered in the experiment. This suggests a feedforward motor strategy that is activated when external disturbances could be predicted. Participants used a feedforward, synergistic motor strategy regardless of the type of external perturbation or the number of digits they could use in the grasping task. When the experiment was replicated with the Shadow hand, it could be observed that a simple stiffness controller was able to reproduce the results obtained from the human counterparts. The resulting force distribution was qualitatively similar between human participants and the robotic hand. A crucial difference, however, was that humans were able to adapt their grasping strategies quickly after a few external disturbances. In the case of the robotic hand, this was not the case.

Since participants could freely choose where they wanted to place their fingers, a large variability of finger placement between participants could be observed. Nevertheless, initial digit position was highly preserved across experiments for individual participants. This could unveil the idiosyncrasy during grasping behaviors [17].

It is important to remark that the goal of the experiment was to observe whether participants controlled different digits' forces independently or in a synergistic fashion, and not whether they used a stiffness or force control. After replicating the experiments with the Shadow hand, it could be confirmed that participants were using a synergistic grasping strategy, using a single controller for the whole hand instead of individual finger force modulation.

The results also showed that the normal forces increased from perturbation to perturbation. This helped reduce the error in net hand torque. This suggests that fingertip forces are adapted across consecutive perturbations. This strategy was sufficient to counterbalance external disturbances on the TACO for both cases in which predictable and unpredictable perturbations were applied. The main difference between these two types of perturbations, however, is that the error in hand net torque decreased faster when perturbations were predictable. An explanation for this would be that precise prediction of the perturbation sequence improves muscle responses, and can therefore reduce latency of muscle activity in relation to the mechanical perturbation [20,21].

The results of the authors in [16] have shown that one force synergy is sufficient to resist different types of external perturbations (either predictable and unpredictable) while achieving a stable grasp. It has been shown that two force synergies accounted for more than 90% of the variability in digit normal force [13,22,12,9]. In our study 1, the first force synergy tends to be voluntary whereas the second one is involuntary. In other words, the first synergy is characterized by full hand synchronous stiffening in order to counteract the external perturbation. The second synergy was produced due to the interaction between the digits and the TACO. Thus, we suggest here that the second synergy is more mechanical. This was confirmed in the shadow robotic hand experiment where the hand was controlled using one synergy (full stiffness closure). In the second study, discussed next, we showed that arbitrary net force and/or torques produced by force closure of a robotic hand satisfy a stable grasp in the absence of friction forces [15,10]. With this work, we showed that one force synergy is sufficient to resist different types of external perturbations (both predictable and unpredictable) while achieving a stable grasp.

7.3.2 Kinematic synergies: human vs. robot

In the second study, we compared the human hand kinematics versus the robotic prosthetic hand. The PCA results of study 2 showed that the first three components of the Hannes hand are quasisimilar to those observed in the human hand. This is in accordance with previous studies on human kinematic synergies that showed that two kinematic synergies accounted for more than 80% of the variability in digit joint angles [3]. The Hannes hand constitutes an excellent example of how synergies can be used to design and control artificial hands, which could make contributions not only to the prosthetics but also to the robotics field. Normally, key aspects including anthropomorphism, biomimicry, and human-like grasping performance are desirable in such artificial devices [23].

The Hannes hand is equipped with one actuation to implement the first human synergy with high fidelity. However, other synergies were obtained and observed in the experimental results of study 2. These additional synergies result from the mechanical intrinsic ability of the prosthetic device to adapt to the shape of the grasped objects. It could be seen that some minor motion discrepancies appeared on the second and third principal components on the Hannes hand, compared to the human hand. This is not an issue, however, as it is well known that only the first synergy requires an assumption of repeatable behavior. Higher-order synergies tend to vary depending on grasping and manipulation tasks requirements.

In this study, we showed that arbitrary net force and/or torques produced by force closure of a Hannes hand achieve a stable grasp in the absence of friction forces. In addition, in the latter work we showed that with single actuation, equivalent to one kinematic synergy, the Hannes was able to achieve stable grasps by grasping different daily objects.

The Hannes hand showcases the usage of kinematic synergistic concepts when designing artificial hands. Research results point to the fact that human hands are built and controlled through controllers that can be approximated and explained by synergies. Therefore, the implementation of such concepts on artificial hands is the logical step to follow when attempting to find a balance between what current technologies are capable of and the highest dexterous performance that can be achieved.

7.4 Conclusions

This chapter has focused on giving the reader a good overview on differences between human and robotic hands. These differences were shown in terms of force and kinematic synergies in order to answer the question of how the human CNS controls the human hand to achieve a stable grasp. Several research works have striven to unveil conclusive answers to this question. Although it is likely that the overall control strategy for human hands is far more complex than a synergistic approach, works such as the ones in [16,15] have helped the research community to gain a deeper understanding of how human hands work on a kinematic and force control level. A synergistic control and design could confirm that controlling each individual DoF of such complex systems as the human hand is greatly exacting, even for the human CNS, which is far more developed than current technological means. This knowledge, if used properly, can have direct implications in one of the open research fields in robotics and prosthetics, the development of dexterous biomimetic artificial hands. In turn, a deeper understanding of robotics and human manipulation will endow the robotics community with resources to expand the robotics field and allow robotics research to develop better robotic hands for both human and robots.

Acknowledgments

This work was funded by the Lighthouse Initiative Geriatronics by StMWi Bayern (Project X, grant no. 5140951), the LongLeif GaPa gGmbH (Project Y, grant no. 5140953), the German Research Foundation (DFG, Deutsche Forschungsgemeinschaft) as part of Germany's Excellence Strategy - EXC 2050/1 - Project ID 390696704 - Cluster of Excellence "Centre for Tactile Internet with Human-in-the-Loop" (CeTI) of Technische Universität Dresden, and the INAIL (Istituto Nazionale per l'Assicurazione contro gli Infortuni sul Lavoro) under grant agreements PPR1A, PPR-AS, and PR19-PAS-P1.

References

[1] Aristotle, Clarendon Aristotle Series: On the Parts of Animals, Oxford University Press, Jan 2002.

[2] Marco Santello, Gabriel Baud-Bovy, Henrik Jörntell, Neural bases of hand synergies, Frontiers in Computational Neuroscience 7 (2013) 23.

[3] Marco Santello, Martha Flanders, John F. Soechting, Postural hand synergies for tool use, Journal of Neuroscience 18 (23) (1998) 10105–10115.

[4] N. Bernstein, The Co-Ordination and Regulation of Movements, Pergamon Press, Oxford, UK, 1967.

[5] Jason Friedman, Tamar Flash, Task-dependent selection of grasp kinematics and stiffness in human object manipulation, Cortex 43 (3) (April 2007) 444–460.

[6] Alexander V. Terekhov, Pesin B. Yakov, Xu Niu, Mark L. Latash, Vladimir M. Zatsiorsky, An analytical approach to the problem of inverse optimization with additive objective functions: an application to human prehension, Journal of Mathematical Biology 61 (3) (2010) 423–453.

[7] E. Todorov, Z. Ghahramani, Analysis of the synergies underlying complex hand manipulation, in: The 26th Annual International Conference of the IEEE Engineering in Medicine and Biology Society, vol. 2, 2004, pp. 4637–4640.

[8] Mark L. Latash, Movements that are both variable and optimal, Journal of Human Kinetics 34 (2012) 5–13.

[9] Vladimir M. Zatsiorsky, Robert W. Gregory, Mark L. Latash, Force and torque production in static multifinger prehension: biomechanics and control. I. Biomechanics, Biological Cybernetics 87 (1) (2002) 1–19.

[10] Antonio Bicchi, Marco Gabiccini, Marco Santello, Modelling natural and artificial hands with synergies, Philosophical Transactions of the Royal Society of London. Series B, Biological Sciences 366 (1581) (2011) 3153–3161.

[11] Michael T. Turvey, Action and perception at the level of synergies, Human Movement Science 26 (4) (2007) 657–697.

[12] Jaebum Park, Vladimir M. Zatsiorsky, Mark L. Latash, Optimality vs. variability: an example of multi-finger redundant tasks, Experimental Brain Research. Experimentelle Hirnforschung. Expérimentation cérébrale 207 (1–2) (Nov 2010) 119–132.

[13] Abdeldjallil Naceri, Marco Santello, Alessandro Moscatelli, Marc O. Ernst, Digit Position and Force Synergies During Unconstrained Grasping, Springer International Publishing, Cham, 2016, pp. 29–40.

[14] C. Piazza, G. Grioli, M.G. Catalano, A. Bicchi, A century of robotic hands, Annual Review of Control, Robotics, and Autonomous Systems 2 (2019) 1–32.

[15] M. Laffranchi, N. Boccardo, S. Traverso, L. Lombardi, M. Canepa, A. Lince, M. Semprini, J.A. Saglia, A. Naceri, R. Sacchetti, et al., The Hannes hand prosthesis replicates the key biological properties of the human hand, Science Robotics 5 (46) (2020).

[16] Abdeldjallil Naceri, Alessandro Moscatelli, Robert Haschke, Helge Ritter, Marco Santello, Marc O. Ernst, Multidigit force control during unconstrained grasping in response to object perturbations, Journal of Neurophysiology 117 (5) (2017) 2025–2036.
[17] Abdeldjallil Naceri, Alessandro Moscatelli, Marco Santello, Marc O. Ernst, Coordination of multi-digit positions and forces during unconstrained grasping in response to object perturbations, in: 2014 IEEE Haptics Symposium (HAPTICS), IEEE, Feb 2014, pp. 35–40.
[18] Frank Rothling, Robert Haschke, Jochen J. Steil, Helge Ritter, Platform portable anthropomorphic grasping with the Bielefeld 20-DOF shadow and 9-DOF TUM hand, in: 2007 IEEE/RSJ International Conference on Intelligent Robots and Systems, IEEE, Oct 2007, pp. 2951–2956.
[19] Cosimo Della Santina, Matteo Bianchi, Giuseppe Averta, Simone Ciotti, Visar Arapi, Simone Fani, Edoardo Battaglia, Manuel Giuseppe Catalano, Marco Santello, Antonio Bicchi, Postural hand synergies during environmental constraint exploitation, Frontiers in Neurorobotics 11 (2017) 41.
[20] J.D. Cooke, S.H. Brown, Movement-related phasic muscle activation, Experimental Brain Research 99 (3) (1994).
[21] K. Akazawa, T.E. Milner, R.B. Stein, Modulation of reflex EMG and stiffness in response to stretch of human finger muscle, Journal of Neurophysiology 49 (1) (1983) 16–27.
[22] Yen-Hsun Wu, Vladimir M. Zatsiorsky, Mark L. Latash, Multi-digit coordination during lifting a horizontally oriented object: synergies control with referent configurations, Experimental Brain Research. Experimentelle Hirnforschung. Expérimentation cérébrale 222 (3) (Oct 2012) 277–290.
[23] Christian Cipriani, Marco Controzzi, Maria Chiara Carrozza, The SmartHand transradial prosthesis, Journal of NeuroEngineering and Rehabilitation 8 (1) (2011) 29.

Chapter 8

Learning form-closure grasping with attractive region in environment

Rui Li[a], Zhenshan Bing[b], and Qi Qi[a]
[a]*School of Automation, Chongqing University, Chongqing, China,* [b]*Department of Informatics, Technical University of Munich, Munich, Germany*

8.1 Background

Robotic manipulation is an emerging area of research in robotic systems. When compared with typical grippers, the robotic manipulator is better suited for complex manipulation tasks owing to its many advantages. The dexterous hand of the robot manipulator is expected to attain the flexibility of the human hand, which is highly desired in home service and flexible manufacturing. Robotic grasping, in this process, is the most important prerequisite for dexterous manipulation.

To grasp *dexterously*, the gripper must first grasp *firmly*. To achieve a *firm* grasp, the closure property of a given object has to be determined first. Force closure and form closure have been proposed for quite a long time. Usually, objects are grasped with force closure, where friction prevents motion in most directions. When objects are grasped by means of form closure, motion in all directions is prevented by frictionless constraints. These methods provide better grasp stability, ensuring good postgrasp manipulation.

With force/form closure, we may define and determine whether a grasp is firm. In the last 30 years, extensive research efforts in this field have led to the development of various technologies, from 2D to 3D and from two-pin grippers to five-fingered dexterous hands. However, we are yet to determine a general method to design the grasp for an arbitrary object. Constructing environmental constraints (ECs), which visualize and utilize the geometric information of the object, the gripper, and the interacting environment in a configuration space, is an effective solution to this issue. The configurations that correspond to force/form closures are characterized as local extrema in such spaces.

We have seen many encouraging results in this direction. Nevertheless, these works still have the following problems. (1) The precision of these works is highly dependent on the geometric models of the gripper and the object, which

Tactile Sensing, Skill Learning, and Robotic Dexterous Manipulation
https://doi.org/10.1016/B978-0-32-390445-2.00016-7

are not always easy to extract in many cases. (2) A difficult trade-off must be made between precision and speed, which causes such methods to fail to work in real-time.

In this chapter, we first review the work on robotic grasping with ECs. Then, to tackle the aforementioned problems, we provide two separate solutions to improve the performance.

8.2 Related work

8.2.1 Closure properties

Closure properties have been under much investigation in the area of robotic grasping ever since form closure and force closure were proposed and introduced [1,2].

Lynch and Park defined the *form closure* state of a body as a set of stationary constraints that prevent all motions of the body. A *form-closure grasp* is a set of constraints provided by the robot's fingers [3]. Meanwhile, the *force closure* state concerns a body and the number of frictional contacts that prevent all motion of the body. In mathematical terms, the force closure state refers to the composite wrench cone, which contains the entire wrench space, so that any external wrench F_{ext} on the body can be balanced by contact forces.

To conservatively determine if a grasp is a form-closure grasp, a first-order form-closure test is conducted in the following manner [4]:

- Check if the $\text{rank}(H) = n$. If not, the columns of H cannot positively span $\mathbb{R}^n$, and thus there is no form closure.
- If the above condition holds, then solve the linear program (LP)

$$\begin{aligned} \text{find} \quad & x \\ \text{minimizing} \quad & \mathbf{1}^\top x \\ \text{such that} \quad & Hx = 0 \\ & Ix - \mathbf{1} \geq 0, \end{aligned} \tag{8.1}$$

where $I \in \mathbb{R}^{k \times k}$ is the identity matrix and $\mathbf{1} \in \mathbb{R}^k$ is a vector with all components of 1. If H is the full rank and the LP of Eq. (8.1) is feasible, then the grasp is a form-closure grasp.

Initially, both form-closure and force-closure grasps were treated as deterministic problems. Given a series of contact points, one can determine whether these points can achieve a form or force closure [5]. However, it is difficult to construct contacts that satisfy an arbitrary closure property. The caging grasp, which was proposed later, provided a looser condition to keep the target object in hand after gripping [6,7]. The concept has been extensively discussed and applied [8], but the closure property still needs to be considered to achieve a stable grasp.

8.2.2 Environmental constraints

ECs are constraints caused by the environment. In real-world scenarios, objects are always placed in an environmental context. Eppner defines robotic grasp strategies that can benefit from contact with the environment as the *exploitation of environmental constraints* [9]. Using ECs, it is possible to design effective manipulation strategies for a robot without the aid of a compliant mechanism or a high-precision sensor. Such principles are possible because humans have quite a similar working mechanism when interacting with objects. We always use tables, walls, etc., to make our grasps easier.

The attractive region in environment (ARIE) is a type of constrained region formed by the environment that exists in the configuration space of the robotic system. This concept was further developed by [10,11] to achieve high-precision sensorless manipulation in production. Through a unique method of formulation and utilization of the attractive region in the configuration space, a strategy to attain high-precision assembly in physical space without a force sensor and flexible wrist was designed, and an approach to achieve 2D and 3D part orientation by sensorless grasping and pushing actions was also proposed.

Based on this theory, several high-precision robotic manipulation tasks have been developed. Several methods have been developed for specific applications. For example, for automobile manufacturing, Su et al. designed an eccentric peg-hole sensorless assembly system with an ARIE-based strategy [12,13]. They also developed a vision-based 3D grasping planning approach using a single image [14]. Liu et al. developed a stable sensorless localization method for 3D objects with a simple pushing mechanism [15]. They further created a vision-based 3D grasping algorithm for grasping 3D objects using a simple 2D gripper [16]. Li and Qiao reported an ARIE-based robotic manipulation strategy for general convex peg-convex hole insertion tasks [17].

In a recent study, Qiao et al. discussed the definition and the generalized conditions of the ARIE [18]. In this work, a general mathematical description of ARIE is presented, and the condition of the existence of the ARIE in different configuration spaces was analyzed. Notably, the relationship of the ARIE in high- and low-dimensional spaces is discussed.

8.2.3 Learning to grasp

Recent researchers have attempted to minimize human participation during robotic manipulation. Instead of *programming* the robot to grasp, they prefer *teaching* it to learn how to grasp. In this process, visual information is provided as the raw input to train the robot for end-to-end, goal-oriented manipulation tasks. To achieve this, researchers first attempted to collect sufficient data for robot manipulation tasks. They studied how to effectively label and preprocess such data. Various datasets for robot task learning have been developed in recent years. For example, Bullock et al. provided a dataset for studying the grasping actions of people. They recorded the grasping actions in home and factory environments, and manually marked different grasp types, objects, and task parameters [19]. Cai et al. designed a UT-grasping dataset. The dataset consisted of four different subjects, who were asked to grasp a series of objects in a controllable environment (in front of the desktop) after telling them how to act [20].

The researchers then studied how to make a robot adapt quickly to task changes. In this scenario, the robot should be able to promptly replan its path to fulfill the new task requirement. Krabbe et al. proposed an interval estimation optimization algorithm based on the fusion of support vector machine (SVM) and principal component analysis (PCA) to train the robot terminal motion parameters and plan the motion trajectory based on the experimental data [21]. Berczi et al. carried out research on autonomous robot grasping based on deep learning. Through the training of grasping modes, the robot could reliably grasp different kinds of routinely used objects, such as a water cup, a key, an eyeglass box, and a book [22]. Duque et al. achieved the robotic assembly operation without hard-coding the process in the algorithm. Using a task parametrized Gaussian mixture model (TP-GMM), the robots learned the motion from human demonstrations [23]. Li et al. implemented a skill acquisition method based on deep reinforcement learning for a low-voltage apparatus assembly [24], where they achieved a success rate of nearly 80%.

In conclusion, the existing robot learning methods require large sample data, and the coverage and size of the sample data have a significant impact on the learning performance. However, the process of acquiring manipulation data is time consuming and laborious.

Current directions of research include learning from smaller datasets [25,26], improving the learning efficiency [27], and training the robot in the simulation and adapting it to the real world without additional adjustments [28]. The improvement of the precision of learning-based robotic assembly tasks while preserving its ability to generalize also remains an open topic of research.

8.3 Learning a form-closure grasp with attractive region in environment

8.3.1 Attractive region in environment for four-pin grasping

The concept of ARIE was initially proposed by Qiao [29]. The concept was inspired by the following phenomenon. Assume that there is a soup bowl on a table and a pea above the bowl. The initial position of the pea is randomly set but within the range of the mouth of the bowl. If we drop the pea, it will fall into the bowl under the effect of gravity. After some time, the pea remains steady at the bottom of the bowl. This concept is formalized mathematically as follows.

Definition 1 (ARIE). Assume that the state of a system can be characterized as

$$\frac{\mathrm{d}x}{\mathrm{d}t} = f(x(t), u(t)), \tag{8.2}$$

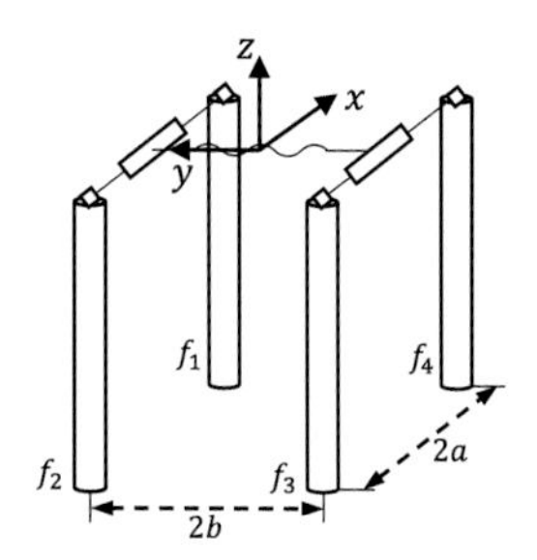

FIGURE 8.1 The four-pin gripper model.

where $x(t) \in X$ is the state of the system and X is the domain of definition. If there exists

1. a region $\Omega \in X$,
2. a self-constructed state-dependent function $g(x)$, and
3. a state-independent input $u(t)$ (other than $u(x)$)

satisfying for all x in Ω that

1. $g(x) > g(x_0)$ when $x \neq x_0$ and $g(x) = g(x_0)$ when $x = x_0$, where $x_0 \in \Omega$, and
2. $g(x)$ has continuous partial derivatives with respect to all components of x,

then the system will be stable in the region Ω, which is defined as the ARIE.

ARIE exists in various robotic manipulation tasks, such as assembly, localization, and grasping. We start the discussion of ARIE in four-pin grasping by introducing the four-pin gripper model as follows.

The four-pin gripper is depicted in Fig. 8.1. The gripper is composed of four parallel column-like fingers (or pins). A gripper frame (expressed by $\{G\}$) is fixed at the top center of the gripper. Each pin can move along the x-axis. The positions of the four pins on the gripper are represented by the coordinates of the end points of the pins as $f_i = [x_i, y_i, z_i]^\top$, $i = 1, 2, 3, 4$. Owing to the symmetry of the mechanical design, the coordinates of the end points are expressed by $\{G\}$ as

$$\begin{cases} f_1^G = [a, b, -h]^\top, \\ f_2^G = [-a, b, -h]^\top, \\ f_3^G = [-a, -b, -h]^\top, \\ f_4^G = [a, -b, -h]^\top, \end{cases} \tag{8.3}$$

where a and b are the displacements along the x-axis and y-axis, respectively, h is the height from the end point of the pin to the origin of $\{G\}$ along the z-axis, and $a, b, h \in \mathbb{R}^+$. The four pins are divided into two groups as $\overline{f_1 f_2}$ and $\overline{f_3 f_4}$. The distance between the pins within a group is defined as the *intragroup* distance $\mathcal{P} = \left\|\overline{f_1 f_2}\right\| = \left\|\overline{f_3 f_4}\right\| = 2a \in [0, \mathcal{P}^+]$, and the distance between the two groups is defined as the *intergroup* distance $\mathcal{Q} = 2b \in [0, \mathcal{Q}^+]$, which is the length of the common normal line between $\overline{f_1 f_2}$ and $\overline{f_3 f_4}$ ($\mathcal{P}^+$ and $\mathcal{Q}^+$ are the upper limits of the distance owing to the mechanical structure). In accordance with the above definitions, the gripper is characterized by two variables, $\mathcal{P}$ and $\mathcal{Q}$.

To form the ARIE, two additional coordinate frames are defined as follows (Fig. 8.2): $\{W\}$ is an inertial frame fixed in the workspace and $\{B\}$ is fixed to the object, the origin of which coincides with the center of mass of the object. The configuration c_{obj} of a rigid body in 3D space is described by six numbers $c_{\text{obj}}(x, y, z, \theta_x, \theta_y, \theta_z) \in \text{SE}(3)$, where (x, y, z) represents the displacement from $\{B\}$ to $\{W\}$ and $(\theta_x, \theta_y, \theta_z)$ are the orientations of $\{B\}$ relative to $\{W\}$, expressed by Euler angles. When finite states exist (let the number be m), where the object stays on a plane stably, the configuration of the object is parametrized as

$$c_{\text{obj}}(S_i) = c_{\text{obj}}\left(x, y, z^{S_i}, \theta_x^{S_i}, \theta_y^{S_i}, \theta_z\right) \tag{8.4}$$

$$= c_{\text{obj}}(x, y, \theta_z | S_i), \ i = 1, 2, \cdots, m, \tag{8.5}$$

where z^{S_i}, $\theta_x^{S_i}$, and $\theta_y^{S_i}$ are constant values determined by the stable state S_i. For example, the gray object shown in Fig. 8.2 has six stable states, $S_1, S_2, \cdots, S_6$. For each stable state, the configuration of the object is described by three variables, (x, y, θ_z).

When the gripper grasps an object, it also creates several ECs on the object. Considering the defined gripper variables and object configuration, the relationship between the gripper and the object (rigid body) is described as follows.

Let $\{G\}$ be fixed relative to $\{W\}$ and let $\{B\}$ be transformed relative to $\{W\}$ by varying x and θ_z in $c_{\text{obj}}(S_i)$. If the value of the gripper variable $\mathcal{P}$ is specified as $\mathcal{P} = \mathcal{P}^*$, then for any given (x, θ_z), there is a unique value of the closing distance d, defined as

$$d(x, \theta_z | S_i) = \left\{\min \mathcal{Q} | \mathcal{P} = \mathcal{P}^*, c_{\text{obj}(S_i)} = \left(x, y^*, \theta_z\right)\right\}, \tag{8.6}$$

where the constant y^* translates $\{B\}$ to the origin of $\{G\}$ along the y-axis.

Fig. 8.3 shows an ARIE of the four-pin gripper grasping a pentagon object. The axes represent the object variables x, θ_z, and the closing distance d, respectively. The three variables form a periodic region in the configuration space. Each point on the surface of the region corresponds to a grasping state for the gripper and the object. According to [30], the region formed by (x, θ_z, d) is an ARIE, and the acute bottom of the region indicates a form-closure grasp configuration.

The terms c_3, c_6, and c_9 correspond to form-closure grasps, but the rest do not because they are not at "acute" bottoms. The trajectory along the surface of the region indicates a continuous transformation of the gripper configuration. If

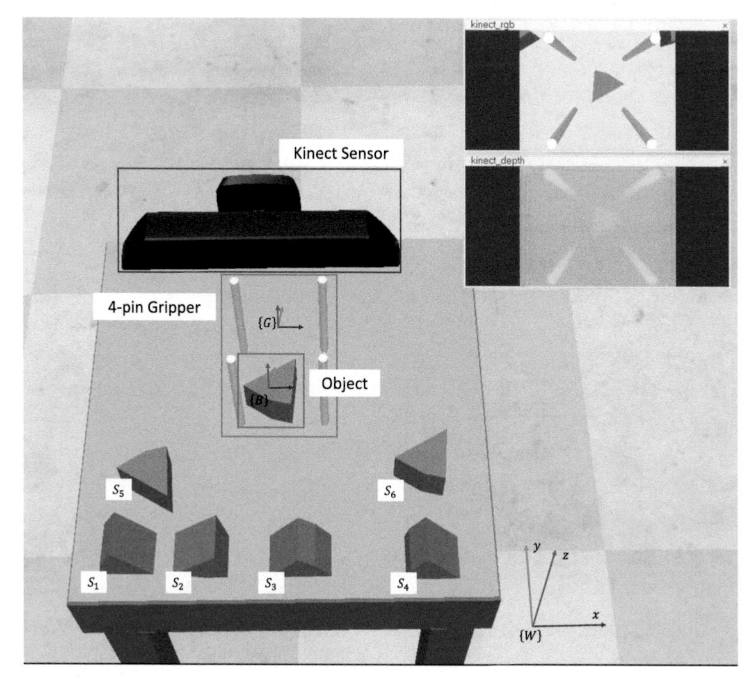

FIGURE 8.2 The illustration of the coordinate frames, the stable states of the object, and the experimental setup.

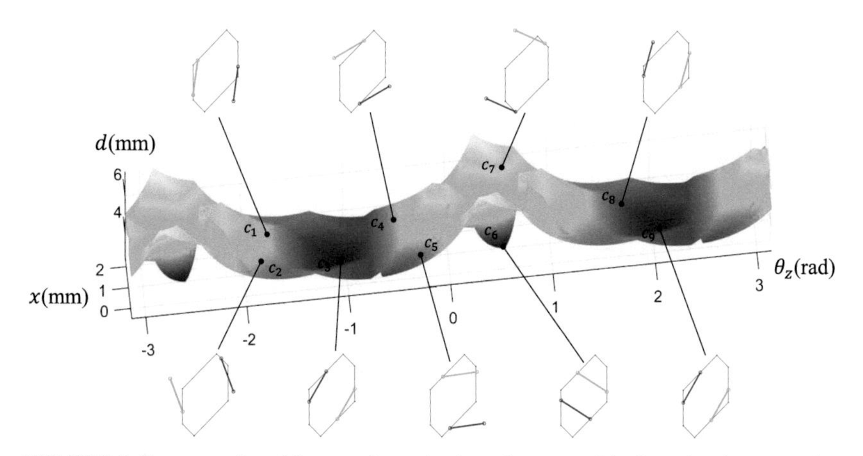

FIGURE 8.3 Representation of the attractive region in environment of the four-pin gripper grasping a pentagon object.

there is a trajectory starting at a certain point and subsequent points are derived by a gradient descent rule, then it represents a gripper-closing action from the initial state. For example, if the initial state of the system is at c_4, then with a decrease in the closing distance d, it gradually converges to the state c_3 and achieves a form-closure grasp. If the state of the system starts at c_5, then it tends to lose the form-closure grasp since such an "acute-bottom" state on its trajectory is nonexistent.

Therefore, the ARIE provides a solution to check if a grasp configuration can converge to a form-closure state. We further extended the process using the watershed algorithm [31] as follows:

1. We establish the ARIE for the given object and gripper.
2. We locate all the local minima (acute bottoms) in the region.
3. We divide the region into several subregions using the watershed algorithm.
4. For each subregion, if there is one acute bottom, then all the grasp configurations within this subregion tend to converge to this local minimum, with the gripper closing action.
5. For any grasp configuration, we only need to check which subregion the configuration belongs to, to know if the configuration converges to a form-closure state.

Finally, to measure the quality of the form-closure grasp configurations, the grasp metric function $\text{Qual}\left(c_{\text{obj}}\left(S\right), \mathcal{P}\right) = \min_i \left(x_i^*\right),\ i = 1, \cdots, k$, is defined to indicate the likelihood that the current grasp configuration escapes from the form closure, where x^* is the solution of Eq. (8.1). After ranking the scores of all the acute-bottom states, an optimal state is selected. The complete scheme of the algorithm is provided in Algorithm 8.1.

8.3.2 Learning to evaluate grasp quality with ARIE

In the previous section, the relationship between form-closure grasps and the local minima in ARIE was established. In practice, we discovered that there is usually more than one form-closure grasp configuration for an arbitrary object and the four-pin gripper. However, not all of the configurations are of the same quality in terms of various disturbances in the real environment. From this perspective, it is vital to select the configuration with the optimal grasping quality to guarantee the success rate of grasping.

To achieve this, a grasp quality measurement (GQM) method is introduced in this section. A GQM function was trained for this purpose. Given the object and grasping points, a score is given by the function as the quality of the grasp. Based on this score, a strategy to generate a robust grasping configuration is proposed.

8.3.2.1 Formation of the grasp quality measurement function

To output the score Γ of a grasp, the following inputs are required: the profile of the object O, pose (configuration) of the object c_{obj}, grasping points G, and parameters of the gripper p_{h}. Then, the function can be defined as

$$\Gamma = f_{\text{GQM}}\left(O, c_{\text{obj}}, G, p_{\text{h}}\right). \tag{8.7}$$

Algorithm 8.1 Calculation of the optimal form-closure grasp configuration.

Require: the shape of the object ∂O,
the range of the gripper parameter $p_i \in [p_{\min}, p_{\max}]$

Ensure: the optimal form-closure grasp configuration $C^* = (x, \theta_z, d)$

```
1: Generate linearly spaced vectors:
   P = linspace(p_min, p_max, N_P)
   X = linspace(x_min, x_max, N_X)
   Θ = linspace(0, 2π, N_Θ)
2: for i ∈ N_P do
3:    Generate potential form-closure grasps:
      Ω^(i) = gen_arie(∂O, X, Θ, P^(i))
      C^(i) = get_local_minima(Ω^(i))
      Ω̂ = water_shed(Ω^(i))
4:    Find all form-closure grasps:
5:    for j_C ∈ C^(i) do
6:       if is_form_closure(j_C) then
7:          for ω_C ∈ Ω̂ do
8:             if is_in_subregion(ω_C, j_C) then
9:                C_fc ⇐ [ω_C, j_C]
10:            end if
11:         end for
12:      end if
13:   end for
14: end for
15: Rank all the form-closure candidates:
    [C_fc, f_fc] = form_closure_measure(C_fc)
    [Ĉ_fc, f̂_fc] = sort(f_fc, [C_fc, f_fc])
    C* = Ĉ_fc[0]
```

In particular, the information of O, c_{obj}, and G can be encoded in a single image $\text{IM}(O, c_{\text{obj}}, G)$. Therefore, the GQM function can be simplified as

$$\Gamma = f_{\text{GQM}}(\text{IM}, p_{\text{h}}). \tag{8.8}$$

In our system, a neural network with learnable parameters is adopted as f_{GQM}. The output of the network is the quality of the grasp with the given contact points (gripper parameters). We implemented a training set to train the network. We collected 9988 samples by dropping 118 object models from the DexNet object dataset [32] to a virtual desk and then taking a picture of it using a top-view RGB-D camera (Fig. 8.2). The corresponding quality scores of these samples were calculated using the method introduced in Section 8.3.1 as the supervision signal for the network training.

8.3.2.2 Uncertainties during grasping process

To select the optimal grasp configuration, we consider three types of uncertainties that are most likely to occur during grasping, as illustrated in Fig. 8.4: (1) the uncertainty in the relative pose between the object and the gripper, (2) the uncertainty in the gripper parameters, and (3) the uncertainty in the object shape.

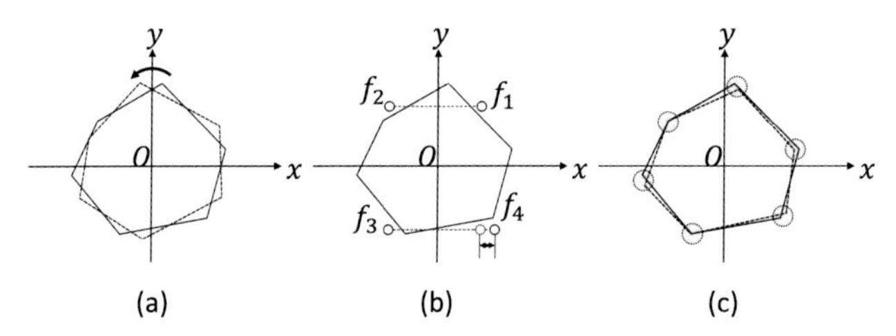

FIGURE 8.4 Three major uncertainty types. (a) Relative pose uncertainty between object and gripper. (b) Gripper parameter uncertainty. (c) Object shape uncertainty.

We model the aforementioned uncertainties using the Gaussian process regression model because it is widely used to estimate continuous functions with sparse and noisy data [33]. We assume that the relative pose uncertainty of the object ν is subject to a Gaussian normal distribution with mean 0 and variance Σ_ν, that is, $\nu \sim \mathcal{N}(0, \Sigma_\nu)$. Then, the pose of the object can be expressed as $c_{\text{obj}} = \hat{c}_{\text{obj}} + \nu$, where $\hat{c}_{\text{obj}}$ is the real pose of the object. Similarly, we assume that the uncertainty of the gripper parameters λ is subject to the Gaussian normal distribution with mean 0 and variance Σ_λ, that is, $\lambda \sim \mathcal{N}(0, \Sigma_\lambda)$, and the uncertainty of the object shape γ is subject to the Gaussian normal distribution with mean 0 and variance Σ_γ, that is, $\gamma \sim \mathcal{N}\left(0, \Sigma_\gamma\right)$. Then, the gripper parameter can be expressed as $p_{\text{h}} = \hat{p}_{\text{h}} + \lambda$, where $\hat{p}_{\text{h}}$ is the real gripper parameter. The object shape can be expressed as $\mathbf{v} = \hat{\mathbf{v}} + \gamma$, where $\mathbf{v} = \mathbf{v}(x_v, y_v)$ is the set of vertex points of the profile of the object O and $\hat{\mathbf{v}}$ is the real value of the vertices.

For each type of uncertainty, we use the *virtual grasping process* to test whether a given grasp configuration is still form-closed when these uncertainties occur. The virtual grasping process is defined as follows.

Definition 2 (Virtual grasping process). For the gripper shown in Fig. 8.1, the virtual grasping process is a process in which the two groups of fingers ($\|\overline{f_1 f_4}\|$ and $\|\overline{f_2 f_3}\|$) close along the gripper's x-axis. The closing of each group stops immediately when any finger within the group touches the object.

In the definition, *virtual* means that the object to be grasped will not move during the entire grasping process. The process can be abstracted as a function $\mathcal{Q} = f_{\text{vgp}}(x, \theta_z)$, where $\mathcal{Q}$ is the intergroup distance defined in Section 8.3.1 and (x, θ_z) is the element that describes the configuration of the object defined in Section 8.3.1.

8.3.2.3 Calculation of the grasp quality score

After determining the uncertainty types for each grasp, we can calculate the grasp quality score for each c_{obj} given p_h. For each uncertainty type described in Section 8.3.2.2, we take $M_j (j = 1, 2, 3)$ samples (in total, $\sum_{j=1}^{3} M_j$ samples) to form the sample set. For each sample, we used the ARIE algorithm to test if it was a form-closure grasp. Separately, there will be K_j $(j = 1, 2, 3)$ samples (in total, $\sum_{j=1}^{3} K_j$ samples) satisfying form closure.

Therefore, for each c_{obj}, given p_h, the score is calculated as

$$\hat{\Gamma} = g\left(c_{\text{obj}} \mid p_h\right) = \frac{\sum_{j=1}^{3} K_j}{\sum_{j=1}^{3} M_j}, \quad \hat{\Gamma} \in [0, 1]. \tag{8.9}$$

The scores range from 0 to 1. A larger score means that the corresponding grasp configuration can achieve form closure in a more robust way, which in this section is referred to as the "grasp quality."

However, it should be mentioned that it is very time consuming to obtain the score for each grasp configuration, because the virtual grasping has to be performed $\sum_{j=1}^{3} M_j$ times to obtain the score. Hence, a neural network is trained to approximate the time-consuming function.

8.3.2.4 Training of the network for grasp quality measurement

Network structure. A convolutional neural network (CNN), called the Grasp Score Regression Network, is designed to evaluate the grasp robustness, as illustrated in Fig. 8.5. The network consists of two branches. The first branch, similar to AlexNet [34], processes the normalized depth images by multiple convolution layers, and the second branch consists of fully connected layers for projecting the gripper width to a high-dimensional space. A dense layer merges these two branches to predict the final grasp score. The output of the network is a float number in the range from 0 to 1, representing the grasping quality. The larger the score, the more robust is the grasping.

Network training. The proposed network was optimized by the supervision of the grasp quality score mentioned in Section 8.3.2.1. Given a set

$$\left\{\left(\text{IM}^1, p_h^1, \hat{\Gamma}^1\right), \ldots, \left(\text{IM}^N, p_h^N, \hat{\Gamma}^N\right)\right\}$$

of N training samples generated from the simulation environment, we calculate the whole regression loss that consists of the squared error between the prediction and the target with regularization loss addition [35] as follows:

$$L\left(\text{IM}, p_h, \Gamma\right) = \sum_{i=1}^{N} \left(\Gamma^i - \hat{\Gamma}^i\right)^2 + \phi * \|W\|, \tag{8.10}$$

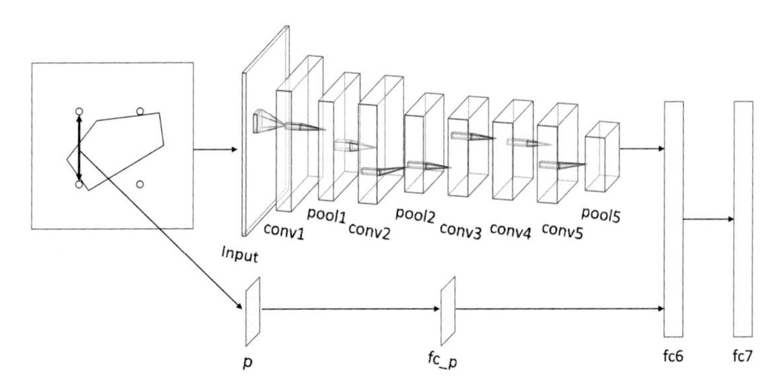

FIGURE 8.5 Structure of the Grasp Quality Measurement Network.

where IM and p_h are the inputs of the network, Γ is the network output, $\hat{\Gamma}^i$ is the corresponding label of the input IM^i and p_h^i, W is the network parameter to be optimized, and ϕ is the regularized factor that guarantees sparsity of parameters, which is helpful in overcoming the network overfitting problem. To minimize the loss, the parameters are updated using stochastic gradient descent (SGD):

$$w_{t+1} = w_t - \eta_{t+1} \nabla L \left(\mathrm{IM}^i, p_h^i, \Gamma^i \right), \tag{8.11}$$

where w_{t+1} and η_{t+1} represent the parameters and learning rate at time $t+1$ and $\nabla L \left(\mathrm{IM}^i, p_h^i, \Gamma^i \right)$ denotes the loss gradient relative to w_t. After sufficient iterations, the network learns to estimate the quality of every candidate grasping configuration facing uncertainties in a real grasping environment.

As mentioned in Section 8.3.2.1, we used 118 objects from the DexNet dataset to generate the dataset to train our Grasp Quality Measurement Network. The sample images were captured by a top-view RGB-D camera, which was mounted over the table and gripper, with a resolution of 640×480 at 30 Hz. For each object, we designed the following process to create a labeled observation:

1. For each of the $N_o = 118$ objects, we sample $N_p = 4$ RGB-D images of the object at random stable poses. The "stable pose" is the state S_i defined in Section 8.3.1.
2. For each pose of the object, we use Algorithm 8.1 to calculate $N_G = \min\left\{ n^{N_p}, 10 \right\}$ candidate form closure grasping configurations, where n^{N_p} is the total number of candidate form-closed configurations for the object and gripper parameters.
3. For each form closure configuration, we generate all three types of uncertainties mentioned in Section 8.3.2.2 and calculate the grasp quality score $\hat{\Gamma}$ by Eq. (8.9)).

Thus, we obtain the labeled observations $\left\{ \left(\mathrm{IM}, p_h, \hat{\Gamma} \right) \right\}$ for supervised learning.

We also tested the proposed grasping methods in the simulation. We conducted an ablation study in the following cases:

- Random ARIE: We used ARIE to find all form-closure grasp configurations and then randomly chose one among them.
- Max-ARIE: We used ARIE to find all form-closure grasp configurations and then chose the one with max $\hat{\Gamma}$ value.
- GQM-ARIE: We used ARIE to find all form-closure grasp configurations and then chose one through the GQM network.

To illustrate the grasping performance, two criteria are considered: (1) the success rate, which is the percentage of successful grasps in all grasping attempts, and (2) the planning time, which is the time interval between the image input and the motion of the gripper.

We chose 40 objects from the DexNet dataset (20 known and 20 unknown) to test the performance of the above methods. Each experiment was conducted 10 times for one object with a random initial pose and Gaussian process noise parameter $\sigma = 0.008$. The experiment results are listed in Table 8.1.

TABLE 8.1 The performance of the grasping methods.

Object type	Test cases		
	Random ARIE	Max-ARIE	GQM-ARIE
Known	82.5%	90.0%	86.1%
Unknown	78.9%	87.5%	87.5%
Planning time	1.21 s	25.83 s	1.14 s

It can be seen that all the three methods achieve a fine success rate for known objects. Max-ARIE has the highest success rate, whereas it has the vital disadvantage of being time consuming, which directly influences the efficiency in reality. In comparison to Max-ARIE, the GQM-ARIE method achieves a comparable success rate for both known and unknown objects, while the planning time is reduced by nearly a factor of 20. For unknown objects, the random ARIE has an unsatisfactory success rate because robustness occupies the dominant position in a successful grasping process. Overall, the GQM-ARIE method achieved a higher success rate during the grasping process for both known and unknown objects, and the average planning time is considerably shorter than that of Max-ARIE, which has the best success rate. Therefore, the proposed GQM-ARIE method is a good trade-off for a stable grasp.

8.3.3 Learning to grasp with ARIE

Reinforcement learning has been used to achieve good grasping in many recent papers. However, few studies have focused on form-closure grasps. Among these works, the reward is often designed to be very sparse. For example, in [36,37], the reward was calculated by gripper on and off tests. In another

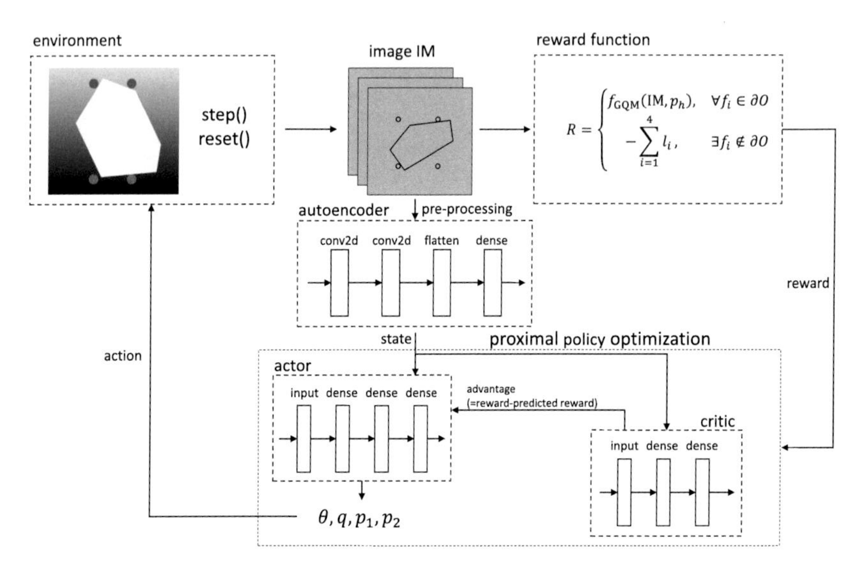

FIGURE 8.6 Full pipeline for training.

example [38], the reward could only have one of three possible values (none, partial, full). Such a sparse reward makes it inefficient to learn anything without providing experience to the trainer. In this section, we introduce a deep reinforcement learning method with a continuous reward. We define how close a grasp configuration is to an optimal form-closure configuration.

8.3.3.1 Formation of the learning pipeline

For these experiments, we used the grasping setup defined in Section 8.3.1. A manipulator (agent) is trained to use the four-pin gripper to grasp the object with form closure. Because the manipulator may block the view of the top-view camera, we used an in-hand camera that would always be able to see the object as well as the gripper pins. The objective of the training process is to teach the agent to achieve a form-closure grasp through trial and error.

The full pipeline of the training process is shown in Fig. 8.6. The input includes an image I_n with dimensions $h \times w \times 3$ and the proprioceptive data of the gripper $(q_n, p1_n, p2_n)$. The output is an action consisting of the gripper's yaw rotation θ_{n+1} around the camera axis and the new gripper parameters $q_{n+1}, p1_{n+1}, p2_{n+1}$, that is,

$$(I_n, q_n, p1_n, p2_n) \rightarrow (\theta_{n+1}, q_{n+1}, p1_{n+1}, p2_{n+1}). \tag{8.12}$$

To focus on the form-closure grasp task, we simplify the task by providing the agent with a set of initial joint values θ. This initial pose was achieved using object detection, camera calibration, and inverse kinematics. We will discuss the details of the learning policy and rewards in the following sections.

8.3.3.2 Learning policy

In this section, we use proximal policy optimization (PPO) as the reinforcement learning policy to train our grasping agent.

In 2017, Schulman et al. [39] from OpenAI proposed a new family of policy gradient methods for reinforcement learning, which is called PPO. These methods have the following three features:

- A surrogate objective function is optimized using stochastic gradient ascent.
- Minibatch updates are performed instead of one gradient update per data sample.
- The surrogate objective is clipped to avoid large policy changes.

Schulman et al. claimed that PPO balances ease of implementation, sample complexity, and ease of tuning. PPO has been applied to learn dexterous in-hand manipulation in some cases [40].

To keep the new policy close to the old policy, PPO uses the technique *clipping*:

$$\mathrm{clip}(r, 1-\epsilon, 1+\epsilon) = \begin{cases} 1-\epsilon, & r < 1-\epsilon, \\ r, & r \in [1-\epsilon, 1+\epsilon], \\ 1+\epsilon, & r > 1+\epsilon, \end{cases} \tag{8.13}$$

so that the agent could benefit from the incoming improvement, without being too far away from the old policy, which may result in worse performance. The hyperparameter ϵ controls the distance that the new policy can move from the old policy. The loss for the actor is then calculated as follows.

Let

$$r = \frac{\pi_\theta\,(a \mid s)}{\pi_{\theta_k}\,(a \mid s)} \tag{8.14}$$

in

$$L\,(s, a, \theta, \theta_k, \epsilon) = \min\left(r A^{\pi_{\theta_k}}\,(s, a)\,, \mathrm{clip}\,(r, 1-\epsilon, 1+\epsilon)\, A^{\pi_{\theta_k}}\,(s, a)\right), \tag{8.15}$$

where s is the state of the agent, a is the predicted action, θ and θ_k are the old and new policies, respectively, and $A^{\pi_{\theta_k}}$ is the advantage function, which indicates the extra reward that can be obtained by the agent by taking the particular action. Also, a small exploratory noise is added to the loss. The implementation was also extended to save the state of the models and parameters for every gradient step. In practice, we use the following simplified version:

$$L\,(s, a, \theta, \theta_k, \epsilon) = \min\left(\frac{\pi_\theta(a \mid s)}{\pi_{\theta_k}(a \mid s)} A^{\pi_{\theta_k}}(s, a), \quad g\left(\epsilon, A^{\pi_{\theta_k}}(s, a)\right)\right), \tag{8.16}$$

where

$$g(\epsilon, A) = \begin{cases} (1+\epsilon)A, & A \geq 0, \\ (1-\epsilon)A, & A < 0. \end{cases} \tag{8.17}$$

The complete algorithm used to train the PPO network is described in Algorithm 8.2.

Algorithm 8.2 High-level algorithm for training actor and critic.

Require: total episodes n_{total}, initial policy parameters θ_0
Ensure: trained models
1: Build actor;
2: Build critic;
3: **if** `previous_run_exists` **then**
4: Load previous run weights W and episode count n;
5: **else**
6: Initialize run weights W and episode count $n = 0$;
7: **end if**
8: **while** $n < n_{\text{total}}$ **do**
9: Collect buffer of observations, actions and rewards by running policy $\pi_k = \pi(\theta_k)$ in the environment;
10: Compute rewards $\hat{R}_t$ using critic and reward function R_t;
11: Compute advantage estimates as $\hat{A}_t = \hat{R}_t - R_t$;
12: Update actor using observations, advantages and actions;
13: Update critic using observations and rewards;
14: **end while**

8.3.3.3 Reward

A well-chosen reward function is crucial for reinforcement learning. It should return a value that defines the performance of the agent for a specific step. In our case, it also returns a done condition that defines whether the current episode is completed after the current step. For the grasping task described in Section 8.3.3, the goal is to achieve a form-closure grasp. The input to the reward function is the encoded image IM, which was introduced in Section 8.3.2.

To establish a continuous reward, we extend the GQM-ARIE method of Section 8.3.2 to obtain a negative value when at least one finger is not on the surface of the object. To achieve this, we use classic computer vision methods to extract the contour of the object ∂O and the coordinates of the pins $F = [f_1, f_2, f_3, f_4]$ with respect to the image frame as follows:

$$\partial O, F = \text{CV}(\text{IM}). \tag{8.18}$$

Then, the pin positions are checked to determine if they are located on the edge of the object. If not, a line is made along the normal direction of the surface of the object from the pin position to the surface of the object. For finger i, the distance of the line is recorded as l_i. Therefore, we define the grasp quality of such cases as

$$\Gamma = f_{\mathrm{GQM}}\left(\mathrm{IM}, p_h\right) = -\sum_{i=1}^{4} l_i, \exists f_i \notin \partial O. \tag{8.19}$$

After the extension to the GQM, we can simply use it as a continuous reward function in the learning to grasp process. The grasp quality scores of three objects are shown in Fig. 8.7. The grasp quality score is obtained from the relationship between the gripper configuration and the object shape.

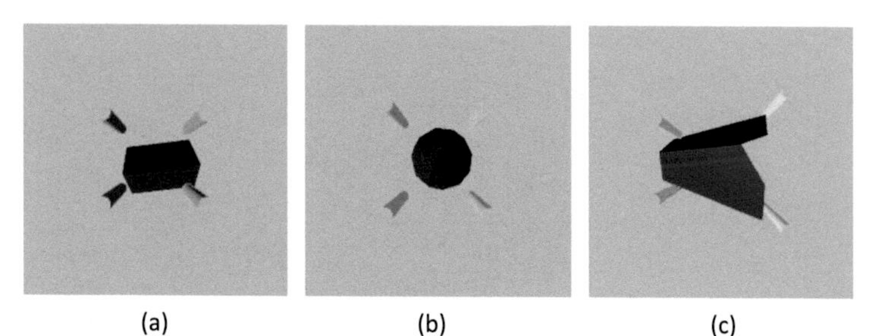

FIGURE 8.7 Example of rewards from the GQM function. (a) Reward: −0.017. (b) Reward: −0.048. (c) Reward: 0.25.

In [38], a reward based on only three form closure values (full, partial, and none) was defined. Such a sparse reward function would not be viable for our case, since initially, almost all cases would fall into the "none" category, a few in the "partial" category, and no cases in the "full" category. This would require a large amount of steps before randomly getting "full" closure cases, since there is no feedback on how to get from the "partial" to the "full" category. Another possible reward function would be to attempt to grasp the object and lift it. Once again, this would also be very sparse, since none of the grasps would be able to lift the object. In our case, a "none" case roughly corresponds to a reward of below 0.3, "partial" to a reward of between 0.3 and 0.7, and "full" to a reward of above 0.7.

We used the parameters listed in Table 8.2 during the training process. The buffer size was decreased to 128 half-way through the process to speed up the startup time after the experiment was interrupted. It was set back to 512 afterwards, which did not seem to have any effect on the performance.

Due to the limitations in computational resources, as mentioned in this section, our experiment can only be run at a real-time speed of 10 s per step, which would result in 115 days for 1,000,000 steps. One consequence of this low training speed is that the model cannot be trained properly in the available time frame. Another important consequence is that it is not possible to test and adjust the training parameters without having to wait 1 week between every test. As a

TABLE 8.2 The performance of the grasping methods.

Variable	Value
Autoencoder	
Image dimensions	128 × 128 × 3
Latent size	128
Type	Convolutional variational
Dataset size	86,000
Object set	Convex objects
PPO	
Buffer size	512 (128)
Batch size	64
Layer size	256
Hidden layers	2
Loss clipping	0.2
Steps per episode	1
Object set	Convex objects

result one set of training parameters, even though suboptimal, are maintained as described in this section.

Finally, the autoencoder was trained based on 86,000 collected images, which were used to encode the input images. Data collection, training, and experiments were performed using CoppeliaSim. The reinforcement learning model was trained over 250 epochs and 38,000 steps, taking 10 real-time days. We observed that the agent could grasp an object after the training process. However, the expected success rate could not be achieved because of the reasons mentioned above. Hence, a powerful computational device will be required to implement this method, as well as some improvements to the current training pipeline.

8.4 Conclusion

In Section 8.2, we first provided a background discussion on *stable grasps* for robotic dexterous manipulation, where we introduced the concept and the current issues of force/form closure in robotic grasping. We then reviewed the related works in this area with respect to closure properties, ECs, and learning to grasp. We introduced the form of the first-order form-closure test, the concept of ECs, and the concept of ARIE.

ARIE reveals the intrinsic relationship between the robot, the object, and the environment. In Section 8.3, we presented the definition of ARIE and its existence in four-pin grasping scenarios. We demonstrated its application by solving a form-closure grasp for the four-pin gripper. First, the relationship between the "acute" bottoms and the form-closure grasps was discussed. Then, an ARIE-based GQM function was established. Later, the function was refactored as a CNN to increase the robustness of the GQM. With the GQM network, we could choose an optimal grasp, given all form-closure grasp candidates.

As a bridge, ARIE also connects synthetic methods with learning methods. We built a PPO network to "teach" a robotic gripper to grasp *in a form-closure way*. We rewarded the agent with an extended GQM function, which was able to output a continuous value even when not every finger gets contacted with the object. Preliminary experiments showed that the proposed PPO network could achieve more form-closure grasps.

In our future work, we will continue to work on a solution to "teach" the robot to grasp *firmly*. The existence of ARIE in more grasping scenarios will be discovered, and more subtle techniques of reinforcement learning will be applied to make the agents learn fast and efficient.

References

[1] B. Dizioğlu, K. Lakshiminarayana, Mechanics of form closure, Acta Mechanica 52 (1–2) (Jul 1984) 107–118.

[2] V.-D. Nguyen, Constructing force- closure grasps, The International Journal of Robotics Research 7 (3) (Jun 1988) 3–16.

[3] K.M. Lynch, F.C. Park, Modern Robotics: Mechanics, Planning, and Control, Cambridge University Press, May 2017.

[4] J. Trinkle, A quantitative test for form closure grasps, in: 1992 IEEE/RSJ International Conference on Intelligent Robots and Systems (IROS), 1992.

[5] J.K. Salisbury, B. Roth, Kinematic and force analysis of articulated mechanical hands, Journal of Mechanisms, Transmissions, and Automation in Design 105 (1) (Mar 1983) 35–41.

[6] A. Sudsang, J. Ponce, N. Srinivasa, Grasping and in-hand manipulation: experiments with a reconfigurable gripper, Advanced Robotics 12 (5) (Jan 1997) 509–533.

[7] A. Sudsang, J. ponce, N. Srinivasa, Grasping and in-hand manipulation: geometry and algorithms, Algorithmica 26 (3–4) (Mar 2000) 466–493.

[8] S. Makita, W. Wan, A survey of robotic caging and its applications, Advanced Robotics 31 (19–20) (Sep 2017) 1071–1085.

[9] C. Eppner, Robot grasping by exploiting compliance and environmental constraints, PhD dissertation, Technische Universität Berlin, 2019.

[10] H. Qiao, Attractive regions formed by the environment in configuration space: the possibility of achieving high precision sensorless manipulation in production, International Journal of Production Research 40 (4) (Jan 2002) 975–1002.

[11] H. Qiao, Two- and three-dimensional part orientation by sensor-less grasping and pushing actions: use of the concept of 'attractive region in environment', International Journal of Production Research 41 (14) (Jan 2003) 3159–3184.

[12] J.H. Su, H. Qiao, C.K. Liu, Z.C. Ou, A new insertion strategy for a peg in an unfixed hole of the piston rod assembly, The International Journal of Advanced Manufacturing Technology 59 (9–12) (2012) 1211–1225.

[13] J.H. Su, H. Qiao, Z.C. Ou, Y.R. Zhang, Sensor-less insertion strategy for an eccentric peg in a hole of the crankshaft and bearing assembly, Assembly Automation 32 (1) (2012) 86–99.

[14] J. Su, H. Qiao, C. Liu, A vision-based 3d grasp planning approach with one single image, in: 2009 International Conference on Mechatronics and Automation (ICMA), IEEE, Changchun, China, Aug 2009, pp. 3281–3286.

[15] C.K. Liu, H. Qiao, B. Zhang, Stable sensorless localization of 3-d objects, IEEE Transactions on Systems, Man, and Cybernetics, Part C (Applications and Reviews) 41 (6) (Nov 2011) 923–941.

[16] C.K. Liu, H. Qiao, J.H. Su, P. Zhang, Vision-based 3-d grasping of 3-d objects with a simple 2-d gripper, IEEE Transactions on Systems, Man, and Cybernetics: Systems 44 (5) (May 2014) 605–620.
[17] R. Li, H. Qiao, Condition and strategy analysis for assembly based on attractive region in environment, IEEE/ASME Transactions on Mechatronics 22 (5) (Oct 2017) 2218–2228.
[18] H. Qiao, M. Wang, J. Su, S. Jia, R. Li, The concept of "attractive region in environment" and its application in high-precision tasks with low-precision systems, IEEE/ASME Transactions on Mechatronics 20 (5) (Oct 2015) 2311–2327.
[19] I.M. Bullock, T. Feix, A.M. Dollar, The Yale human grasping dataset: grasp, object, and task data in household and machine shop environments, The International Journal of Robotics Research 34 (3) (2015) 251–255.
[20] M. Cai, K.M. Kitani, Y. Sato, A scalable approach for understanding the visual structures of hand grasps, in: 2015 IEEE International Conference on Robotics and Automation (ICRA), 2015, pp. 1360–1366.
[21] E. Krabbe, E. Kristiansen, L. Hansen, D. Bourne, Autonomous optimization of fine motions for robotic assembly, in: 2014 IEEE International Conference on Robotics and Automation (ICRA), IEEE, Hong Kong, China, May 2014, pp. 4168–4175.
[22] L.P. Berczi, I. Posner, T.D. Barfoot, Learning to assess terrain from human demonstration using an introspective Gaussian-process classifier, in: 2015 IEEE International Conference on Robotics and Automation (ICRA), 2015, pp. 3178–3185.
[23] D.A. Duque, F.A. Prieto, J.G. Hoyos, Trajectory generation for robotic assembly operations using learning by demonstration, Robotics and Computer-Integrated Manufacturing 57 (Jun 2019) 292–302.
[24] F. Li, Q. Jiang, S. Zhang, M. Wei, R. Song, Robot skill acquisition in assembly process using deep reinforcement learning, Neurocomputing (Feb 2019).
[25] D. Xu, S. Nair, Y. Zhu, J. Gao, A. Garg, L. Fei-Fei, S. Savarese, Neural task programming: learning to generalize across hierarchical tasks, in: 2018 IEEE International Conference on Robotics and Automation (ICRA), IEEE, May 2018.
[26] Y. Duan, M. Andrychowicz, B. Stadie, O. Jonathan Ho, J. Schneider, I. Sutskever, P. Abbeel, W. Zaremba, One-shot imitation learning, in: I. Guyon, U.V. Luxburg, S. Bengio, H. Wallach, R. Fergus, S. Vishwanathan, R. Garnett (Eds.), Advances in Neural Information Processing Systems 30, Curran Associates, Inc., 2017, pp. 1087–1098 [online], available: http://papers.nips.cc/paper/6709-one-shot-imitation-learning.pdf.
[27] S. Levine, P. Pastor, A. Krizhevsky, J. Ibarz, D. Quillen, Learning hand-eye coordination for robotic grasping with deep learning and large-scale data collection, The International Journal of Robotics Research 37 (4–5) (Jun 2017) 421–436.
[28] Y. Chebotar, A. Handa, V. Makoviychuk, M. Macklin, J. Issac, N.D. Ratliff, D. Fox, Closing the sim-to-real loop: adapting simulation randomization with real world experience, CoRR, arXiv:1810.05687 [abs], 2018.
[29] H. Qiao, Attractive regions in the environment [motion planning], in: 2000 IEEE International Conference on Robotics and Automation (ICRA), IEEE, San Francisco, CA, USA, 2000, pp. 1420–1427.
[30] Z. Ou, H. Qiao, Analysis of stable grasping for one-parameter four-pin gripper, in: Intelligent Robotics and Applications, Springer-Verlag GmbH, 2008, pp. 630–639.
[31] J. Cousty, G. Bertrand, L. Najman, M. Couprie, Watershed cuts: minimum spanning forests and the drop of water principle, IEEE Transactions on Pattern Analysis and Machine Intelligence 31 (8) (2009) 1362–1374.
[32] J. Mahler, J. Liang, S. Niyaz, M. Laskey, R. Doan, X. Liu, J.A. Ojea, K. Goldberg, Dex-net 2.0: deep learning to plan robust grasps with synthetic point clouds and analytic grasp metrics, in: Robotics: Science and Systems (RSS), Robotics: Science and Systems Foundation, Cambridge, Massachusetts, USA, Jul 2017.
[33] S.T. Ounpraseuth, Gaussian processes for machine learning, Journal of the American Statistical Association 103 (481) (Mar 2008) 429.

[34] A. Krizhevsky, I. Sutskever, G.E. Hinton, ImageNet classification with deep convolutional neural networks, Communications of the ACM 60 (6) (May 2017) 84–90.

[35] A.E. Hoerl, R.W. Kennard, Ridge regression: biased estimation for nonorthogonal problems, Technometrics 42 (1) (Feb 2000) 80–86.

[36] Y. Jiang, S. Moseson, A. Saxena, Efficient grasping from RGBD images: learning using a new rectangle representation, in: 2011 IEEE International Conference on Robotics and Automation, IEEE, May 2011.

[37] Y. Jiang, J.R. Amend, H. Lipson, A. Saxena, Learning hardware agnostic grasps for a universal jamming gripper, in: 2012 IEEE International Conference on Robotics and Automation, IEEE, May 2012.

[38] A. Elahibakhsh, M. Ahmadabadi, F. Sharifi, B. Araabi, Distributed form closure for convex planar objects through reinforcement learning with local information, in: 2004 IEEE/RSJ International Conference on Intelligent Robots and Systems (IROS), IEEE, 2004.

[39] J. Schulman, F. Wolski, P. Dhariwal, A. Radford, O. Klimov, Proximal policy optimization algorithms, ArXiv e-prints, Jul 2017.

[40] M. Andrychowicz, B. Baker, M. Chociej, R. Józefowicz, B. McGrew, J. Pachocki, A. Petron, M. Plappert, G. Powell, A. Ray, J. Schneider, S. Sidor, J. Tobin, P. Welinder, L. Weng, W. Zaremba, Learning dexterous in-hand manipulation, The International Journal of Robotics Research 39 (1) (Nov 2019) 3–20.

Chapter 9

Learning hierarchical control for robust in-hand manipulation

Tingguang Li
Tencent Robotics X, Shenzhen, China

9.1 Introduction

In-hand manipulation is defined as moving a grasped object to any other pose within the workspace of a robot hand [1–3]. It is traditionally achieved by moving the objects with fingers and possibly assisted by using the other hand parts such as the palm as support. In this chapter, we are particularly concerned with the robustness of in-hand manipulation, where the object is continuously grasped within the hand without dropping. This is an important ability in human manipulation, e.g., when pouring water, a human grasps the cup within the hand and adjusts its pose to pour the water. While tremendous progress has been made for pick-and-place tasks using simple grippers and suction cups [4–6], robots are surprisingly limited when using multifingered dexterous hands [7–9].

There are two main challenges in dexterous manipulation: (1) modeling and controlling a complex hand with many degrees of freedom (DoFs) and (2) coordinating multiple fingers during a long manipulation sequence to achieve an object goal pose. Traditional methods towards dexterous manipulation mostly rely on modeling the geometry and dynamics of the hand and the object in contact [1,2,10]. They can typically control a single manipulation primitive that locally adjusts the object pose in-hand by either moving the fingers coherently or changing the contact locations of the fingers. However, these methods are not suitable for generating manipulation sequences that need to chain multiple manipulation primitives together for moving an object to a more distant pose. The small noises in the model measurement and interaction with the environment also cause the error in the control system, which can easily cause failure.

In recent years, deep reinforcement learning (DRL) has been applied to the problem of dexterous manipulation as it has proved effective for systems with complex dynamics [11]. For example, Kumar et al. [12] and Andrychowicz et al. [13] successfully applied DRL to enable in-hand manipulation of different objects using a complex anthropomorphic robot hand. Zhu et al. [14] applied DRL and demonstration learning to rotate valves with different hands. However, the learned policies in most DRL methods do not generalize well over

Tactile Sensing, Skill Learning, and Robotic Dexterous Manipulation
https://doi.org/10.1016/B978-0-32-390445-2.00017-9

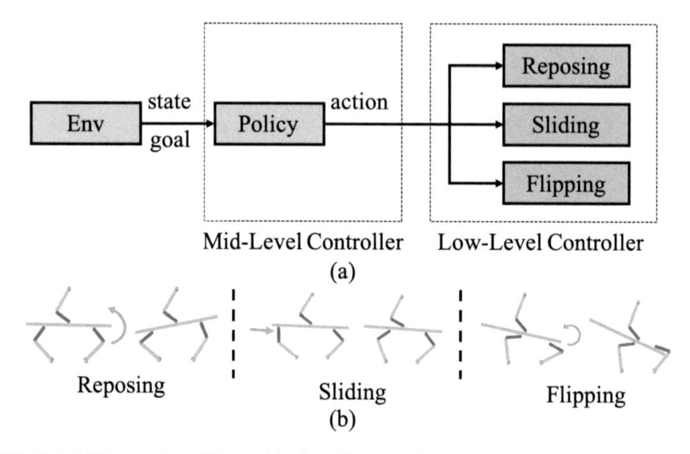

FIGURE 9.1 (A) The two-level hierarchical architecture for in-hand manipulation. We define three manipulation primitives at the low level, which are executed with torque controllers. At the mid-level, we train a DRL policy to pick a manipulation primitive and its parameterization. (B) Demonstration of the three manipulation primitives.

different object geometries, leading to per-object policies. These works are typically demonstrated on manipulation tasks that do not require the object to be grasped firmly by the hand. For example, the palm may be facing upwards and thereby prevent the object from dropping [12,13], or the hand is manipulating an articulated object that is rigidly attached in the world [14]. Learning to robustly hold the object would likely require DRL methods, an even larger number of training episodes, and a very carefully designed reward function. Nevertheless, this is an important ability in many real-world manipulation tasks.

In this chapter, we introduce a hierarchical control structure that enables in-hand manipulation of objects within the entire workspace of the robotic hand (see Fig. 9.1 (A)). We leverage both traditional controllers for robust, low-level torque control of a single manipulation primitive and DRL for mid-level policy sequencing of manipulation primitives. Thereby, we can ensure that the object does not fall out of the hand even if the palm of the hand is facing down. We also show that the learned policies are robust against noise in the object pose and inaccuracy of the dynamics model.

In our method, we define three manipulation primitives: reposing, sliding, and flipping, as shown in Fig. 9.1 (B). Reposing, or in-grasp manipulation [15], refers to changing the pose of an object while maintaining the same contact locations. Sliding changes contact locations by sliding fingers along the object surface. Flipping means releasing one finger from one side and making a new contact on a different side. All the three manipulation primitives are implemented with torque controllers so that they can stably grasp the object. At the mid-level, DRL is trained to select a sequence of the three manipulation primitives and plan a complicated action trajectory to reach the distant goal.

We test our approach from three aspects: how well does our approach conduct complex in-hand manipulation; how robust is our approach to noise; and whether our approach can generalize to different shapes. We first compare our method with an end-to-end method and a graph search-based method [16]. The experiments show our method achieves a significantly higher success rate of reaching a target object pose. Then we test our model in environments with observation noise and model inaccuracy. We also show the generalization result by manipulating a cube. All the experiments are conducted in simulated environments with a three-fingered robot hand.

9.2 Related work

Model-based approaches to dexterous manipulation: Traditionally, researchers try to build the kinematics, dynamics, and contact model of the controller and objects, and then control the hands to perform predefined manipulation motions. Those motions typically include reposing/reorientation, which refers to moving the objects by moving the fingers while keeping the contact, and finger sliding or gaiting, which refers to moving the fingers on the objects to change the contact. Rus [10,17] developed different methods of controlling the fingers to generate rotational motions of a grasped object in a plane. Okada [18] and Kerr and Roth [19] applied sliding and rolling of robot fingers to manipulate objects of round and unspecified shapes by applying inverse kinematics. A typical challenge for the model-based approaches is to plan the long sequence of manipulation that involves switching between different primitives, in order to reach a large workspace of object poses. Mordatch et al. [20] and Fan et al. [21] developed different motions to access a larger workspace, one by a trajectory optimization method that uses a contact-invariant cost and another by leveraging a two-staged planner. It plans finger gaits first and then uses a controller to move the object, which is robust to external perturbations and modeling inaccuracies. However, these methods are not robust under either the noise of the system dynamics or the object model. Apart from the manipulation with fully actuated hands with a high number of DoFs, Odhner and Dollar [22] and Calli and Dollar [23] developed a method to roll or slide in-hand objects using two-finger underactuated grippers and demonstrated this by applying model predictive control within a visual-servoing scheme. However, the limitation of the motion limits the workspace of object motion.

Model-free approaches to dexterous manipulation: Another class of methods applied to tasks in dexterous manipulation are model-free; these use limited or no prior model of the gripper and object and attempt to learn to control the object through interaction. These approaches, which we refer to as end-to-end learning, try to directly learn a mapping between the observations of object pose to the operational space of the manipulator. Kumar et al. [12] proposed a method that learns to rotate a cylinder with an anthropological hand by optimizing a linear Gaussian controller via iterative LQR, and showed that their model could adapt to learn the task in few iterations. Recently, DRL has been successfully used to learn some dexterous manipulation skills using a model-free approach, such as reorienting a cube or a pen in the hand. Andrychowicz et al. [13] developed a way to train a DRL policy of using domain and physics randomization in simulation to enable sim-to-real transfer of a policy that can reconfigure a cube in-hand. The resulting policy acquired skills for specific grasp types, controlling the use of gravity and using finger gaiting, while also doing state estimation from vision. However, their setup does not require robust grasping of the object in order to manipulate the object, which can be problematic when the object must be controlled being held in a position. Another recent method [24] used tactile sensing as input to a reinforcement learning algorithm, and was able to learn a policy that could roll and reorient cylindrical objects on a table, robust to changes in the size and weight of the objects. In order to address the issues of sample complexity with DRL, Rajeswaran et al. [25] and Zhu et al. [14] developed a method combining imitation learning and reinforcement learning, and therefore reduced training time. However, it is worth noting that the tasks were constructed in a way where the objects are supported and cannot be dropped. This limitation is inherent to end-to-end model-free approaches, as policy learning from scratch will not immediately learn (and may never learn) to manipulate the object with full force closure.

Combining deep reinforcement learning with models: Our work looks to synthesize these different approaches to dexterous manipulation into a hybrid framework that aims to highlight the advantages of both traditional motion-planning methods and more recent DRL methods. There have been recent efforts in a similar direction, as explored in Silver et al. [26] and Johannink et al. [27]. Both works combined traditional hand-designed controllers with DRL methods that learned to output residual joint-commands added to the controller's output to account for inaccuracies that may exist in the model and the controller and environmental noise. Our method is distinguished from theirs by being hierarchical in structure, where the reinforcement learning policy directly selects the corresponding manipulation primitive to be executed by the low-level controller, as opposed to being compositional, where the outputs of the policy and the controller are directly composed together.

9.3 Methodology

We aim at conducting in-hand manipulation with a fully actuated hand, i.e., to move the grasped object from one pose $\boldsymbol{X}_0$ to the target pose $\boldsymbol{X}_g$. For simplicity, we assume using precision grasps, that the hand grasps objects using only fingertips, where the hard-finger model [28] applies to the contact. The hard-finger model means point contact, while both normal force and frictional force exist. We demonstrate our method in a simulated three-finger hand in a 2D vertical plane with gravity.

We describe the manipulation state with $(\boldsymbol{X}, \boldsymbol{C})$, where $\boldsymbol{X}$ is the object pose and $\boldsymbol{C}$ is the contact configuration of the hand. We assume the state is

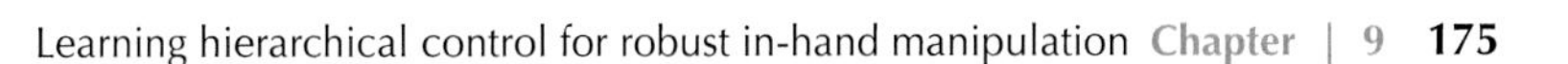

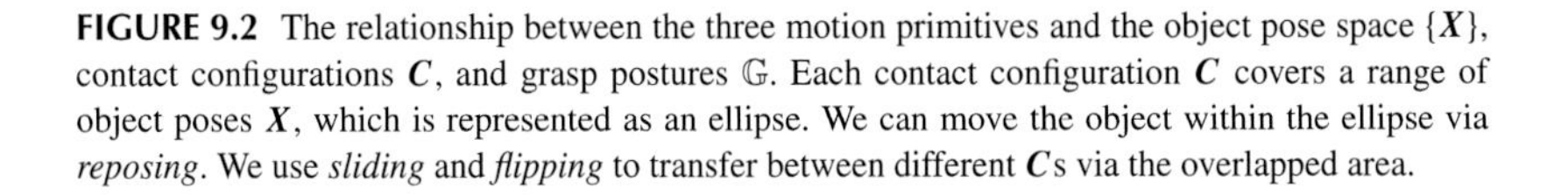

FIGURE 9.2 The relationship between the three motion primitives and the object pose space $\{X\}$, contact configurations C, and grasp postures $\mathbb{G}$. Each contact configuration C covers a range of object poses X, which is represented as an ellipse. We can move the object within the ellipse via *reposing*. We use *sliding* and *flipping* to transfer between different Cs via the overlapped area.

measurable, and the shape, mass, and surface friction of the robot and objects are known. In reality, the pose and contact configuration can be measured by robot vision or motion tracking systems. In our setting, $X = (x, y, \theta)$ and C is represented by the contact location of each fingertip on the object. During manipulation, the hand could either keep the contact configuration C while moving the object to change X or move fingers on the object to change C.

Each contact configuration C covers a range of object poses X that can be reached, and different Cs may overlap on $\{X\}$ space. Additionally, the hand could grasp an object with different grasp postures $\mathbb{G}$, which we define as how the fingers are arranged on different sides of the object. In the three-fingered hand and pole-shaped object case, the relationship between C, X, and $\mathbb{G}$ is demonstrated in Fig. 9.2. Under this setting, an in-hand manipulation task is analogous to finding a path from X_0 to X_g in the X space, while the state could move from one point X_1 to X_2 via reposing if they are in the range of same C, via jumping between overlapping Cs via sliding, or via flipping if the two Cs belong to different $\mathbb{G}$s. We therefore design a hierarchical structure that uses a low-level dynamic controller to conduct that basic primitive motion and a DRL mid-level controller to choose the primitive motion.

9.3.1 Hierarchical structure for in-hand manipulation

We discretize the in-hand manipulation sequence into three basic motion primitives: reposing, sliding, and flipping. Reposing refers to moving the object with the current contact configuration C, so that the object follows the motion of the fingertips. Sliding and flipping refers to changing the contact configuration C, by either sliding a fingertip on the object surface or making a fingertip jump to a distant contact location. In the example of manipulating a pole, flipping means one finger leaves the pole and quickly jumps to the other side of the pole. Fig. 9.1 (B) shows examples of the three motion primitives.

To achieve a longer sequence of in-hand manipulation, we design a hierarchical control method, which combines a low-level dynamic controller to perform the motion primitives and a mid-level controller to choose the current motion primitive and subgoal parameters. This allows the robot hand to perform a long and complicated sequence motion by combining different motion primitives, thus being able to move the object anywhere in a large range. The definition of low-, mid-, and high-level controllers was formalized by [1]. The structure is demonstrated in Fig. 9.1 (A). We use a traditional dynamic controller [29] for the low-level control part and a neural network for the mid-level part.

9.3.2 Low-level controller

Motion primitives are implemented by low-level torque controllers. The controllers model the dynamics of both the object and the hand and calculate the actuation torque on the hand in order to achieve a specific $\boldsymbol{X}$ or $\boldsymbol{C}$. When the hand holds the object firmly, with effective contact points on all the fingertips, the dynamics for the robot hand is

$$\boldsymbol{M}_{hnd}(\boldsymbol{q})\ddot{\boldsymbol{q}} + V(\boldsymbol{q}, \dot{\boldsymbol{q}}) + \boldsymbol{J}^T\boldsymbol{\lambda} = \boldsymbol{\tau}, \tag{9.1}$$

where $\boldsymbol{q} \in \mathbb{R}^n$ represents joint angles, $\boldsymbol{M}_{hnd}(\boldsymbol{q})$ is the mass matrix of the robot hand, V computes the centrifugal and Coriolis terms, $\boldsymbol{J}$ is the stacked Jacobian matrix for each finger tip, $\boldsymbol{\lambda}$ contains the contact force and moment applied at each contact point, and $\boldsymbol{\tau} \in \mathbb{R}^n$ is the torque applied at each joint. Similarly, the dynamics of the object is

$$\boldsymbol{M}_{obj}(\boldsymbol{X})\dot{\boldsymbol{\nu}}_{obj} - \boldsymbol{G}\boldsymbol{\lambda} = \boldsymbol{g}, \tag{9.2}$$

where $\boldsymbol{M}_{obj}(\boldsymbol{X})$ is the mass matrix of the object, $\boldsymbol{\nu}_{obj}$ is the twist of the object, $\boldsymbol{G}$ is the grasp matrix, which describes how contact forces applied to the object result in a wrench on the object, and $\boldsymbol{g}$ contains the external forces applied on the object, including gravity. Then, by plugging Eq. (9.2) into Eq. (9.1), we have

$$\boldsymbol{M}_{hnd}(\boldsymbol{q})\ddot{\boldsymbol{q}} + V(\boldsymbol{q}, \dot{\boldsymbol{q}}) + \boldsymbol{J}^T\boldsymbol{G}^{-1}(\boldsymbol{M}_{obj}(\boldsymbol{X})\dot{\boldsymbol{\nu}}_{obj} - \boldsymbol{g}) = \boldsymbol{\tau}. \tag{9.3}$$

Note that the solution to the dynamic systems is not unique, and there are extra constraints on the contact force $\boldsymbol{\lambda}$, that the normal force components $\boldsymbol{\lambda}_N$ should be larger than 0 and the shear force components should be smaller than $\mu\boldsymbol{\lambda}_N$, where μ is the friction coefficient. To address the problems, we calculate the null space of $\boldsymbol{G}$ and denote it as $\mathcal{N}(\boldsymbol{G})$, so that for any $\boldsymbol{\lambda}_n \in \mathcal{N}(\boldsymbol{G})$, there is $\boldsymbol{G}\boldsymbol{\lambda}_n = \boldsymbol{0}$. We choose some $\boldsymbol{\lambda}_n$ with a large enough normal force component to ensure $\boldsymbol{\lambda} = \boldsymbol{\lambda}_p + \boldsymbol{\lambda}_n$, where $\boldsymbol{\lambda}_p$ is a particular solution in (9.2), the normal force component is always large enough, and the shear force is within the friction cone range.

Reposing: Reposing changes the pose of an object without changing the contact locations. We assume no slip at the fingertips, so the twists of the fingertips are the same as those of contact points on the objects. This yields

$$\boldsymbol{J}\dot{\boldsymbol{q}} = \boldsymbol{v}_{ftips} = \boldsymbol{v}_{contact} = \boldsymbol{G}^T \boldsymbol{v}_{obj}. \tag{9.4}$$

Considering the definition of the Jacobian $\boldsymbol{v}_{ftips} = \boldsymbol{J}\dot{\boldsymbol{q}}$, we differentiate it and obtain $\dot{\boldsymbol{v}}_{ftips} = \dot{\boldsymbol{J}}\dot{\boldsymbol{q}} + \boldsymbol{J}\ddot{\boldsymbol{q}}$. Therefore, the joint-space acceleration can be computed as $\ddot{\boldsymbol{q}} = \boldsymbol{J}^{-1}\dot{\boldsymbol{v}}_{ftips} - \boldsymbol{J}^{-1}\dot{\boldsymbol{J}}\dot{\boldsymbol{q}}$. Substituting this into (9.3), we have

$$\begin{aligned} &(\boldsymbol{M}_{hnd}(\boldsymbol{q})\boldsymbol{J}^{-1} + \boldsymbol{J}^T\boldsymbol{G}^{-1}\boldsymbol{M}_{obj}(\boldsymbol{X})\boldsymbol{G}^{-T})\dot{\boldsymbol{v}}_{ftips} + \\ &(V(\boldsymbol{q}, \dot{\boldsymbol{q}}) - \boldsymbol{M}_{hnd}(\boldsymbol{q})\boldsymbol{J}^{-1}\dot{\boldsymbol{J}}\dot{\boldsymbol{q}}) - \boldsymbol{J}^T\boldsymbol{G}^{-1}\boldsymbol{g} = \boldsymbol{\tau}. \end{aligned} \tag{9.5}$$

Given a target object state $(\boldsymbol{X}_d, \dot{\boldsymbol{X}}_d, \ddot{\boldsymbol{X}}_d)$, the desired fingertip acceleration $\dot{\boldsymbol{v}}_{ftips}$ can be computed using a PID controller: $\boldsymbol{u}_{ftips} = \ddot{\boldsymbol{X}}_d + \boldsymbol{K}_d(\dot{\boldsymbol{X}}_d - \dot{\boldsymbol{X}}) + \boldsymbol{K}_p(\boldsymbol{X}_d - \boldsymbol{X})$. It can be proved that the error dynamics of this controller are exponentially stable with suitable gain matrices. Substituting $\boldsymbol{u}_{ftips}$ for $\dot{\boldsymbol{v}}_{ftips}$ in Eq. (9.5) gives the actuation torque.

Sliding: Action sliding changes the contact configuration $\boldsymbol{C}$ by sliding one finger on the object surface while keeping the other finger contacts fixed. Given a target contact configuration $\boldsymbol{C}_d$, the desired joint angles $\boldsymbol{q}_d$ can be computed using inverse kinematics. We then use inverse dynamics to compute the joint torque

$$\boldsymbol{\tau} = \boldsymbol{M}_{hnd}(\boldsymbol{q})\boldsymbol{u}_{joint} + V(\boldsymbol{q}, \dot{\boldsymbol{q}}) + \boldsymbol{J}^T\boldsymbol{G}^{-1}(\boldsymbol{M}_{obj}(\boldsymbol{X})\dot{\boldsymbol{v}}_{obj} - \boldsymbol{g}), \tag{9.6}$$

where the control input is $\boldsymbol{u}_{joint} = -\boldsymbol{K}_d\dot{\boldsymbol{q}} + \boldsymbol{K}_p(\boldsymbol{q}_d - \boldsymbol{q})$.

Flipping: Action flipping, also called finger gaiting [15] or pivoting [3], refers to releasing one finger off the object and making a new contact on a different side. The moving finger may temporarily leave the object and recontact on the new location. Specifically, when manipulating a pole-shaped object, we apply flipping motion when moving a finger from one side of the pole to the other side by designing a hand-crafted trajectory $\{\boldsymbol{x}_0, ..., \boldsymbol{x}_N, \boldsymbol{x}_d\}$ for the moving fingertip, where $\boldsymbol{x}_d$ is the destination point located on the other side of the pole, which is chosen heuristically.

9.3.3 Mid-level controller

In the mid-level controller, DRL learns to sequence the aforementioned motion primitives. At each step, given the current state $\boldsymbol{S}$, the DRL policy selects an action $\boldsymbol{A}$, which is a parameterized motion primitive. We build a discrete motion set as shown in Table 9.1, where each action $\boldsymbol{A}$ contains the motion primitive and direction, with a fixed motion magnitude Δ. For example, when manipulating a pole-shaped object, the action could be moving the object using reposing motion for Δ or $-\Delta$ on any of the directions, or sliding any one of the fingers on the pole

TABLE 9.1 Discretization of manipulation primitives into 14 possible actions.

Reposing	$+\Delta X^x, -\Delta X^x, +\Delta X^y, -\Delta X^y, +\Delta X^\theta, -\Delta X^\theta$
Sliding	$+\Delta C^1, -\Delta C^1, +\Delta C^2, -\Delta C^2, +\Delta C^3, -\Delta C^3$
Flipping	Flip either left or right finger to the other side

in either the clockwise or the counterclockwise direction, or flipping a finger. After the DRL policy generates a specific action $\boldsymbol{A}$, the low-level controller will conduct the closed-loop torque control to reach the temporary goal.

Note that some motion is not valid for any finger configuration $\boldsymbol{C}$. For example, flipping can only succeed when the two still fingers are close by on the opposite sides of the object; otherwise the object will be rotated when the other finger is released. Apart from those necessary conditions, the frequency of the flipping motion is much lower than those of reposing and sliding. As a result, the DRL policy would always try to avoid the flipping motion. To address this challenge, we particularly design a reward function for the flipping motion. If the flipping motion is successful and useful, the agent will receive a positive reward; if the flipping motion fails, the agent will receive a small negative reward. The "usefulness" of the flipping motion is defined by whether the new configuration $\boldsymbol{C}$ adds some extra moving freedom to the object towards the goal. In this case, the DRL network tends to conduct flipping motion when it is necessary and likely to succeed. Additionally, the flipping reward will also make the network learn to approach the configuration that can make a successful flipping motion when necessary. Meanwhile, another common cause for manipulation failure is losing contact, which commonly happens when the target pose is beyond the workspace boundary. These invalid motions are not desirable, and therefore, we designed a feasibility filter according to the system's kinematics. The feasibility filter also considers the heuristic conditions for flipping, which requires the two still fingers to be within a distance d to each other. The workflow of our architecture is shown in Fig. 9.3. The DRL model picks one action, and then the low-level controller conducts the corresponding motion primitive in a closed-loop pattern for n steps.

9.4 Experiments

In this section, we report experimental results of manipulating a pole-shaped object and test the robustness of our model by adding noise to the measurement of object pose. We also test the generality of the method by experimenting with other kinds of objects.

We implement our approach in PyBullet [30], where a three-fingered hand has to move an object within a 2D vertical plane with downward gravity. The robot hand has to continuously hold the object. Each finger has two independent joints. In the first two experiments, the hand manipulates a pole with a length of

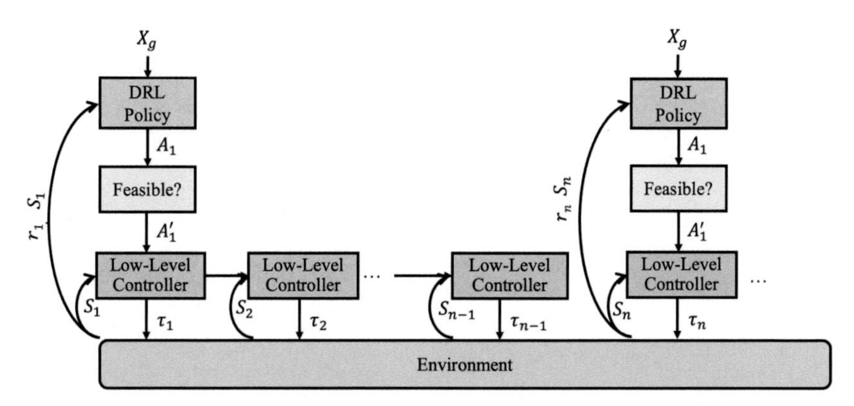

FIGURE 9.3 The workflow of our architecture. Given a state S_0 and a goal pose X_g, the DRL planner outputs an action A_0. The feasible action filter determines whether A_0 is valid and generates a valid action A'_0. Then the chosen low-level controller will conduct the closed-loop torque control for n steps until the DRL planner makes a new action A_1.

50 cm and a width of 2 cm. In the last experiment, we use a cube with a length of 10 cm.

9.4.1 Training mid-level policies and baseline

We train the hierarchical policy using proximal policy optimization (PPO) [31] with 16 parallel environments. The agent can observe joint angles $\boldsymbol{q}$, joint velocities $\dot{\boldsymbol{q}}$, the current object pose $\boldsymbol{X}_t$, and the goal pose $\boldsymbol{X}_g$, where $\boldsymbol{q}, \dot{\boldsymbol{q}} \in \mathbb{R}^6$, $\boldsymbol{X}_t, \boldsymbol{X}_g \in \mathbb{R}^3$. The agent receives a reward of $+5$ if it reaches the goal $\boldsymbol{X}_g$ and a penalty of -0.01 if DRL generates an invalid action. Besides, to encourage flipping, an additional reward of $+1$ is given when the flipping action is necessary and successful. We say an episode is successful if the object is close to the goal, with a translational error of less than 1 cm and a rotational error of less than 5.7 degrees (0.1 rad). Since continuously grasping the object is our particular concern, we consider dropping the object a failure case, where dropping is defined as losing contact with the pole at one or both sides of it. Another failure case is when the hand fails to reach the goal pose within a time limit.

We test the effectiveness and robustness of our method with an end-to-end policy trained using deep deterministic policy gradient (DDPG) [32] and a search-based baseline [16], respectively. DDPG is an off-policy RL algorithm that works well for continuous control tasks, in particular for complicated dexterous manipulation tasks [14,25]. The DDPG agent operates in the same environment with a continuous policy that outputs joint torques for each of the joints with an identical network structure with three fully connected layers with 256 units. The reward function is sparse with $+5$ when it succeeds; otherwise it is 0.

TABLE 9.2 Success rate and drop rate (with standard deviation) for 500 episodes of manipulating a pole averaged over three random seeds. Average time represents the real-world time for successful manipulation sequences. Experiments are conducted in Easy, Medium, and Hard groups.

	Success rate	Drop rate	Avg. time (s)
Ours – Easy	95.4%±0.7%	3.8%±0.7%	7.1±0.2
Ours – Mid	94.6%±0.4%	3.7%±1.3%	8.9±0.6
Ours – Hard	79.3%±0.6%	15.7%±1.4%	13.4±0.3
DDPG – Easy	72.5%±13.5%	19.3%±14.2%	1.82±0.1
DDPG – Mid	61.7%±6.6%	26.8%±10.2%	1.97±0.1
DDPG – Hard	11.1%±6.8%	84.0%±10.8%	2.35±0.1
Search – Easy	91.4%	5.2%	8.1
Search – Mid	79.8%	13.6%	8.1
Search – Hard	54.4%	34.8%	8.25

For the search-based baseline, we first discretize all the feasible object poses $\boldsymbol{X}$ and contact configurations $\boldsymbol{C}$ with a resolution of 2 cm and 0.2 rad for $\boldsymbol{X}$ and 5 cm for $\boldsymbol{C}$. Then we build a graph where the grid cells are feasible object poses and the edges connecting neighboring nodes correspond to the manipulation primitives. In this way, the task of manipulating an object is equivalent to finding a path from the current pose to the goal pose. In our experiment, we use Dijkstra's algorithm [33] to find the shortest path.

9.4.2 Dataset

The overall motion range of the pole object is set to be $[-30, 30]$ (cm) in the x-direction and $[-20, 20]$ (cm) in the y-direction. Because not all the poses are reachable, we collect a comprehensive dataset to train the DRL controller, which covers the entire range of reachable object poses. Further, the dataset is divided into three groups: Easy Goal, Medium Goal, and Hard Goal. In the Easy Goal group, the goal pose $\boldsymbol{X}_g$ can be reached without changing the contact configuration $\boldsymbol{C}$, which means the robot hand can manipulate the object using reposing only. The Medium Goal group contains pairs where the initial pose $\boldsymbol{X}_0$ and goal pose $\boldsymbol{X}_g$ are in the same grasp posture $\mathbb{G}$, so sliding might be required. Similarly, in the Hard Goal group, flipping might be necessary for some goal poses.

9.4.3 Reaching desired object poses

In this part, we report the experiment results of moving a pole-shaped object to desired goal poses. We compare our approach to an end-to-end and search-based baseline. The results are reported in Table 9.2. It is obvious that our method has a significantly higher success rate and lower drop rate on all the three groups,

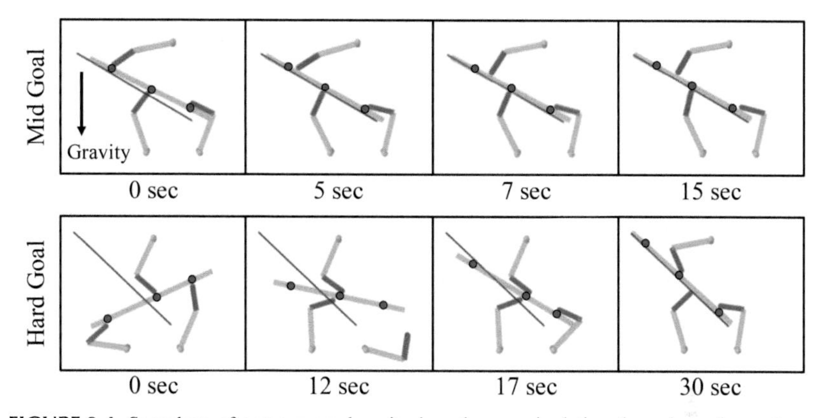

FIGURE 9.4 Snapshots of some example episodes when manipulating the pole to the goal pose (red line). To show the change of contact points, the initial contact points are indicated in blue.

especially for the hard goals. Particularly, DDPG fails in almost all cases in the Hard Goal group. The results show that our approach is advantageous for the goal poses that require complex manipulation sequences. Fig. 9.4 gives some episode examples.

Compared with our approach, the end-to-end policies generate much faster manipulation sequences. It frequently pushes the object so that the object moves quickly to the goal pose. However, it is hard to use these dynamic manipulation behaviors for they are never robust when implemented on a real robotic hand. It is more likely that the object will be dropped frequently. In contrast, our approach uses motion primitives which are implemented with torque controllers and do not learn these fast behaviors. Therefore, it will grasp the object firmly and drop it less often. Besides, as we will show in the next section, our approach is more robust when there are observation noise, external perturbations, and modeling errors. We believe such robustness can help to reduce the sim-to-real gap when transferring our model to the real world.

Compared with our method, the success rate of search-based baseline decreases faster when the goal pose gets difficult. One problem of the search-based method is its computational complexity. For example, the number of graph nodes will reach 10^9 when the position resolution is 1 cm and the angle resolution is 0.1 rad for the Medium Goal tasks. The resolutions have a significant influence on the planning performance, where a lower resolution results in a lower accuracy. Taking the flipping primitive, for example, since this primitive is conditioned on the distance of the other two finger contact points, a low-resolution graph will result in a low success rate.

9.4.4 Robustness analysis

In this part, we evaluate the robustness of our method against the measurement error in object dimension and pose. Traditional model-based manipulation meth-

TABLE 9.3 Success rate for inaccurate object models (geometry and dynamics) and/or observation noise. Experiments are done with the Medium Goal group. The first row shows results for the model that is trained and tested without observation noise. The second row reports results for the model that is trained and tested with observation noise.

	Trained Obj.	Heavier	Lighter	Thicker	Thinner	Longer	Shorter
No $\epsilon_{x,y,\theta}$	94.4%	94.4%	95.8%	91.8%	94.2%	91.8%	95.2%
$\epsilon_{x,y,\theta}$	93.2%	92.4%	94.6%	94.4%	91%	89.2%	92.6%

ods highly depend on the accuracy of the object and system model, and the end-to-end methods require large numbers of training examples for a single object model. However, in reality, the pose estimation system and the dynamic model of hands and objects are both noisy. We evaluate our model in the presence of measurement noises.

We evaluate the robustness from two aspects: observation noise and parameter inaccuracy. We use the model with the same parameters as in Section 9.4.3 (Trained Obj.) and test it under different conditions. The geometry and inertia parameters of the object are changed, i.e., heavier (1.1× mass), lighter (0.9× mass), thicker (1.5× width), thinner (0.5× width), longer (1.1× length), and shorter (0.9× length). For observation noise $\epsilon_{x,y,\theta}$, we add a translational error uniformly distributed in $[-0.5, 0.5]$ (cm) and a rotational error uniformly distributed in $[-2.86, 2.86]$ (degrees) to the object pose $\boldsymbol{X}_t$. The result is reported in Table 9.3, which shows the averaged success rates over 500 evaluation trials. It is obvious that our model is very robust to the inaccuracy of the object model. The performance is almost the same when the pole's mass or dimension changes. Besides, our model can still perform well even under observation noise. We observe in the experiments that the noises in the system or observation may cause inaccuracy of the low-level dynamic controller and that both $\boldsymbol{C}$ and $\boldsymbol{X}$ may deviate from the subgoals of the low-level controller; however, the mid-level controller is able to fix the inaccuracy of motion and choose the motion primitive according to the current states. On the other hand, the force-closure design of the low-level dynamic controller can keep the grasp stability regardless of the noise in the object measurement.

9.4.5 Manipulating a cube

In this part, we test the generality of our method over different objects. We want to show our method for the hierarchical structure of both learning and dynamic methods and the design of motion primitives can be generalized to other objects. We consider the task of manipulating a cube (Fig. 9.5). The edges of a cube are much shorter. Therefore, we disable the sliding primitive in the experiment. Besides, we decompose flipping into two stages: releasing and landing.

We tested our model for 500 episodes. It achieved a success rate of 71.4% and a dropping rate of 20.8%, and the average run-time was 7.8 s. The re-

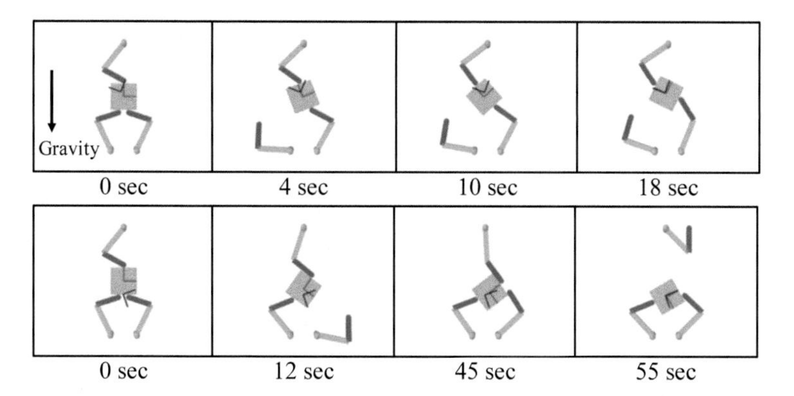

FIGURE 9.5 Example episodes for manipulating a small cube. The goal pose is indicated in red.

sults show our hierarchical architecture can be generalized to objects with other shapes, with minor modifications to the low-level controllers. Actually, our method can be extended to other shapes as long as the grasped object is in force closure.

9.5 Conclusion

In this chapter, we introduce a hierarchical approach with low-level torque controllers and mid-level reinforcement learning policy. We test our method in a simulation environment with a three-fingered hand. Our low-level torque controller uses the geometry and dynamics of the hand and object and controls the hand to conduct some predefined motion primitives; for the mid-level controller, we apply a DRL policy to generate actions for a sequence of different motion primitives to reach a desired goal position. We show our method could generate a complicated manipulation sequence that moves the object to any possible pose in the workspace. Experiments show that our method has a significantly higher success rate and lower drop rate, compared with an end-to-end and search-based baseline. We also show the robustness of our method under perturbations of the object shape, mass, or pose measurement, and it can be generalized to objects with other shapes. We believe our method can be generalized to other in-hand manipulation tasks and other hand configurations as well, and help robots to perform more dexterously in real-world tasks.

References

[1] A.M. Okamura, N. Smaby, M.R. Cutkosky, An overview of dexterous manipulation, in: Proceedings 2000 ICRA. Millennium Conference. IEEE International Conference on Robotics and Automation. Symposia Proceedings (Cat. No. 00CH37065), vol. 1, IEEE, 2000, pp. 255–262.

[2] A. Bicchi, Hands for dexterous manipulation and robust grasping: a difficult road toward simplicity, IEEE Transactions on Robotics and Automation 16 (6) (2000) 652–662.

[3] R.R. Ma, A.M. Dollar, On dexterity and dexterous manipulation, in: 2011 15th International Conference on Advanced Robotics (ICAR), IEEE, 2011, pp. 1–7.
[4] J. Mahler, M. Matl, V. Satish, M. Danielczuk, B. DeRose, S. McKinley, K. Goldberg, Learning ambidextrous robot grasping policies, Science Robotics 4 (26) (2019) eaau4984.
[5] D. Kalashnikov, A. Irpan, P. Pastor, J. Ibarz, A. Herzog, E. Jang, D. Quillen, E. Holly, M. Kalakrishnan, V. Vanhoucke, S. Levine, Scalable deep reinforcement learning for vision-based robotic manipulation, in: A. Billard, A. Dragan, J. Peters, J. Morimoto (Eds.), Proceedings of the 2nd Conference on Robot Learning, in: Proceedings of Machine Learning Research, PMLR, vol. 87, 2018, pp. 651–673.
[6] A. Zeng, S. Song, K.-T. Yu, E. Donlon, F.R. Hogan, M. Bauza, D. Ma, O. Taylor, M. Liu, E. Romo, N. Fazeli, F. Alet, N.C. Dafle, R. Holladay, I. Morona, P.Q. Nair, D. Green, I. Taylor, W. Liu, T. Funkhouser, A. Rodriguez, Robotic pick-and-place of novel objects in clutter with multi-affordance grasping and cross-domain image matching, in: Proceedings of the IEEE International Conference on Robotics and Automation, 2018.
[7] S. Jacobsen, E. Iversen, D. Knutti, R. Johnson, K. Biggers, Design of the Utah/MIT dextrous hand, in: Proceedings. 1986 IEEE International Conference on Robotics and Automation, vol. 3, IEEE, 1986, pp. 1520–1532.
[8] H. Kawasaki, T. Komatsu, K. Uchiyama, Dexterous anthropomorphic robot hand with distributed tactile sensor: Gifu hand ii, IEEE/ASME Transactions on Mechatronics 7 (3) (2002) 296–303.
[9] D.-H. Lee, J.-H. Park, S.-W. Park, M.-H. Baeg, J.-H. Bae, Kitech-hand: a highly dexterous and modularized robotic hand, IEEE/ASME Transactions on Mechatronics 22 (2) (2016) 876–887.
[10] D. Rus, Dexterous rotations of polyhedra, in: Proceedings 1992 IEEE International Conference on Robotics and Automation, IEEE, 1992, pp. 2758–2763.
[11] J. Lee, J. Hwangbo, M. Hutter, Robust recovery controller for a quadrupedal robot using deep reinforcement learning, arXiv:1901.07517 [abs], 2019.
[12] V. Kumar, E. Todorov, S. Levine, Optimal control with learned local models: application to dexterous manipulation, in: 2016 IEEE International Conference on Robotics and Automation (ICRA), IEEE, 2016, pp. 378–383.
[13] M. Andrychowicz, B. Baker, M. Chociej, R. Jozefowicz, B. McGrew, J. Pachocki, A. Petron, M. Plappert, G. Powell, A. Ray, et al., Learning dexterous in-hand manipulation, arXiv preprint, arXiv:1808.00177, 2018.
[14] H. Zhu, A. Gupta, A. Rajeswaran, S. Levine, V. Kumar, Dexterous manipulation with deep reinforcement learning: efficient, general, and low-cost, in: 2019 International Conference on Robotics and Automation (ICRA), IEEE, 2019, pp. 3651–3657.
[15] B. Sundaralingam, T. Hermans, Geometric in-hand regrasp planning: alternating optimization of finger gaits and in-grasp manipulation, in: 2018 IEEE International Conference on Robotics and Automation (ICRA), IEEE, 2018, pp. 231–238.
[16] S. Cruciani, K. Hang, C. Smith, D. Kragic, Dual-arm in-hand manipulation and regrasping using dexterous manipulation graphs, arXiv preprint, arXiv:1904.11382, 2019.
[17] D. Rus, Coordinated manipulation of objects in a plane, Algorithmica 19 (1–2) (1997) 129–147.
[18] T. Okada, Computer control of multijointed finger system for precise object-handling, IEEE Transactions on Systems, Man, and Cybernetics 12 (3) (1982) 289–299.
[19] J. Kerr, B. Roth, Analysis of multifingered hands, The International Journal of Robotics Research 4 (4) (1986) 3–17.
[20] I. Mordatch, Z. Popović, E. Todorov, Contact-invariant optimization for hand manipulation, in: Proceedings of the ACM SIGGRAPH/Eurographics Symposium on Computer Animation, Eurographics Association, 2012, pp. 137–144.
[21] Y. Fan, T. Tang, H.-C. Lin, Y. Zhao, M. Tomizuka, Real-time robust finger gaits planning under object shape and dynamics uncertainties, in: 2017 IEEE/RSJ International Conference on Intelligent Robots and Systems (IROS), IEEE, 2017, pp. 1267–1273.

[22] L.U. Odhner, A.M. Dollar, Stable, open-loop precision manipulation with underactuated hands, The International Journal of Robotics Research 34 (11) (2015) 1347–1360.
[23] B. Calli, A.M. Dollar, Robust precision manipulation with simple process models using visual servoing techniques with disturbance rejection, IEEE Transactions on Automation Science and Engineering (99) (2018) 1–14.
[24] H. Van Hoof, T. Hermans, G. Neumann, J. Peters, Learning robot in-hand manipulation with tactile features, in: 2015 IEEE-RAS 15th International Conference on Humanoid Robots (Humanoids), IEEE, 2015, pp. 121–127.
[25] A. Rajeswaran, V. Kumar, A. Gupta, G. Vezzani, J. Schulman, E. Todorov, S. Levine, Learning complex dexterous manipulation with deep reinforcement learning and demonstrations, in: Proceedings of Robotics: Science and Systems (RSS), 2018.
[26] T. Silver, K. Allen, J. Tenenbaum, L. Kaelbling, Residual policy learning, arXiv preprint, arXiv:1812.06298, 2018.
[27] T. Johannink, S. Bahl, A. Nair, J. Luo, A. Kumar, M. Loskyll, J.A. Ojea, E. Solowjow, S. Levine, Residual reinforcement learning for robot control, in: 2019 International Conference on Robotics and Automation (ICRA), IEEE, 2019, pp. 6023–6029.
[28] J.K. Salisbury, B. Roth, Kinematic and force analysis of articulated mechanical hands, Journal of Mechanisms, Transmissions, and Automation in Design 105 (1) (1983) 35–41.
[29] C.C. de Wit, B. Siciliano, G. Bastin, Theory of Robot Control, Springer Science & Business Media, 2012.
[30] E. Coumans, Y. Bai, Pybullet, a python module for physics simulation for games, robotics and machine learning, http://pybullet.org, 2016–2019.
[31] J. Schulman, F. Wolski, P. Dhariwal, A. Radford, O. Klimov, Proximal policy optimization algorithms, arXiv preprint, arXiv:1707.06347, 2017.
[32] T.P. Lillicrap, J.J. Hunt, A. Pritzel, N. Heess, T. Erez, Y. Tassa, D. Silver, D. Wierstra, Continuous control with deep reinforcement learning, arXiv preprint, arXiv:1509.02971, 2015.
[33] E.W. Dijkstra, A note on two problems in connexion with graphs, Numerische Mathematik 1 (1) (1959) 269–271.

Chapter 10

Learning industrial assembly by guided-DDPG

Yongxiang Fan
FANUC Advanced Research Laboratory, FANUC America Corporation, Union City, CA, United States

10.1 Introduction

Industrial robots have been applied in automation assembly for decades. They are programmed and integrated into different systems. They are rigid, reliable, precise, and highly efficient in traditional mass production. With dedicated manual tuning, industrial robots can provide considerable reliability for specific tasks. The recent popularity of mass customization, however, presents considerable challenges to dedicated tuning methods. The personalized supply chains with numerous workpieces introduce considerable variations and uncertainties, which exponentially complicate the assembly of products. To reduce human involvement and increase the robustness to uncertainties, more researches are focusing on learning of the assembly skills.

There are two major learning techniques in the community: supervised learning and reinforcement learning (RL). While it is ideal to implement a supervised learning algorithm in case of sufficient training data, collecting data in practice is inefficient under various environment uncertainties. Besides supervised learning with popular deep neural networks, supervised learning can include other learning formats. For example, the human demonstration is used to train a Gaussian mixture model in [1] for peg-hole insertion skill learning. To simplify the task complexity, the policy is constrained to explore in a plane, and it is not clear whether the trained policy can be applied to different environments.

On the other hand, RL techniques enable the robot to explore environments and learn optimal actions to maximize the expected rewards. Multiple types of RL methods have been proposed, including direct policy gradients such as REINFORCE [2], Q-learning-based methods such as deep Q-learning (DQN) [3], and actor-critic frameworks such as deep deterministic policy gradient (DDPG) [4] or proximal policy optimization (PPO) [5]. Since the dynamics model of the system is not used in policy learning, these methods are called model-free RL. Model-free RL methods have been introduced to assembly tasks [6,7] successfully despite a lack of dynamics. To explore the state/action

Tactile Sensing, Skill Learning, and Robotic Dexterous Manipulation
https://doi.org/10.1016/B978-0-32-390445-2.00018-0

space and reconstruct the transitions of the states, the model-free RL needs to collect considerable data. Therefore, the model-free RL is not time-efficient and data-efficient.

Model-based RL [8,9] is introduced to address the efficiency issue of the model-free RL. Model-based RL applies optimal control such as iterative linear quadratic regulator (iLQR) or iterative linear quadratic Gaussian (iLQG) [10] on a fitted local model to compute the optimal trajectories. Random noise is added to the action space to enable certain exploration. A neural network policy is then trained from the optimized trajectory in a supervised manner. The model-based RL is generally faster than the model-free RL since the trajectory is constrained locally without broader exploration. However, it is worth pointing out that the performance of the model-based RL depends on the behavior of the optimal controller (i.e., supervisor), which in turn is affected by the accuracy of the local dynamics model. For the rigid industrial manipulators with force/torque as states, the dynamics model is less smooth, which means that the dynamics change dramatically as the trajectory slightly changes. This implies that the dynamics fitting may not be accurate. Consequently, the model-based RL cannot converge consistently. In practice, model-based RL is often used in robot systems with soft dynamics (e.g., with serial elastic actuators) such as Baxter, PR2 [9] with position/velocity feedback by ignoring the force/torque states.

This chapter proposes a learning framework to incorporate both positional signals with force/torque feedbacks to train a more natural assembly policy. To learn a precise assembly task, we combine the model-free actor-critic and model-based RL. To learn tasks efficiently, we adopt the trajectory optimization from the model-based RL to compute the optimal trajectories and train the policy network. The trajectory optimization takes both positional and force/torque signals as feedback. The performance of the trained policy network is affected by the smoothness of the local fitted dynamics model. To prevent inconsistency, instability, or tedious case-by-case parameter tuning, we introduce a critic network to learn the correct critic value (Q-value). Of note, we train the policy network by combining the supervised learning (trajectory optimization) with policy gradient (actor-critic method) instead of training the policy network by pure supervision. We also employ the Q-value from trajectory optimization to further accelerate the efficiency of the critic network training.

The contributions of this work are as follows. First, the proposed method with trajectory optimization is more data-efficient compared with actor-critic methods. By exploring in a narrower space and solving for the optimal trajectories mathematically, the trajectory optimization constrains the exploration space in a safe region and guides the initial training of the networks. In comparison, actor-critic methods explore space randomly at the first iterations. Secondly, the proposed method with the combined critic network is more stable and consistent in rigid robotic systems. By building up a ground truth critic for the policy network, the proposed method is able to address the potential inconsistency and instability issues of the trajectory optimization caused by the rigid dynamics and the force/torque feedback.

10.2 From model-free RL to model-based RL

The objective of an assembly task is to learn an optimal policy $\pi_\theta(a_t|o_t)$ to choose an action a_t based on the current observation o_t in order to minimize an expected loss:

$$\min_{\pi_\theta} E_{\tau\sim\pi_\theta}(l(\tau)), \tag{10.1}$$

where θ represents policy parameterization, $\tau = \{s_0, a_0, s_1, a_1, ..., s_T, a_T\}$ denotes a trajectory, $\pi_\theta(\tau) = p(s_0)\prod_1^T p(s_t|s_{t-1}, a_{t-1})\pi_\theta(a_t|s_t)$ represents the likelihood of the trajectory τ, and $l()$ is the loss of τ.

Optimization can be used to solve Eq. (10.1) if the global dynamics $p(x_t|x_{t-1}, u_{t-1})$ is known. In practice, the global dynamics model is extremely difficult to model especially for a contact-rich complex manipulation task. Learning-based approaches are proposed to address the issue, and the dynamics are either avoided [7] or fitted into local linear models [1,8,9].

Without dynamics, it generally requires excessively data for an RL algorithm to explore the space and locate the optimal policy, especially in learning the assembly tasks with high-dimensional action spaces. On the other hand, the locally fitted dynamics or globally learned dynamics may not be stable once the robotic system is rigid or the force/torque feedback is included in the optimal controller. Consequently, the performance of the model-based RL can be downgraded.

A learning framework is proposed in this chapter to address the aforementioned issues. It combines the actor-critic framework and optimal control for efficient high-accuracy assembly. The actor-critic framework is modified from the DDPG algorithm, while the optimal controller is adapted from the model-based RL [9]. These two algorithms will be briefly introduced below.

10.2.1 Guided policy search

With guiding distribution $p(\tau)$, the problem defined by Eq. (10.1) can be rewritten as

$$\min_{\pi_\theta, p} E_p(l(\tau)), \quad s.t. \;\; p(\tau) = \pi_\theta(\tau). \tag{10.2}$$

Guided policy search (GPS) solves Eq. (10.2) by alternatively minimizing the augmented Lagrangian with respect to primal variables p, π_θ and updating the Lagrangian multipliers λ. The augmented Lagrangian for θ and p optimization are

$$\begin{aligned} L_p(p,\theta) &= E_p(l(\tau)) + \lambda\left(\pi_\theta(\tau) - p(\tau)\right) + \nu D_{KL}\left(p(\tau)\|\pi_\theta(\tau)\right), \\ L_\theta(p,\theta) &= E_p(l(\tau)) + \lambda\left(\pi_\theta(\tau) - p(\tau)\right) + \nu D_{KL}\left(\pi_\theta(\tau)\|p(\tau)\right), \end{aligned} \tag{10.3}$$

where λ is the Lagrangian multiplier, ν is the penalty parameter for the violation of the equality constraint, and D_{KL} represents the Kullback–Leibler (KL) divergence. The optimization of primal variable p is called trajectory optimization. It optimizes the guiding distribution p with learned local dynamics. To ensure the accuracy of dynamics fitting, the optimization is constrained within the trust region ϵ:

$$\min_{p} L_p(p, \theta), \quad s.t.\ D_{KL}(p(\tau) \| \hat{p}(\tau)) \leq \epsilon, \tag{10.4}$$

where $\hat{p}$ is the guiding distribution of the previous iteration. The Lagrangian of Eq. (10.4) is $\mathcal{L}(p) = L_p(p, \theta) + \eta(D_{KL}(p(\tau) \| \hat{p}(\tau)) - \epsilon)$, where η is the Lagrangian multiplier for the constraint optimization. With the Gaussian assumption of the dynamics, $\mathcal{L}(p)$ is solved by iLQG [11]. To avoid large derivation from the fitted dynamics, η is adapted by comparing the predicted KL divergence with the actual one.

Supervised learning is used to optimize the policy parameters θ. With the Gaussian policy $\pi_\theta(a_t|o_t) = \mathcal{N}(u_\theta(o_t), \Sigma_t^\pi)$, $L_\theta(p, \theta)$ in Eq. (10.3) can be rewritten as

$$\begin{aligned} L_\theta(\theta, p) = \frac{1}{2N_b} \sum_{i,t=1}^{N_b, T} E_{p_i(s_t, o_t)}[\operatorname{tr}\left(C_{ti}^{-1} \Sigma_t^\pi\right) - \log|\Sigma_t^\pi| + \\ \left(u_\theta(o_t) - u_{ti}^p(s_t)\right)^T C_{ti}^{-1} \left(u_\theta(o_t) - u_{ti}^p(s_t)\right) + 2\lambda_t^T u_\theta(o_t)], \end{aligned} \tag{10.5}$$

where $p_i(u_t|s_t) \sim \mathcal{N}(u_{ti}^p(s_t), C_{ti})$ is the guiding distribution. Eq. (10.5) contains the decoupled form of the variance optimization and policy optimization. The reader may refer to [9] for more details.

10.2.2 Deep deterministic policy gradient

The DDPG algorithm trains a critic network Q_ϕ and actor network u_θ parametrized by ϕ and θ by collecting sample data (s_j, a_j, s_{j+1}, r_j) from the replay buffer R.

The policy network is updated by

$$\theta \leftarrow \operatorname*{argmax}_{\theta} \frac{1}{N_{dd}} \sum_{j=1}^{N_{dd}} Q_{\hat{\phi}}(s_j, u_\theta(s_j)), \tag{10.6}$$

where θ is the parameter for the policy network to be optimized and $\hat{\phi}$ is the parameter of the target critic network. Policy gradient is applied to update the parameters of the actor network:

$$\theta \leftarrow \theta + \alpha \frac{1}{N_{dd}} \sum_{j=1}^{N_{dd}} \nabla_a \hat{Q}(s, a)|_{s=s_j, a=a_j} \nabla_\theta u_\theta(s)|_{s=s_j}, \tag{10.7}$$

where α is the learning rate of the actor network.

The critic network is updated by

$$\begin{aligned}\phi &\leftarrow \underset{\phi}{\operatorname{argmin}} \frac{1}{N_{dd}} \sum_{j=1}^{N_{dd}} \left(y_j - Q_\phi(s_j, a_j)\right)^2, \\ y_j &= r_j + \gamma Q_{\hat{\phi}}(s_{j+1}, u_{\hat{\theta}}(s_{j+1})),\end{aligned} \tag{10.8}$$

where N_{dd} is the batch size for DDPG, $\hat{\theta}$ denotes the parameters of the target actor network, and γ is the discount for future reward.

The target networks are updated by

$$\begin{aligned}\hat{\theta} &\leftarrow \delta\theta + (1-\delta)\hat{\theta}, \\ \hat{\phi} &\leftarrow \delta\phi + (1-\delta)\hat{\phi},\end{aligned} \tag{10.9}$$

where δ is the target update rate and set to be a small value ($\delta \approx 0.01$).

10.2.3 Comparison of DDPG and GPS

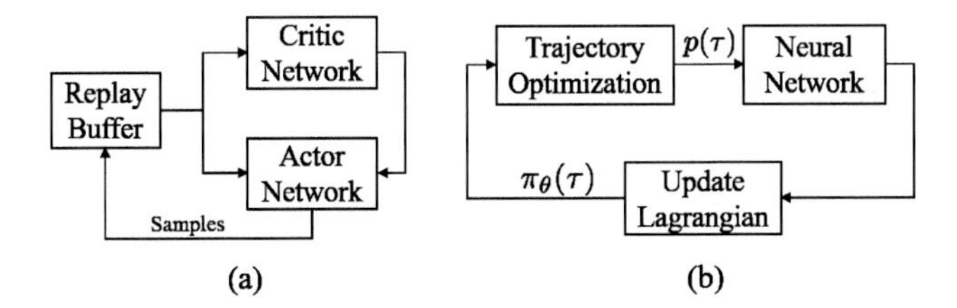

FIGURE 10.1

DDPG jointly trains the critic network and the actor (policy) network using the rollout samples from a replay buffer, as shown in Fig. 10.1 (A). The Q-value is gradually built from rollouts by the critic network and is applied to train the actor network based on policy gradient. The actor-critic method provides a more stable policy in the tasks with nonsmooth dynamics. These tasks are common in a high-precision industrial assembly where the system has higher stiffness and contains force/torque feedback in the states. However, DDPG is less data-efficient due to the intensive exploration, which is usually unnecessary since assembly tasks only require exploration in a narrow trajectory space.

In comparison, GPS consists of trajectory optimization and policy network training, as shown in Fig. 10.1 (B). The performance of the policy network after training relies on the quality of the trajectory optimization supervision. By fitting the dynamics from sampling data and training the network with the computed optimal trajectory, GPS is more efficient than the DDPG and many other model-free RL algorithms. However, the performance of the policy would be compromised if the system has high stiffness and force/torque feedbacks as states due to the less smooth dynamics and the smaller trust region.

10.3 Guided deep deterministic policy gradient

Industrial manipulators have high system stiffness in order to resist external disturbances and achieve precise tracking performance. With high stiffness, small clearance, and force/torque feedback, both the model-free and the model-based RL methods cannot accomplish assembly tasks efficiently and stably. In this chapter, we propose a learning framework called guided deep deterministic policy gradient (guided-DDPG) for precise industrial assembly. Guided-DDPG combines the actor-critic with the model-based RL for high-precision industrial assembly. It behaves more efficiently than the actor-critic and is more stable/reliable than the model-based RL.

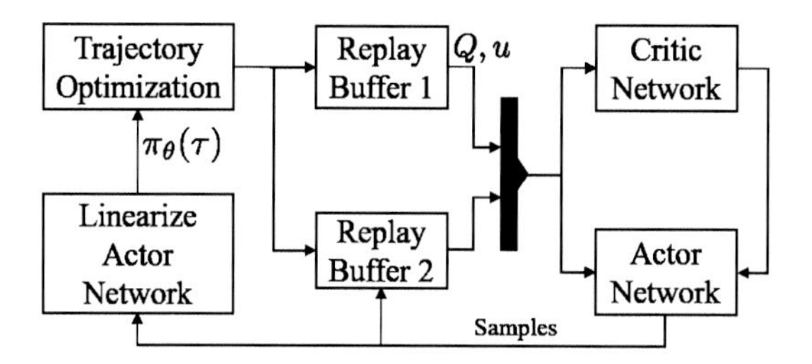

FIGURE 10.2

Fig. 10.2 illustrates the proposed guided-DDPG algorithm. Due to the low efficiency and large exploration space of the actor-critic algorithm, trajectory optimization is introduced to guide the initial exploration. Since the trajectory optimization may have unstable or inconsistent behavior and may require case-by-case dedicated parameter tuning in rigid assembly systems, the involvement of the supervision from the trajectory optimization will be reduced as the training progresses and the critic network becomes more accurate. In other words, trajectory optimization serves as a semisupervisor to train the actor-critic to establish the initial critic and constrain the network in a narrow task space. The supervision reduces as the critic gradually builds up an evaluation system for the assembly task. Eventually, the actor-critic should exhibit a superior performance compared with the trajectory optimization.

To be more specific, the trajectory optimization has

$$\min_{p} E_p(l(\tau)), \quad s.t. \;\; D_{KL}(p(\tau)\|\hat{p} * \theta(\tau)) \leq \epsilon, \tag{10.10}$$

where $\hat{p} * \theta$ is set as the trajectory distribution generated by actor policy at the first subiteration and is set as the previous trajectory distribution $\hat{p}$ for the successive $N_{trajopt} - 1$ subiterations. Eq. (10.10) is optimized by the dual:

$$\max_{\eta}\{\min_{p} E_p(l(\tau)) + \eta(D_{KL}(p(\tau)\|\hat{p}_\theta(\tau)) - \epsilon)\}. \tag{10.11}$$

The optimization of p is solved by LQG with fixed η and dynamics, and the optimization of η is done heuristically: decrease η if $D_{KL}(p(\tau)\|\hat{p} * \theta(\tau)) < \epsilon$, otherwise increase η. The trust region ϵ varies based on the expected improvement and the actual one; ϵ would be reduced once the actual improvement is far smaller than the expected one, thus the network focuses on penalizing the KL divergence from $\hat{p} * \theta(\tau)$.

We collect the trajectory after $N_{trajopt}$ subiterations to replay buffer R_1 for supervised training of actor-critic nets and feed all the sample data during $N_{trajopt}$ executions to replay buffer R_2. With the supervision from R_1, the critic is trained by

$$\begin{aligned}\phi \leftarrow \operatorname*{argmin}_{\phi} & \frac{1}{N_{dd}}\sum_{j=1}^{N_{dd}}\left(y_j - Q * \phi(s_j, a_j)\right)^2 \\ & + w * to\frac{1}{N_{to}}\sum_{i=1}^{N_{to}}\|Q * \phi(s_i, a_i) - Q_i^{to}\|^2,\end{aligned} \tag{10.12}$$

where w_{to}, N_{to} are the weight and batch size of the semisupervisor and y_j has the same form as Eq. (10.8); (s_i, a_i, Q_i^{to}) is the supervision data from R_1, and (s_j, a_j, r_j, s_{j+1}) is the sample data from R_2.

The actor is trained by

$$\theta \leftarrow \operatorname*{argmax}_{\theta} \frac{1}{N_{dd}}\sum_{j=1}^{N_{dd}} Q * \hat{\phi}(s_j, u * \theta(s_j)) + w_{to}\frac{1}{N_{to}}\sum_{i=1}^{N_{to}}\|u_\theta(s_i) - a_i\|^2. \tag{10.13}$$

The supervision weight w_{to} decays as the number of training rollouts N_{roll} increases. We use $w_{to} = \frac{c}{N_{roll}+c}$, where c is a constant to control the decay speed.

We summarize the guided-DDPG algorithm in Algorithm 10.1. The critic and actor are initialized in Line 2. Guided-DDPG runs for EP epochs in total. In each epoch, a semisupervisor is first executed to update the trajectories for supervision. With the high stiffness, small clearance, and force/torque feedback, the fitted dynamics (Line 7) is discontinuous and has a small trust region. Therefore, the trajectories generated from the semisupervisor might be suboptimal. Nevertheless, they are sufficient to guide the initial training of the actor-critic. The actor-critic is trained in Lines (12–22) following the standard procedure of DDPG with the modified objective function (Line 19). The supervision weight w_{to} is decreased as the training progresses due to the superior performance of the actor-critic compared with the semisupervisor.

Algorithm 10.1 Guided-DDPG.

1: **input:** $EP, N_{ddpg}, N_{inc}, N_{trajopt}, N_{roll} = 0, R_{1/2} \leftarrow \Phi$
2: **init:** $Q_\phi(s,a), u_\theta(s)$, set target nets $\hat{\phi} \leftarrow \phi, \hat{\theta} \leftarrow \theta$
3: **for** $epoch = 0 : EP$ **do**
4: $\quad p_{prev} \leftarrow u_\theta$
5: $\quad$ **for** $it = 0 : N_{trajopt}$ **do**
6: $\quad\quad \mathcal{S} \leftarrow sampleData(p_{prev}), R_2 \leftarrow R_2 \cup \mathcal{S}$
7: $\quad\quad f_{dy} \leftarrow fitDynamics(\mathcal{S})$
8: $\quad\quad \hat{p} * \theta \leftarrow linearizePolicy(p * prev, \mathcal{S})$
9: $\quad\quad p \leftarrow updateTrajectory(f_{dy}, \hat{p} * \theta), p * prev \leftarrow p$
10: $\quad$ **end for**
11: $\quad S \leftarrow sampleData(p), R_1 \leftarrow R_1 \cup \mathcal{S}, R_2 \leftarrow R_2 \cup \mathcal{S}$
12: $\quad$ **for** $it = 0 : N_{ddpg}$ **do**
13: $\quad\quad \mathcal{N} * ex \leftarrow sampleExplorationNoise()$
14: $\quad\quad s_0 \leftarrow observeState(), w * to = \frac{c}{c + N_{roll}++}$
15: $\quad\quad$ **for** $t = 0 : T$ **do**
16: $\quad\quad\quad a_t = u_\theta(s_t) + \mathcal{N} * ex(t)$, observe $s * t + 1, r_t$
17: $\quad\quad\quad R_2 \leftarrow R_2 \cup (s_t, a_t, s_{t+1}, r_t)$
18: $\quad\quad\quad$ sample N_{to}, N_{dd} transitions from R_1, R_2
19: $\quad\quad\quad$ update critic and actor nets by Eq. (10.12) and Eq. (10.13)
20: $\quad\quad\quad$ update target nets by Eq. (10.9)
21: $\quad\quad$ **end for**
22: $\quad$ **end for**
23: $\quad N_{ddpg} \leftarrow N_{ddpg} + N_{inc}$
24: **end for**

10.4 Simulations and experiments

This section presents both the simulation and experimental results of the guided-DDPG to verify the effectiveness of the proposed learning framework. The videos are available at [12].

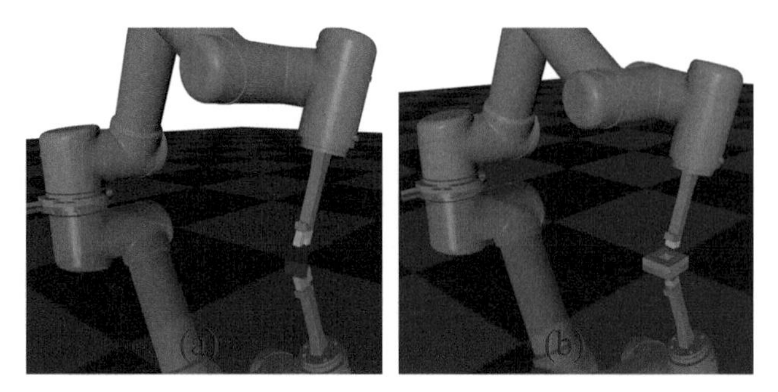

FIGURE 10.3

We built up a simulation model using the Mujoco physics engine [13] in order to compare the performance of the guided-DDPG with other state-of-the-art RL algorithms. The host computer we used was a desktop with 32 GB RAM, 4.0 GHz CPU and GTX 1070 GPU. A six-axis UR5 robot model from Universal Robots was used to perform the tasks. Two different assembly tasks were simulated, the first one was the U-shape joint assembly, as shown in Fig. 10.3 (A), and the second one was the Lego brick insertion, as shown in Fig. 10.3 (B).

10.4.1 Parameter lists

The number of the maximum epoch is set to $EP = 100$, and the initial number of rollouts for DDPG and trajectory optimization was $N_{ddpg} = 21$ and $N_{trajopt} = 3$, respectively. To ensure fewer visits of trajectory optimization as the training progresses, we increased the number of rollouts by $N_{inc} = 15$ for each DDPG iteration. The sizes of the replay buffers R_1, R_2 were 2000 and 1E6, respectively. The soft update rate $\gamma = 0.001$ in Eq. (10.9). The batch sizes for trajectory optimization N_{to} and DDPG N_{dd} were both 64. The algorithm used a cost function $l(s, a) = 0.0001\|a\|_2 + \|FK(s) - p_{tgt}(s)\|_2$, where FK represents the forward kinematics and p_{tgt} are the target end-effector points.

10.4.2 Simulation results

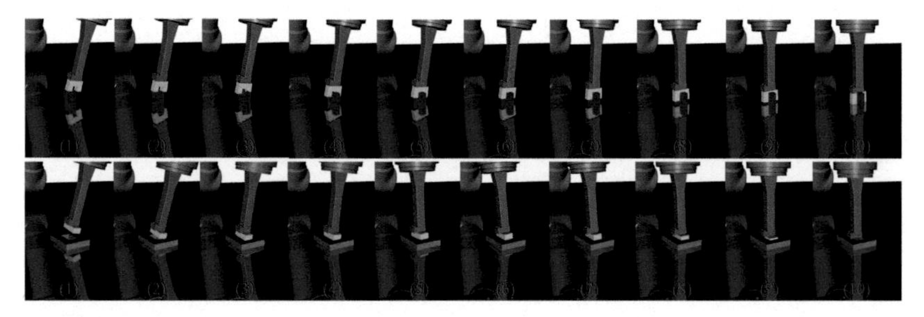

FIGURE 10.4

The simulation results on U-shape joint assembly and Lego brick insertion are shown in Fig. 10.4. Both simulations were trained with assembly clearance as 0.1 mm. Guided-DDPG takes poses and force/torque measurements of the end-effector as the states and generates joint torques as action to drive the robot. The U-shape joint has a more complicated surface than the Lego brick, and a successful assembly requires matching the shapes twice, as shown in Fig. 10.4 (top). Despite the difficulties, the proposed algorithm was able to train the policy within 1000 rollouts. We also visualized the adaptability of the trained policy on the Lego brick insertion task, as shown in Fig. 10.4 (bottom). The policy was trained with a brick of size 2×2 and clearance 0.1 mm and tested with a brick of size 4×2 and clearance 1 μm. Moreover, the brick position had an unknown offset (1.4 mm) to the network. The proposed network was able to address these uncertainties and successfully inserted the brick into a tighter hole with uncertain positions.

10.4.2.1 Comparison of different supervision methods

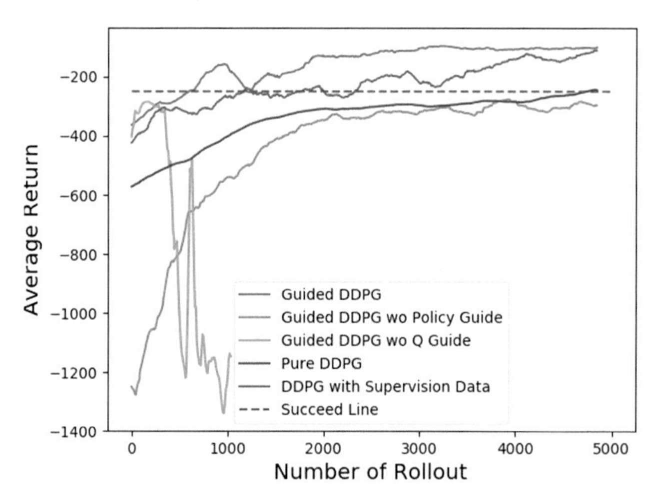

FIGURE 10.5

The proposed learning framework guides both the critic and the actor. To illustrate the necessity of the proposed guidance, we compared the results of guided-DDPG with several other supervision methods, including guided-DDPG with partial guidance, pure-DDPG with supervision data to replay buffer (no supervision on objective function), and pure-DDPG. The result is shown in Fig. 10.5. The proposed guided-DDPG achieved the best performance. The partial guidance without critic (Fig. 10.5, green) was able to guide the actor and realized safe exploration at the beginning. However, the actor network behaved worse as the involvement of the semisupervisor reduced and the weight of the critic increased, since the critic is trained purely by the contaminated target actor (Eq. (10.8)). In contrast, the partial guidance without actor (Fig. 10.5, orange) had a poorly behaved actor since the actor was trained purely by the policy gradient using the contaminated critic (Eq. (10.6)). The pure-DDPG with supervision data (Fig. 10.5, purple) achieved better performance than pure-DDPG, since the trajectories obtained from the semisupervisor were better behaved than the initial rollouts of DDPG. This kind of supervision is similar to the human demonstration in [6].

10.4.2.2 Effects of the supervision weight w_{to}

The supervision weight w_{to} balances the model-based supervision and model-free policy gradient in actor/critic updates, as shown in Eq. (10.13) and

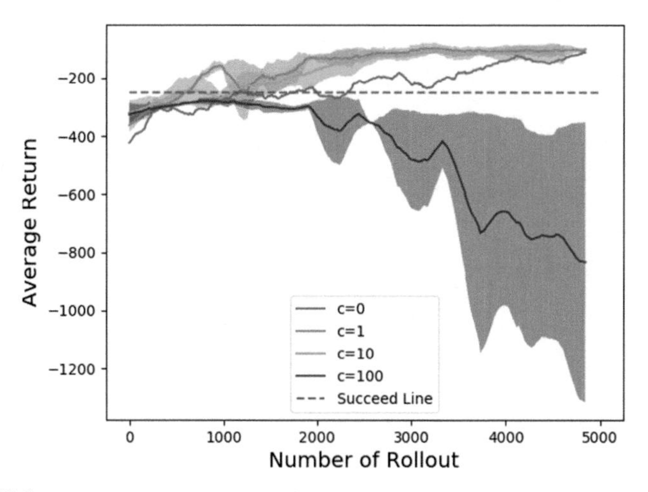

FIGURE 10.6

Eq. (10.12). The results of different weights on Lego brick insertion are shown in Fig. 10.6. With $c = 1$, the supervision weight is $w_{to} = \frac{1}{1+N_{roll}}$. The weight starts with 1 and decays to 0.001 as $N_{roll} = 1000$, while $c = 100$ makes w_{to} decay to 0.1 as $N_{roll} = 1000$. Slower decay provides excessive guidance by the semisupervisor and contaminates the original policy gradient and makes the DDPG unstable. Empirically, $c = 1 \sim 10$ achieves comparable results.

10.4.2.3 Comparison of different algorithms

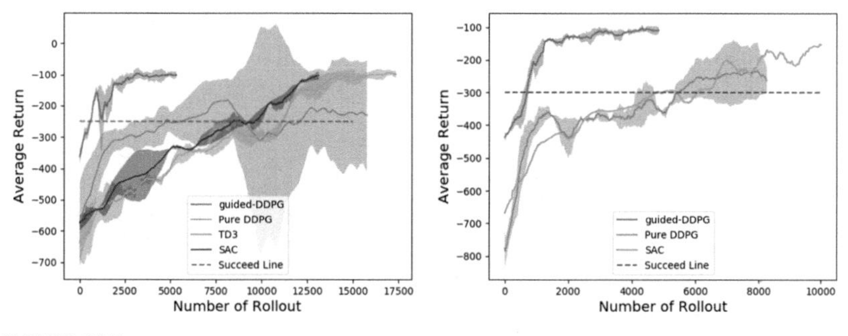

FIGURE 10.7

The proposed learning framework was compared with other state-of-the-art algorithms, including pure-DDPG, twin delayed DDPGs (TD3) [14], and the soft actor-critic (SAC) [15]. Default parameters were used for TD3, as shown in [16]. As for SAC, we used the default parameters in [16] with tuned reward scale as 10. The comparison result on the Lego brick insertion task is shown in Fig. 10.7 (left). The proposed guided-DDPG passed the success threshold (shaded purple line) at 800 rollouts and consistently succeeded the task after

TABLE 10.1 Comparison of DDPG and guided-DDPG.

Items	DDPG	Guided-DDPG
Time (min)	83	**37.3**
Data (rollouts)	7000	**1500**

2000 rollouts. In comparison, pure-DDPG passed the success threshold at 5000 rollouts and collapsed around 10,000 rollouts. The performance of pure-DDPG was inconsistent in seven different trials. TD3 and SAC had an efficiency that was similar to that of pure-DDPG. The comparison of the algorithms on U-shape joint assembly is shown in Fig. 10.7 (right). Similar with Lego brick insertion, guided-DDPG achieved a more stable and efficient learning performance. The time efficiency and data efficiency of DDPG and guided-DDPG are compared in Table 10.1.

10.4.2.4 Adaptability of the learned policy

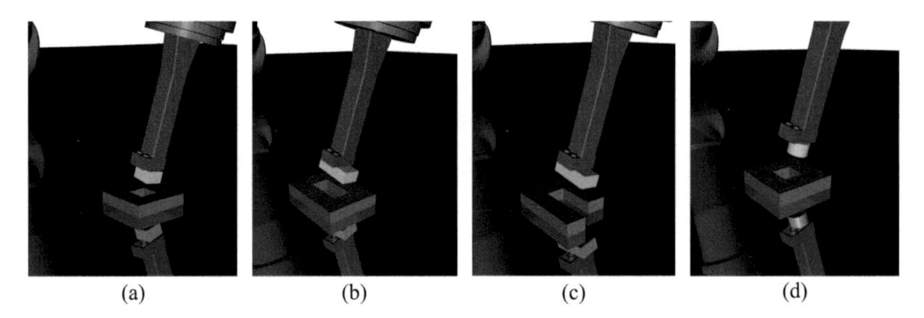

FIGURE 10.8

Three different types of uncertainties were considered. The first type was the unknown hole position. The learned policy was able to successfully insert the brick when moving the hole to an uncalibrated position (maximum offset is 5 mm, hole has a width of 16 mm). The second type of uncertainty was the shapes of peg/hole. We found that the learned policy is robust to different shapes shown in Fig. 10.8. Fig. 10.8 (A) shows the 2×2 brick used in training, Fig. 10.8 (B) shows the 4×2 brick scenario, Fig. 10.8 (C) shows the 4×2 brick with incomplete hole, and Fig. 10.8 (D) shows a cylinder brick. The third type was the different clearance. The policy was trained with clearance 0.1 mm and tested successfully on insertion tasks with clearance 10 μm, 1 μ, and 0. The simulation videos are available at [12].

10.4.3 Experimental results

Experimental results are presented in this section. The Lego brick was attached to a 3D printed stick at the end-effector of the Universal Robot UR5. A Robotiq

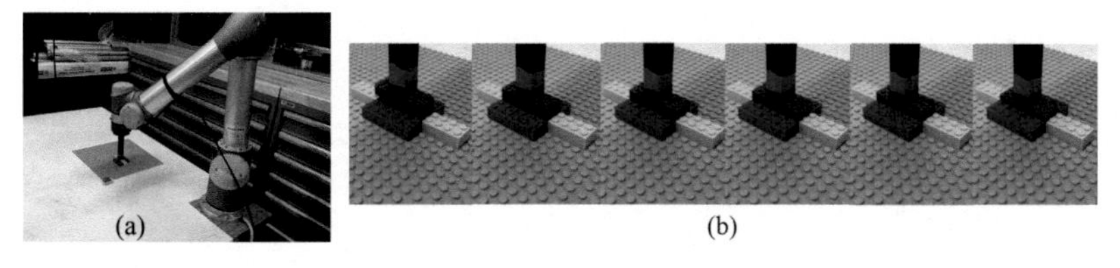

FIGURE 10.9

FT 300 force torque sensor was used to collect the force/torque signal at the wrist. The experimental setup is shown in Fig. 10.9 (A). The policy took the estimated hole position and the force/torque reading as inputs and generated transitional velocities for the end-effector. The velocity was tracked by a low-level tracking controller. The clearance of the Lego brick is less than 0.2 mm. The target position of the hole had 0.5 mm uncertainty, yet the policy was able to successfully locate the hole and insert the brick, as shown in Fig. 10.9 (B). It took 2 hours for pure-DDPG to find a policy in the exploration space bounded within 1 mm around the hole, and it took 1.5 hours for guided-DDPG to find a policy in a larger exploration space bounded within 3 mm around the hole. The experimental videos are shown in [12].

10.5 Chapter summary

The performance of the automatic assembly has been improved dramatically by deep learning [7,9,17]. On the one hand, supervised learning is reliable when the training data are sufficient. Practically, collecting data under various uncertainties is inefficient. On the other hand, RL learns a sequence of optimal actions by exploring the environment to maximize the expected reward. With the abstract reward, RL also suffers from data shortage and the low efficiency issue. Meanwhile, both types of learning methods and policies are embedded into the trained environment, causing overfitting and challenges to adapt to different environments. Model-based RL [8,9] has been proposed to address the data shortage by fitting models from data. The model fitting would be erroneous and the performance would be compromised if the contact surface is stiff or jerky. Moreover, defining a physically safe training environment is very challenging for different tasks on the end-user side. Domain randomization [18] improves robustness by randomizing uncertain states and training exhaustively in simulation. The force responses in the physical world, however, are extremely difficult to be captured accurately, especially for those tasks with stiff and jerky contacts. The contact dynamics become more different when constraining the workspace in precise assembly.

A recent successful peg-hole insertion experiment with a simple proportional controller[1] revealed the strength of compliance control. It can be seen that the

[1] Experiment with FANUC LRMate 200iD/7L manipulators and H7/h7 Aluminum peg and hole. P controller is tuned trivially. Experiment credit: Shiyu Jin in Mechanical Systems Control Lab, UC Berkeley.

human or soft robotic systems usually have better performance because of the system compliance. Therefore, a group of assembly learning methods with compliance control and impedance control [1,19] have been successfully applied. The compliance is provided by the controller to the closed-loop system in order to compensate for the jerky and high stiffness of the contact surfaces.

This chapter proposed a learning framework for automatic assembly tasks. The framework contains a trajectory optimization and an actor-critic structure. The trajectory optimization served as a semisupervisor to provide initial guidance to actor-critic, and the critic network established the ground truth quality of the policy by both learning from the semisupervisor and exploring with policy gradient. The actor network learned from both the supervision of the semisupervisor and the policy gradient of the critic. The involvement of the critic network successfully addressed the stability issue of the trajectory optimization caused by the high stiffness and the force/torque feedback. The proposed learning framework constrained the exploration in a safe narrow space, improved the consistency and reliability of the model-based RL, and reduced the data requirements to train a policy. Simulation and experimental results verified the effectiveness of the proposed learning framework.

The method, however, still requires excessive time to train. A promising direction of future research is to integrate the impedance/compliance control into the guided-DDPG to compensate for the effect of the discontinuous dynamical model. Another direction is to evaluate the algorithm on more realistic industrial applications such as connector insertion, furniture assembly, and tight peg-in-hole tasks.

References

[1] T. Tang, H.-C. Lin, Y. Zhao, Y. Fan, W. Chen, M. Tomizuka, Teach industrial robots peg-hole-insertion by human demonstration, in: Advanced Intelligent Mechatronics (AIM), 2016 IEEE International Conference on, IEEE, 2016, pp. 488–494.

[2] R.J. Williams, Simple statistical gradient-following algorithms for connectionist reinforcement learning, Machine Learning 8 (3–4) (1992) 229–256.

[3] V. Mnih, K. Kavukcuoglu, D. Silver, A.A. Rusu, J. Veness, M.G. Bellemare, A. Graves, M. Riedmiller, A.K. Fidjeland, G. Ostrovski, et al., Human-level control through deep reinforcement learning, Nature 518 (7540) (2015) 529.

[4] T.P. Lillicrap, J.J. Hunt, A. Pritzel, N. Heess, T. Erez, Y. Tassa, D. Silver, D. Wierstra, Continuous control with deep reinforcement learning, arXiv preprint, arXiv:1509.02971, 2015.

[5] J. Schulman, F. Wolski, P. Dhariwal, A. Radford, O. Klimov, Proximal policy optimization algorithms, arXiv preprint, arXiv:1707.06347, 2017.

[6] M. Vecerík, T. Hester, J. Scholz, F. Wang, O. Pietquin, B. Piot, N. Heess, T. Rothörl, T. Lampe, M.A. Riedmiller, Leveraging demonstrations for deep reinforcement learning on robotics problems with sparse rewards, CoRR, arXiv:1707.08817 [abs], 2017.

[7] T. Inoue, G. De Magistris, A. Munawar, T. Yokoya, R. Tachibana, Deep reinforcement learning for high precision assembly tasks, in: Intelligent Robots and Systems (IROS), 2017 IEEE/RSJ International Conference on, IEEE, 2017, pp. 819–825.

[8] S. Levine, V. Koltun, Guided policy search, in: International Conference on Machine Learning, 2013, pp. 1–9.

[9] S. Levine, C. Finn, T. Darrell, P. Abbeel, End-to-end training of deep visuomotor policies, The Journal of Machine Learning Research 17 (1) (2016) 1334–1373.
[10] Y. Tassa, T. Erez, E. Todorov, Synthesis and stabilization of complex behaviors through online trajectory optimization, in: Intelligent Robots and Systems (IROS), 2012 IEEE/RSJ International Conference on, IEEE, 2012, pp. 4906–4913.
[11] E. Todorov, W. Li, A generalized iterative lqg method for locally-optimal feedback control of constrained nonlinear stochastic systems, in: Proceedings of the 2005, American Control Conference, 2005, IEEE, 2005, pp. 300–306.
[12] Experimental Videos for Learning Industrial Assembly by Guided-DDPG, https://yongxf.github.io/ICRA2019/guidedddpg.html, 2019.
[13] E. Todorov, T. Erez, Y. Tassa, Mujoco: a physics engine for model-based control, in: 2012 IEEE/RSJ International Conference on Intelligent Robots and Systems, IEEE, 2012, pp. 5026–5033.
[14] S. Fujimoto, H. van Hoof, D. Meger, Addressing function approximation error in actor-critic methods, arXiv preprint, arXiv:1802.09477, 2018.
[15] T. Haarnoja, A. Zhou, P. Abbeel, S. Levine, Soft actor-critic: off-policy maximum entropy deep reinforcement learning with a stochastic actor, arXiv preprint, arXiv:1801.01290, 2018.
[16] V. Pong, rlkit: reinforcement learning framework and algorithms implemented in pytorch, https://github.com/vitchyr/rlkit.git, 2018.
[17] D. Martínez, G. Alenya, C. Torras, Relational reinforcement learning with guided demonstrations, Artificial Intelligence 247 (2017) 295–312.
[18] J. Tobin, R. Fong, A. Ray, J. Schneider, W. Zaremba, P. Abbeel, Domain randomization for transferring deep neural networks from simulation to the real world, in: 2017 IEEE/RSJ International Conference on Intelligent Robots and Systems (IROS), IEEE, 2017, pp. 23–30.
[19] L. Peternel, T. Petrič, J. Babič, Robotic assembly solution by human-in-the-loop teaching method based on real-time stiffness modulation, Autonomous Robots 42 (1) (2018) 1–17.

Part III

Robotic hand adaptive control

Chapter 11

Clinical evaluation of Hannes: measuring the usability of a novel polyarticulated prosthetic hand

Marianna Semprini[a], Nicolò Boccardo[a], Andrea Lince[a], Simone Traverso[a], Lorenzo Lombardi[a], Antonio Succi[a], Michele Canepa[a], Valentina Squeri[a], Jody A. Saglia[a], Paolo Ariano[b], Luigi Reale[c], Pericle Randi[d], Simona Castellano[d], Emanuele Gruppioni[d], Matteo Laffranchi[a,e], and Lorenzo De Michieli[a,e]

[a]*Rehab Technologies Lab, Istituto Italiano di Tecnologia, Genova, Italy,* [b]*Center for Sustainable Future Technologies, Istituto Italiano di Tecnologia, Torino, Italy,* [c]*Area sanità e salute, ISTUD Foundation, Milano, Italy,* [d]*Centro Protesi INAIL, Vigorso di Budrio (BO), Italy*

11.1 Introduction

The human hand is characterized by complexity, dexterity, and adaptability that are not yet replicated by any polyarticulated hand [28,4,17]. Indeed, myoelectric polyarticulated hand prostheses are basically robotic devices, relying on feedback control for the regulation of force and position of each finger [12,15,16,3]. The feedback control regulates the adaptability of the prosthesis to the surroundings or to grasped objects [10,18], whereas their design is characterized by rather approximate kinematics and low anthropomorphism [2]. The existing gap between state-of-the-art devices and the human hand strongly affects the perception of these prostheses as tools rather than replacements of the missing limb [28,4].

The purpose of this study is twofold. First, we aimed to draft proper requirements for the development of a novel polyarticulated prosthetic hand to match the needs of end users. This was achieved following a preliminary study in which upper limb amputees were interviewed through questionnaires and in a focus group. This preliminary study led to the development of the Hannes hand

[e] Senior equal contribution author.

Tactile Sensing, Skill Learning, and Robotic Dexterous Manipulation
https://doi.org/10.1016/B978-0-32-390445-2.00020-9

[17] and of novel electromyographic (EMG) sensors for myoelectric hand control [5,19,24,8]. The second goal of this work was to evaluate the usability of these tools through clinical validation.

The usability of the Hannes hand was assessed in a clinical study which enrolled three transradial amputees that were tested with standardized protocols before and after a short training with an occupational therapist and a period of autonomous usage of the Hannes hand at home. The same tests were also executed with their currently used hand, which was used as reference. Results confirmed that, although the training and usage period was relatively short, Hannes exhibits highly promising results, i.e., high levels of effectiveness and usability as compared to advanced state-of-the-art prostheses.

The EMG control of prosthetic devices through novel custom developed EMG sensors was addressed in a further study. We enrolled 10 patients and tested them in a manual task both with standard EMG sensors and with those developed by us. This study confirmed the equivalence of our newly developed sensors with their golden standard counterpart.

The chapter is organized as follows. Section 11.2 details the preliminary study, while Section 11.3 describes the design and implementation of the Hannes system, including both the Hannes hand and the EMG sensors. The following sections describe the clinical studies: Section 11.4 presents the pilot study aimed at validating the Hannes hand, whereas Section 11.5 addresses the description of the cross-over study for assessing the validity of the custom EMG sensors. Following this, in Section 11.6 we discuss the implementation choices that led to the design of the Hannes system and the outcomes of the clinical studies. We conclude the chapter by providing our perspective on future directions that upper limb prosthetic should take in order to narrow the gap between robotic hands and their biological counterpart.

11.2 Preliminary study

We performed a preliminary study aimed at understanding the real requirements of upper limb amputees on prosthetic hands in daily life activities (i.e., domestic, professional, and free time). The study was carried out in collaboration with the ISTUD foundation[1] and the Prosthetic Center of INAIL.[2]

Specifically, through interaction with upper limb amputees, the objective was (1) to identify the main requirements for the functionality of hand prostheses in different contexts of daily life and (2) to understand if and how currently available prostheses fulfill these needs. The results of this study were then used

[1] ISTUD: Health and Safety section, research institute accredited by the Italian Ministry of University and Research.

[2] INAIL: Istituto Nazionale per l'Assicurazione contro gli Infortuni sul Lavoro, Italian workers' compensation system.

to shape the design and development of a novel polyarticulated prosthetic hand, the IIT-INAIL Hannes hand, and novel EMG sensors.

11.2.1 Data collection

The preliminary study consisted of (1) a questionnaire sent to 65 upper limb amputees and (2) a focus group involving a subset of eight patients.

11.2.1.1 Questionnaire

The questionnaire included a structured form with multiple-choice questions, a set of open questions, and space available for general comments and suggestions. Filling the questionnaire required the authorization to treat sensitive data, in accordance with the related national laws (D.Lvo. 196/2003).

Questions regarded the following contents:

- sociodemographic information;
- amputation: year, cause, type, and place of occurrence;
- prosthetic use: first and current prostheses used, modality and duration of usage of each device, prosthesis choice process, satisfaction;
- daily life: ability in different tasks with or without the prosthesis, tactile and grasping requirements, and related satisfaction;
- work/school: ability in different tasks with or without the prosthesis, tactile and grasping requirements, and related satisfaction;
- free time, hobby, and sport: ability in different tasks with or without the prosthesis, tactile and grasping requirements, and related satisfaction;
- phantom limb pain: pain occurrence and intensity;
- ideas for new prosthesis: suggestions for improving currently available devices.

11.2.1.2 Focus group

The goal of the focus group was to deepen the themes covered by the questionnaire. A selected group of eight amputees participated in a session chaired by a moderator, proposing specific topics focusing on how upper limb prosthetics might and should fulfill different movement requirements. Participants were first requested to introduce themselves, their habits, and the usage settings of the prosthetic device. They were then asked to describe their ideal prosthesis, the environments it should be applicable to, and the functionalities that it should be equipped with.

11.2.2 Outcomes

This study provided us with an overview on the autonomy of upper limb amputees in performing various activities of daily life (ADLs). Information was collected from 65 volunteers located around Italy and although this sample is not representative of the entire amputees' realm, it allowed understanding the use modalities and habits of users and current prosthetic devices available on the market, specifically related to grasp and movement needs in different contexts: home, work, and everyday life.

The observed population was composed of 92% males, with an average age of 54 years. In 64% of cases, amputation occurred before the age of 35 and in 95% it was due to a traumatic event, which occurred in 78% of cases at work. The most used device is the tridigital myoelectric hand Ottobock Variplus [23] (49% of cases), in most cases (91%) provided with a passive pronosupination wrist. The study revealed a massive use of the prosthesis, on average used 6 days a week and in 77% of cases for more than 9 hours a day. The context of use is diverse: work (65%), recreational (88%), free time (69%), and domestic (71%). Users declared to be efficiently autonomous in most activities at home. The major difficulties consist in the use of cutlery, washing, writing, and the use of a computer. At work the most challenging tasks are keyboard use, browsing pages, and precision activities. Of all participants, 63% also own a second prosthesis, mainly cosmetic, which is used on average twice a week for specific activities as a substitute of the principal one (such as certain domestic work) or in specific contexts (e.g., during free time activities, such as at the beach). Of the interviewed people, 65% feel the prosthesis as a part of their own body, very important for interacting with others. The level of satisfaction is good on average, whereas complaints are mainly related to grasping naturalness (score 5.8/10), wrist (mostly nonelectric) functionality (score 5.68/10), and functionality for precision activities (score 5.14/10). Main issues are represented by glove ripping (43% of cases) and mechanical breaks (15%). Most people (72%) indicated functionality over esthetics as an improvement priority, but we collected suggestions for improving both aspects:

1. *Technical and functional indications:* lighter prosthesis, electric rotation of the wrist, autonomous movement and control of the fingers, socket comfort, complete hand closure, regulation of force, reduction of noise, removable and interchangeable battery, use of index finger for pointing, led panel for visualizing device functions (e.g., battery status).
2. *Esthetic indications:* naturalness of the grasp, reduction of the hand dimension, natural posture of the hand when not actuated, improved robustness to dust and stains, glove more easily washable, harness removal, restoration of sensation, solving the sweating problem during summer with the consequent deterioration of the prosthesis.

In conclusion, this study clarified how currently available devices allow upper limb amputees to reach an acceptable level of autonomy in most activities of everyday life. However, there is room for improvement in both the functional and esthetic domains, in order to increase the level of satisfaction and to respond to the users' demands.

11.3 The Hannes system

11.3.1 Analysis of survey study and definition of requirements

Analyzing the outcomes of the survey study, a few aspects emerge which are relevant to the overall layout of the prosthesis. First of all, although embodiment is indicated as a priority, it appears clear that the percentage of amputees who feel the prosthesis as a part of their own body is still relatively low (65%). Secondly, leaving issues that can be solved with proper engineering practice such as robustness, maintenance, and management aside, both technical and appearance/esthetics-related aspects tend to point to seamless use and human-like behavior (grasp naturalness, rotation of the wrist, autonomous movement of the fingers, complete hand closure, regulation of force, natural posture of the hand, and restoration of sensation). From this, it emerges that the ideal prosthesis would need to be very much human-like in its function, esthetics, physical properties, and performance. We believe that the satisfaction of such requirements can additionally increase embodiment and it could ultimately have an impact on the percentage of abandonment of the device. Therefore, for the development of the Hannes system, we conceived a biomimetic design, which takes these features into consideration.

11.3.2 System architecture

The Hannes system was detailed in [17]. Briefly, it consists of (1) a myoelectric polyarticulated prosthetic hand that exploits a differential underactuated mechanism as inspired by the work in [6] and (2) a passive wrist flexion-extension module (Fig. 11.1). Muscular activity from the residual limb of the amputee is detected by two dry surface EMG sensors placed in the socket. The signals are fed to a dedicated board module, the EMG Master, to be conditioned and converted to digital form. These are then used to synthetize a reference position of the hand and sent to the "Motor Board" module. The "Motor Board" controls accordingly the hand actuation system, which consists of a DC motor coupled to a custom hypocycloidal gearbox. The drive train tensions a dyneema cable routed within the palm that transmits the motion to the fingers.

A lithium battery pack placed inside the socket provides energy to the entire Hannes system for up to 1 day of usage. A battery management system controls the state of charge of the pack and ensures overall safety. The batteries can then be recharged connecting the dedicated charger to the magnetic plug connector when the system is not in use.

The flexion-extension module can be added at the proximal side of the Hannes system. It relies on a spring-based mechanism that allows passive wrist movement (flexion/extension) as well as the capability of locking the wrist in five different wrist postures.

An electromechanical connector system that replicates the (de facto standard) system developed by Ottobock connects the prosthesis to the socket. This system allows to relay power and control signals from the residual limb to the hand, while also providing the quick connect/disconnect capabilities.

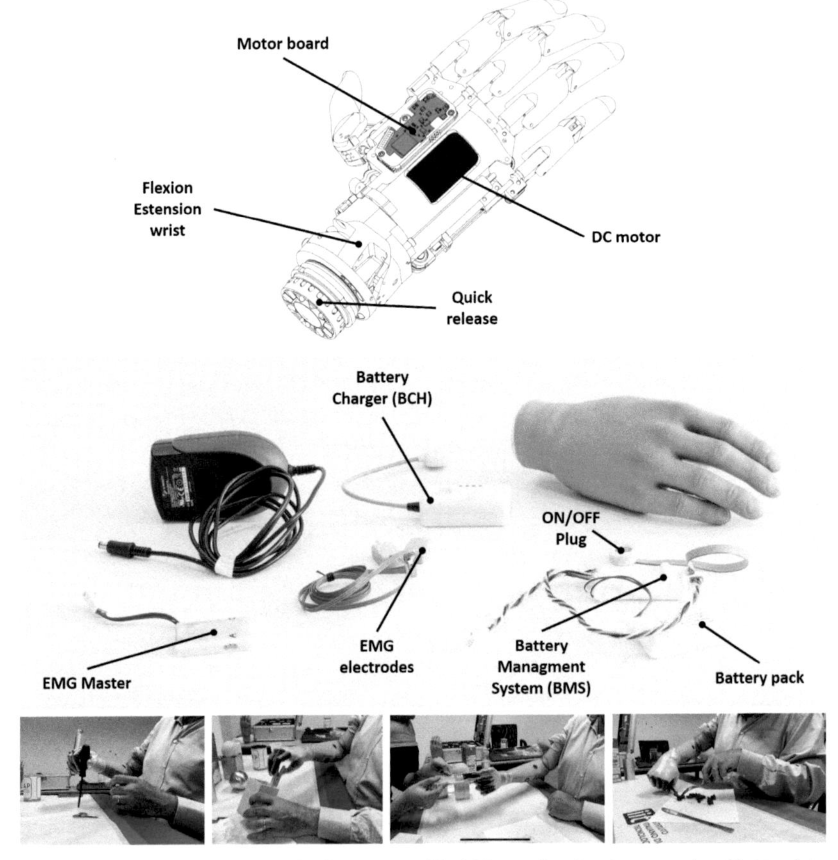

FIGURE 11.1 The Hannes prosthetic system. (Top) Hannes hand and accessories. **(Bottom)** An amputee manipulating objects with Hannes.

11.3.2.1 The Hannes hand

We considered the following specifications for the design of the hand: (1) adaptability and human-like coordination of movement and force; (2) anthropomorphism; (3) biomimetic performance (velocity and force); (4) biomimetic physical properties (kinematics, weight); and (5) intuitive control. These guidelines led to the development of a system that relies on a synergetic coordination of the finger motions and forces. This has been shown to be the most effective method to obtain a stable, robust, and biomimetic grasp [3,10,18]. Synergistic behavior has been implemented in several polyarticulated hands, either through multiple motors or through a mechanical design that implements specific hand movements [2,8,6,25]. Hannes was designed with the aim of directly implementing this adaptability and coordination of motion and force. This differential drive, actuated by a single motor, allows to adapt its posture to the grasped object whilst equally distributing forces over all five fingers. The relative timing between the distal and proximal phalanges is also implemented mechanically. Hannes' control relies on typical myoelectric direct control, which sends opening or closure commands based on the intensity of the sensed EMG signal. The patient can thus move the prosthesis by suitably contracting the flexor or extensor muscle. This results in an overall behavior that closely mimics the human hand [10].

Weight of the prosthesis and overall range of motion (RoM) are also among the considered metrics of anthropomorphism. The prosthesis weighs 480 g, close to its biological counterpart. As for the RoMs, with the exception of the distal interphalangeal joint (DIP) for the long fingers and the interphalangeal (IP) of the thumb, all other degrees of freedom were designed to match those of the human hand. Another important metric is the delay introduced between muscle activation and hand actuation, which has been shown to be crucial for embodiment [27]. Introducing only 10 ms delay, the Hannes system compares well with the physiological electromechanical counterpart [7].

Speed and grasp force are also key parameters. The train drive developed achieves the performance required to perform typical daily activities, whilst also minimizing weight and volume. This led to a high-density actuation that achieves the optimal compromise between grasp force and speed (150 N peak and 3–4 rad/s) without compromising anthropomorphism [28,26].

11.3.2.2 Custom EMG sensors

Superficial EMG sensors have been used for over 50 years to control prostheses. These are noninvasive devices that contain two (monopolar) or three (bipolar) conductive elements that can capture the electric potential differences generated by the contraction of the muscles in the stump. The potentials are typically in the order of tens of μV up to 1 mV and thus very sensitive to electromagnetic noise. A small analog front-end amplifies, filters, and demodulates the raw signal, transforming it into a more manageable signal for A/D conversion. In case of transradial amputation, in the most used configuration, two sensors are placed inside the socket on an antagonistic muscle pair (typically wrist flexor and extensor). An external electronic board receives the processed signals, performs the A/D conversion, and transforms them into appropriate commands for the prosthesis.

Technological advances have led to the development of small sensors with high signal-to-noise ratio that can be placed inside the socket. Therefore, there is no need to further increase the functionality of these devices; rather, there is room for improving more technical/applicative features that affect the reliability of the sensors and the prosthesis and their cost. We thus designed a novel type of sensor, inspired by the commercial golden standard provided by the Ottobock sensors [1], and added some new feature, detailed hereafter.

The sensor fabrication followed a procedure aimed at guaranteeing waterproofness, in order to obtain a device more robust to malfunctions due to sweat (which can penetrate inside the sensor, damaging its electronics). Specifically, we made use of a rotational molding technique that allows to completely cover the integrated circuits (ICs) with the polymer, so as not to leave air gaps.

Moreover, our sensors can be wirelessly programmed via a near field communication (NFC) programmer. In this way, working parameters, such as cutoff frequency (50 or 60 Hz) and gain (1÷7), can be remotely adjusted by the user with no need of a certified prosthetist and orthotist (CPO) intervention.

Other improvements with respect to the state of the art are related to the prosthesis fitting procedure. Prosthesis fitting is crucial and is mainly devoted to the calibration of the EMG sensors. Indeed, if sensors are not correctly calibrated, the patient is unable to control the prosthesis. We thus developed a calibration device which consists in an electronic board that continuously receives signals from the EMG sensors and is able to visualize it on an external monitor. This allows to assess the optimal positioning of the EMG sensors and this information is then used for the socket manufacturing and fitting procedure.

The developed EMG sensor and its CAD model are presented in Fig. 11.2.

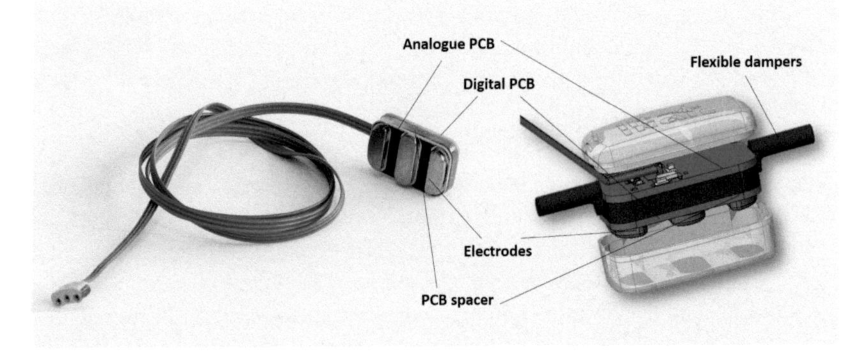

FIGURE 11.2 The IIT-INAIL EMG sensor. **(Left)** a picture of the device. **(Right)** CAD model.

11.4 Pilot study for evaluating the Hannes hand

We performed a clinical evaluation of the Hannes hand with the primary goal of defining its functionality, safety, and reliability. Moreover, we aimed at assessing the potential benefits that the Hannes hand might add to the quality of life of amputees. The clinical study was performed in collaboration with the Prosthetic Center of INAIL in Vigorso di Budrio (Italy) in accordance with the Helsinki declaration and with the guidelines of good clinical practice (GCP), and was approved by the Ethical Committee of Bologna and Imola (CE-BI, proto-

TABLE 11.1 Subjects information. * Subject 2 also uses the Michelangelo prosthesis during weekends. Modified from [17].

	Subject #1	Subject #2	Subject #3
Age	62	29	55
Age at amputation	14	23	34
Gender	male	male	male
Missing hand	right	right	right
Level of amputation	medial	medial	distal
Dominant hand	right	right	right
Years of prosthesis use	48	6	20
Years of myoelectric prosthesis use	48	5	19
Commonly used prosthesis	Michelangelo	VariPlus*	VariPlus

col number 16051). Personal data were treated according to the related national laws (D.Lvo. 196/2003).

11.4.1 Materials and methods

11.4.1.1 Subjects

Three male right-handed subjects (aged 62, 29, and 55 years, respectively) participated in the study. All subjects presented a transradial amputation of the right hand and the level of amputation was medial for subjects #1 and #2 and distal for subject #3. None of them presented psychological comorbidity related to hand loss. All subjects had residual muscles efficiently active and were expert in the use of a myoelectrically controlled hand. The currently used myoelectric hand served as reference hand in the study: an Ottobock Michelangelo [22] hand for patient #1 and an Ottobock Variplus hand [23] for patients #2 and #3. Table 11.1 summarizes information about the recruited subjects. All subjects signed an informed consent form before enrollment in the study.

11.4.1.2 Study protocol

We first tested subjects with their commonly used myoelectric prosthesis in order to assess their baseline (TB, see Fig. 11.3) and we performed a myometric exam to evaluate the functional state of the residual muscles, the amount of EMG signals, and the optimal positions of Hannes EMG sensors. It followed a training procedure, during which subjects became acquainted with the Hannes prosthetic system. Subjects were then dismissed and provided with the Hannes prosthetic hand for a period of 15 days for domestic use and daily utilization. The same tests used for baseline assessment (TB) were repeated with the Hannes hand before the training procedure (T0), at the end of the training period (T1), and at the end of the study (T2). Questionnaires (see Section 11.4.1.3) were presented at TB and T2.

Fig. 11.3 depicts the experimental protocol and its various phases. The training with the occupational therapist was repeated for 4 days and each session lasted 4 hours, during which subjects performed a set of tasks combining generic manipulation and execution of several ADLs. Specifically, the tasks consisted of:

- full and partial opening/closing of the prosthetic hand;
- grasping and manipulating wooden objects of different shapes (cubes, cylinders, etc.);
- grasping and manipulating small objects such as buttons or coins;
- grasping, manipulating, and releasing cups made of different materials (i.e., plastic or glass);
- laying and clearing the table;
- tidying up a room;
- utilizing cutlery;
- cutting an apple and spreading butter on a toast;
- utilizing toothbrush and toothpaste;
- opening and closing doors using handles and locking/unlocking doors;
- simulated driving (turn the steering wheel, shifting gears);
- utilizing occupational therapy panels with zips and strings;
- tying laces;
- dressing (shirts, jackets);
- writing with a pen on a sheet of paper;
- using a computer's keyboard.

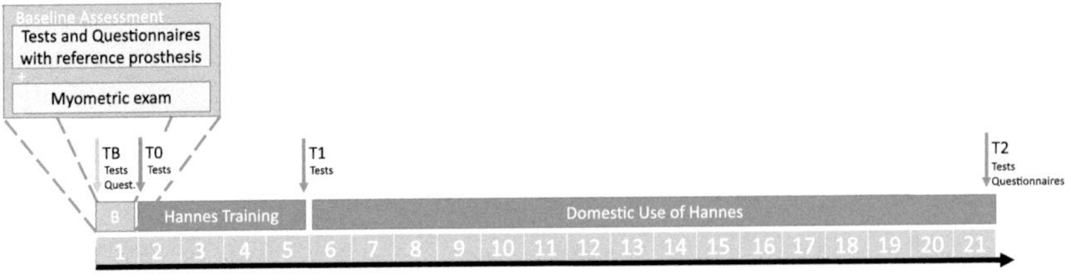

FIGURE 11.3 Outline of the experimental protocol. "B" (day 1) refers to the baseline phase, during which tests and questionnaires are executed with the reference prosthetic hand (TB) and then the subject receives myometric exams for setting EMG electrode positions of the Hannes hand. Following is Hannes training (days 2–5), which begins with tests of Hannes (T0) and terminates with other tests (T1). Finally, the subject brings Hannes home for domestic use (days 6–21) and undergoes a final test and questionnaire phase (T2). Modified from [17].

11.4.1.3 Clinical evaluation measures

During the different phases of the clinical study, we monitored parameters collected from tests and questionnaires. Specifically, the tests used were the Minnesota Manual Dexterity Test, Placing only (MMDT-P) and the Southampton Hand Assessment Procedure (SHAP). The first assesses the ability to move small objects and the score is computed by recording the total seconds required to complete the chosen number of test trials [9]. The second scores the ability of a prosthetic hand with values ranging from 0 to 100 (100 being the functionality typical of a healthy limb [20]), and is calculated for overall hand function (index of function) or for specific postures (i.e., spherical, tripod, power, lateral, tip, and extension). Questionnaires provide a qualitative evaluation on the independence of the user in the execution of ADLs. The employed questionnaires were (1) the Orthotics and Prosthetics User Survey-Upper Extremity Functional Status module (OPUS-UEFS) [13], (2) the Disabilities of the Arm, Shoulder, and Hand (DASH) [14], (3) the Trinity Amputation and Prosthesis Experience Scales (TAPES) [11], and (4) a final evaluation questionnaire. Specifically, the OPUS-UEFS evaluates the efficacy in carrying out daily tasks, such as general self-care and usage of domestic tools; the scores indicate how easily subjects perform the tasks and if they use the prosthesis for these activities. The DASH returns measurements on the functional activities of the upper limb in ADLs, work, and sport and produces a score indicating the level of disability, ranging from 0 (performance as for a healthy limb) to 100 (full disability). Finally, the TAPES questionnaire considers the degree of satisfaction in the utilization of the prosthesis in various tasks. As mentioned before, these tools were used to assess the behavioral performance of subjects using a prosthetic hand and thus were performed both when subjects used the Hannes hand and when they used the commonly used myoelectric hand.

11.4.2 Results

The scores of the tests and questionnaires (see Section 11.4.1.3) executed on the three subjects with the reference hand (only during baseline assessment TB) and with the Hannes hand along the whole study are reported in Table 11.2. In order to highlight the difference between scores recorded at T2 with Hannes and scores recorded at TB with the reference hand, Table 11.3 summarizes the changes in test and questionnaire scores from baseline to end of the study with values indicating improvement highlighted in green (note that for the MMDT-P and DASH tests, negative values indicate improvement). The results obtained for the Minnesota tests with Hannes are on average better than those obtained with the reference prosthesis and all subjects improved the Minnesota test with Hannes during the training. Indeed, subjects #1 and #2 performed better with the Hannes hand with a decrease of time needed to perform the task (Table 11.3), while subject #3 performed better with the reference hand, showing slower task execution during T0, T1, and T2 with respect to TB (Table 11.2). Subject #3 had only experience with a tridigital hand, and was not familiar with polyarticulated devices and its related grasp as the other subjects. Indeed, novice users of polyarticulated devices tend to drift away the ring and little fingers more than in tridigital grasp, with the consequence that while they grasp an object, they also move away the other nearby objects used during the Minnesota test and

TABLE 11.2 Scores of tests and questionnaires in each phase of the study. Modified from [17].

	Subject #1				Subject #2				Subject #3			
	TB	**T0**	**T1**	**T2**	**TB**	**T0**	**T1**	**T2**	**TB**	**T0**	**T1**	**T2**
Minnesota Test												
Score ([s])	137.33 ±15.31	132 ±6	125.33 ±8.08	123.67 ±4.04	171.67 ±10.07	181.67 ±20.50	142.33 ±6.43	133.67 ±6.03	140.67 ±10.50	210 ±20.30	171 ±38.74	166.33 ±10.07
SHAP Test												
IoF ([%])	76	78	71	74	72	66	77	76	58	43	62	61
Spherical ([%])	79	82	77	77	74	82	90	88	75	57	76	76
Tripod ([%])	63	70	60	70	44	35	44	53	29	24	49	50
Power ([%])	75	77	65	71	59	53	59	66	52	42	57	63
Lateral ([%])	85	75	73	72	77	68	80	83	70	34	58	55
Tip ([%])	60	68	65	63	70	58	70	65	45	30	55	51
Extension ([%])	75	78	70	76	71	66	78	80	67	44	61	69
OPUS-UEFS Questionnaire												
Score ([%])	96.05	x	x	80.36	86.78	x	x	82.5	67.5	x	x	88.75
Usage ([%])	67.85	x	x	60.71	60.71	x	x	71.4	71.42	x	x	71.42
DASH Questionnaire												
ADLs ([%])	0.08	x	x	2.5	2.5	x	x	2.5	4.3	x	x	10
WORK ([%])	0	x	x	0	-	x	x	-	0	x	x	18.8
SPORT ([%])	0	x	x	0	0	x	x	0	-	x	x	-
TAPES Questionnaire												
Score ([%])	98.3	x	x	93.1	88.3	x	x	90.8	74.1	x	X	76.6

TABLE 11.3 Difference between scores recorded at T2 and TB. Modified from [17].

	Subject #1	Subject #2	Subject #3
Minnesota Test			
Score ([s])	−13.66	−38	26.66
SHAP Test			
IoF ([%])	−2	4	3
Spherical ([%])l	−2	14	1
Tripod ([%])	7	9	21
Power ([%])	−4	7	11
Lateral ([%])	−13	6	−15
Tip ([%])	3	−5	6
Extension ([%])	1	9	2
OPUS-UEFS Questionnaire			
Score ([%])	−15.69	−4.26	21.25
Usage ([%])	−17.86	10.72	−2.62
DASH Questionnaire			
ADLs ([%])	1.7	0	6.7
WORK ([%])	0	-	18.8
SPORT ([%])	0	0	-
TAPES Questionnaire			
Score ([%])	−5.2	2.5	2.5

thus execution time increases. Subject #1 performed better with Hannes since T1, possibly because of his expertise in the use of a polyarticulated hand. The SHAP test indicated that the scores obtained with the reference hand at TB and with Hannes at T2 are on average higher for all subjects and in many cases, there was an increase from TB to T2 for all subjects (see highlighted values in Table 11.3). Overall tests results suggest that the Hannes prosthetic system has a quick learning curve and its task execution performance is on average better than that of commercial state-of-the-art polyarticulated prostheses.

Questionnaires were used to qualitatively evaluate the independence of the user in the execution of ADLs. Functional activities executed with the prosthesis are scored with the OPUS-UEFS. According to this questionnaire, subject #1 preferred the reference hand, while the others preferred Hannes (Table 11.3). More in detail, subject #2 increased the use of Hannes, while subject #3 perceived the usage of Hannes as identical to that of the reference hand. The level of impairment in functional activities was measured with the DASH questionnaire, and according to it, subjects #1 and #3 performed better with the reference prosthesis, whereas for subject #2 there was no difference (see Table 11.3). Instead, the Hannes system was preferred over the reference hand in subjects #2 and #3 according to the TAPES questionnaire (Table 11.3), which scores the degree of satisfaction in the utilization of the prosthesis.

11.5 Validation of custom EMG sensors

We performed a cross-over study to assess the equivalence of our custom EMG sensors (NS) to the commercial golden standard by Ottobock (GS) [1].

11.5.1 Materials and methods

We enrolled 10 subjects who performed the study protocol first using the prosthesis equipped with one type of sensors and then the other ones. Patients were randomly assigned to one of two groups of five people each: group 1 first performed the tests with the NS and then with the GS, while group 2 first performed the tests with the GS and then with the NS. The test used was the MMDT-P, as in the pilot study. The study protocol was conducted in accordance with the Helsinki Declaration and with the standards of GCP and was approved by the Ethical Committee of Bologna and Imola (CE-BI, protocol number 15033). Personal data were treated according to the related national laws (D.Lvo. 196/2003.

11.5.1.1 Subjects

Ten subjects were enrolled at the INAIL Prosthetic Center in Vigorso di Budrio and provided informed consent. The study was single blind, so subjects were randomly assigned to one of two possible groups, each testing the two types of sensors in different order. Subjects were unaware of the sensors in use because they were opportunely covered such that they looked identical. As for the pilot study, patients had their dominant limb amputated and were expert users of the myoelectric prosthetic hand. Table 11.4 reports subjects' information.

11.5.1.2 Study protocol

Fig. 11.4 details the study protocol. Enrolled patients were first instructed with the objectives and modalities of the study. Once they signed the informed consent, a myometric exam was conducted in order to assess the functionality of the residual muscles, the amount of EMG production, and the optimal position of the new sensors. After the prosthesis was prepared, subjects were provided in single blind fashion with one of the two sensors. They thus underwent a training procedure followed by testing with the MMDT-P protocol. Subjects were thus provided with the other prosthetic system and underwent a second set of training and testing. Finally, we provided a short questionnaire (15' duration) where patients were requested to highlight the differences between the two systems.

The myometric test lasted 30' and was aimed at identifying the optimal location for the two sensors inside the socket, corresponding to the highest amplitude of the recorded signals that could be independently elicited by the patients (lowest cross-talk). These myometric exams were different from standard clinical

TABLE 11.4 Subjects information. MICH = Michelangelo hand, developed by Ottobock; TRI = Variplus Tridigital hand, developed by Ottobock.

Subject ID	#1	#2	#3	#4	#5	#6	#7	#8	#9	#10
Age	62	46	30	50	64	40	47	56	22	61
Age at amputation	13	24	23	29	17	36	21	34	19	56
Gender	M	F	M	M	M	M	M	M	M	M
Missing hand	R	R	R	R/L	R	R	R	R	L	R
Level of amputation	MID	DIST	MID	MID	PROX	MID	MID	DIST	MID	PROX
Dominant hand	R	R	R	R	R	R	R	R	L	R
Years of prosthesis use	>10	>10	7	>10	>10	3	>10	>10	1	4
Years of myoelectric use	>10	>10	6	>10	>10	3	>10	>10	1	3
Commonly used device	MICH	TRI	TRI	TRI	TRI	TRI	MICH	MICH	TRI	TRI

tests, as here the purpose was to compare prosthesis control with clinical evaluation of the muscles. These tests were thus conducted by a CPO.

The training protocols (T1, T3) involved the accomplishment of a set of tasks combining generic manipulation and execution of several ADLs, as in Section 11.4.1.2:

- full and partial opening/closing of the prosthetic hand;
- grasping and manipulating wooden objects of different shapes (cubes, cylinders, etc.);
- grasping and manipulating small objects such as buttons or coins;
- grasping, manipulating, and releasing cups made of different materials (i.e., plastic or glass);
- laying and clearing the table;
- tidying up a room;
- utilizing cutlery;
- cutting an apple and spreading butter on a toast;
- utilizing toothbrush and toothpaste;
- opening and closing doors using handles and locking/unlocking doors;
- utilizing occupational therapy panels with zips and strings;
- tying laces;
- dressing (shirts, jackets);
- writing with a pen on a sheet of paper;
- using a computer's keyboard.

For testing (T2, T4), we used a modified version of the MMDT-P test and we evaluated the execution time while moving 60 plastic disks into a specific grid, using first the real and then the prosthetic hand. During each testing phase, the test was repeated three times, followed by a resting time of 3' to 5', and overall it lasted 30'.

11.5.1.3 Analysis

The purpose of the study was to evaluate whether the NS maintained the same functionality and efficacy of the GS sensors. To this end, we assessed the equivalence of the two systems by performing a Wilcoxon test of the MDDT-P scores obtained with NS and with GS. The Kolmogorov–Smirnov test was used to assess the normal distribution of the data. Statistical analysis was performed with MATLAB® (the Mathworks).

11.5.2 Results

In this study, each patient wore two prosthetic devices (one with the GS and one with the NS) and with each of them the patient went through a training and a testing phase with the MDDT-P protocol. We report in Table 11.5 the execution time of such test in three different trials for both devices.

For each subject we then calculated the average value for the MDDT-P test with the GS and the NS. Fig. 11.5 shows the linear relationship between the

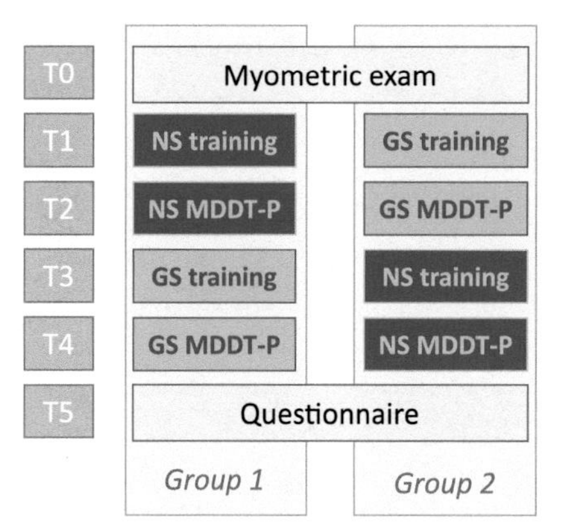

FIGURE 11.4 Study protocol. All subjects received a myometric exam (T0) and were then divided in two groups. Group 1 was first provided with the prosthesis equipped with the NS, received training (T1), and was tested with the MDDT-P protocol (T2). Subjects of group 1 were then provided with the prosthesis equipped with the GS, received training (T3), and were tested with the MDDT-P protocol (T4). On the contrary, group 2 was first provided with the prosthesis equipped with the GS, received training (T1), and was tested with the MDDT-P protocol (T2). Subjects of group 2 were then provided with the prosthesis equipped with the NS, received training (T3), and were tested with the MDDT-P protocol (T4). At the end of the procedure all subjects filled a questionnaire (T5).

TABLE 11.5 Outcome of the MDDT-P in the various phases of the study protocol.

Subject ID	Execution time with NS			Execution time with GS		
	Trial #1	Trial #2	Trial #3	Trial #1	Trial #1	Trial #3
#1	152	151	138	142	140	138
#2	197	182	174	203	216	131
#3	155	146	147	149	142	139
#4	108	104	108	120	107	108
#5	182	199	165	198	192	191
#6	145	141	141	135	135	141
#7	141	145	127	134	142	145
#8	210	200	197	228	174	192
#9	116	113	118	126	126	119
#10	212	215	200	232	207	213

execution times with both sensors, i.e., subjects that were faster/slower with the GS with respect to others were also faster/slower when using the NS sensors. Finally, to assess the equivalence of the two systems, we tested whether the timings obtained with the two different sensors were statistically significant. Because data were not normally distributed (as assessed by the Kolmogorov–Smirnov test), we run a Wilcoxon test, which confirmed that there were no significant differences between the data obtained with the two types of sensors ($P > 0.5$).

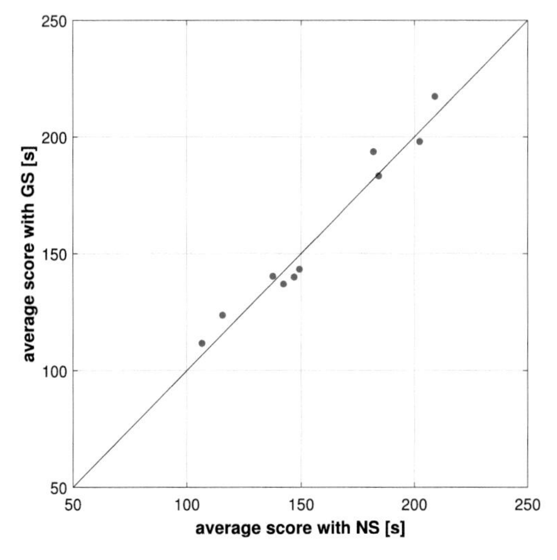

FIGURE 11.5 Average MDDT-P scores with NS vs. scores obtained with GS. Blue dots represent single subjects' data.

11.6 Discussion and conclusions

This work focused on a survey study that was carried out to draft the requirements for the development and on the clinical evaluation of a novel polyarticulated myoelectric prosthetic hand (Hannes) conceived with the goal of narrowing the gap between the robotic prosthesis and their biological counterpart and of novel EMG sensors.

For designing an effective system, we first collected information from a set of patients of the INAIL prosthetic center. This sample is not fully representative of the whole population of hand amputees of the Italian area, but offers an important overview of the needs, desires, and expectations of users for the development of prosthetic devices of the future. Using this information as driving force, we thus decided to focus the innovative features of our system on anthropomorphism, in terms of both functional control and esthetics.

The mechatronic design of Hannes, already detailed in [17], was here presented and analyzed to show how human-likeness, performance, and physical properties were implemented into the device. We believe that the main strength of Hannes is that it naturally and seamlessly adapts itself in a very human-like manner when interacting with the environment and/or when grasping objects. This feature offers an extremely high degree of adaptability to the shape of the grasped object, during both hold and manipulation, thus resulting in effective and natural grasps characterized by high stability and robustness. This enables

Hannes to reach performance that is comparable with, and in some cases even better than, that of high-end state-of-the-art devices, such as the Michelangelo hand. We further confirmed this with the pilot study conducted on three subjects, which showed how the synergetic implementation of such characteristics into one device could be beneficial to the capability of the amputees to execute a variety of tasks in an efficient way. It is worth noting that trials were performed by expert users of myoelectric hands with several years of experience, while the training with a prototypical version of the Hannes system was relatively short. We therefore expect to obtain much better results with a refined version of the device and following a longer usage. Interestingly, we observed a stronger improvement during the training in subjects that normally use the tridigital hand (i.e., #2 and #3), although they also showed slower learning rates with respect to subject #1, who was already expert in the use of a polyarticulated hand and thus showed good performance since the beginning of the training. Subjects #2 and #3 instead had to cope with the different grasping modality of the polyarticulated hand, which involves also the movement of the ring and little fingers. More in detail, subject #2 obtained higher scores with respect to subject #3 and this can be explained both by the fact that he occasionally used the polyarticulated hand and was thus not completely novice to its grasp configuration and also because of his younger age. Overall, amputees who underwent the trials explained in this work were mostly satisfied with the Hannes system and its general usability and performance. However, we believe that improvements can be made to increase the range of functionalities that this system can perform. Indeed, advancements will be introduced by raising the number of degrees of freedom, either passive or actuated, to realize more human-like capabilities and potentially reach higher levels of embodiment. This will in turn open new control-related issues and scenarios that will generate the need for the development of novel control paradigms aimed at realizing a synergetic combination of human-like functionalities and seamless ease of use.

The custom EMG sensors were efficiently used for myoelectrically controlling a hand prosthesis and they presented some novel features with respect to those available on the market, such as waterproofness and wireless programming. We are also currently working on a novel, miniaturized version of the sensors (round shape, 1.8 cm diameter) with the final goal of enabling more advanced control strategies that make use of more than two EMG sensors [21]. As additional outcome of the cross-over study, we found that reducing sensor size is more desirable than, for example, waterproofness, given that robotic hands are not yet waterproof and thus use in water is still prevented. On the contrary, reducing sensor size would allow keeping at minimum the possible access points for dangerous agents and, consequently, would augment the overall reliability and robustness of the system.

This chapter clearly shows how a "patient-centered" approach is crucial in order to achieve quick development time and high satisfaction from the users. We therefore want to stress once more the importance of continuously dealing with the final users, whose opinion is fundamental to understand which future improvements represent an urgency and how to efficaciously shape their implementation.

References

[1] MyoBock Sensors, http://www.ottobock.com/.

[2] J.T. Belter, et al., Mechanical design and performance specifications of anthropomorphic prosthetic hands: a review, Journal of Rehabilitation Research and Development (ISSN 0748-7711) 50 (5) (2013) 599–618.

[3] L. Biagiotti, et al., How far is the human hand? A review on anthropomorphic robotic end-effectors, University of Bologna, 2008.

[4] Elaine A. Biddiss, Tom T. Chau, Upper limb prosthesis use and abandonment: a survey of the last 25 years, Prosthetics and Orthotics International (ISSN 0309-3646) 31 (3) (2007) 236–257, https://doi.org/10.1080/03093640600994581.

[5] Hanneke Bouwsema, Corry K. van der Sluis, Raoul M. Bongers, Learning to control opening and closing a myoelectric hand, Archives of Physical Medicine and Rehabilitation (ISSN 0003-9993) 91 (9) (2010) 1442–1446, https://doi.org/10.1016/j.apmr.2010.06.025, http://www.sciencedirect.com/science/article/pii/S000399931000362X.

[6] M.G. Catalano, et al., Adaptive synergies for the design and control of the Pisa/IIT SoftHand, The International Journal of Robotics Research (ISSN 0278-3649) 33 (5) (2014) 768–782, https://doi.org/10.1177/0278364913518998.

[7] P.R. Cavanagh, P. Komi, Electromechanical delay in human skeletal muscle under concentric and eccentric contractions, Journal of Applied Physiology 42 (159) (1979).

[8] Christian Cipriani, Marco Controzzi, Maria Chiara Carrozza, The SmartHand transradial prosthesis, Journal of NeuroEngineering and Rehabilitation (ISSN 1743-0003) 8 (1) (2011) 29, https://doi.org/10.1186/1743-0003-8-29.

[9] Company, L.I. Catalog, 1969.

[10] Marco Controzzi, Christian Cipriani, Maria Chiara Carrozza, Design of artificial hands: a review, in: Ravi Balasubramanian, Veronica J. Santos (Eds.), The Human Hand as an Inspiration for Robot Hand Development, Springer, Cham, 2014.

[11] Pamela Gallagher, Malcolm MacLachlan, The Trinity amputation and prosthesis experience scales and quality of life in people with lower-limb amputation ¹, Archives of Physical Medicine and Rehabilitation (ISSN 0003-9993) 85 (5) (2004) 730–736, https://doi.org/10.1016/j.apmr.2003.07.009.

[12] M. Grebenstein, et al., The DLR hand arm system, in: IEEE International Conference on Robotics and Automation (ICRA), 2011.

[13] A.W. Heinemann, R.K. Bode, C. O'Reilly, Development and measurement properties of the Orthotics and Prosthetics User's Survey (OPUS): a comprehensive set of clinical outcome instruments, Prosthetics and Orthotics International (ISSN 0309-3646) 27 (3) (2003) 191–206, https://doi.org/10.1080/03093640308726682.

[14] Pamela L. Hudak, et al., Development of an upper extremity outcome measure: the DASH (disabilities of the arm, shoulder, and head), American Journal of Industrial Medicine (ISSN 0271-3586) 29 (6) (1996) 602–608, https://doi.org/10.1002/(SICI)1097-0274(199606)29:6<602::AID-AJIM4>3.0.CO;2-L.

[15] Eun-Hye Kim, Seok-Won Lee, Yong-Kwun Lee, A dexterous robot hand with a bio-mimetic mechanism, International Journal of Precision Engineering and Manufacturing (ISSN 2005-4602) 12 (2) (2011) 227–235, https://doi.org/10.1007/s12541-011-0031-x.

[16] Yuichi Kurita, et al., Human-sized anthropomorphic robot hand with detachable mechanism at the wrist, Mechanism and Machine Theory (ISSN 0094-114X) 46 (1) (2011) 53–66, https://doi.org/10.1016/j.mechmachtheory.2010.08.011.

[17] M. Laffranchi, et al., The Hannes hand prosthesis replicates the key biological properties of the human hand, Sciences Robotics (ISSN 2470-9476) 5 (46) (2020), https://doi.org/10.1126/scirobotics.abb0467, https://www.ncbi.nlm.nih.gov/pubmed/32967990.
[18] Thierry Lalibertfè, Lionel Birglen, Clément M. Gosselin, Underactuation in robotic grasping hands, Machine Intelligence & Robotic Control 4 (3) (2002) 1–11.
[19] Shunchong Li, et al., Design of a myoelectric prosthetic hand implementing postural synergy mechanically, Industrial Robot: An International Journal 41 (5) (2014) 447–455, https://doi.org/10.1108/IR-03-2014-0312.
[20] Colin M. Light, Paul H. Chappell, Peter J. Kyberd, Establishing a standardized clinical assessment tool of pathologic and prosthetic hand function: normative data, reliability, and validity, Archives of Physical Medicine and Rehabilitation (ISSN 0003-9993) 83 (6) (2002) 776–783, https://doi.org/10.1053/apmr.2002.32737.
[21] A. Marinelli, et al., Miniature EMG sensors for prosthetic applications, in: 2021 IEEE Neural Engineering Conference (NER), 2021.
[22] Michelangelo prosthetic hand, https://www.ottobockus.com/prosthetics/upper-limb-prosthetics/solution-overview/michelangelo-prosthetic-hand/, 2017.
[23] MyoHand VariPlus speed, https://professionals.ottobock.com.au/Products/Prosthetics/Prosthetics-Upper-Limb/Adult-Terminal-Devices/8E38-9-MyoHand-VariPlus-Speed/p/8E38~59, 2019.
[24] Christian Pylatiuk, Artem Kargov, Stefan Schulz, Design and evaluation of a low-cost force feedback system for myoelectric prosthetic hands, JPO: Journal of Prosthetics and Orthotics (ISSN 1040-8800) 18 (2) (2006) 57–61, https://journals.lww.com/jpojournal/Fulltext/2006/04000/Design_and_Evaluation_of_a_Low_Cost_Force_Feedback.7.aspx.
[25] Gionata Salvietti, Replicating human hand synergies onto robotic hands: a review on software and hardware strategies, Frontiers in Neurorobotics (ISSN 1662-5218) 12 (2018) 27, https://doi.org/10.3389/fnbot.2018.00027, http://www.ncbi.nlm.nih.gov/pmc/articles/PMC6001282/.
[26] H. Atakan Varol, et al., Biomimicry and the design of multigrasp transradial prostheses, in: Ravi Balasubramanian, Veronica J. Santos (Eds.), The Human Hand as an Inspiration for Robot Hand Development, Springer International Publishing, Cham, ISBN 978-3-319-03017-3, 2014, pp. 431–451.
[27] Frédérique de Vignemont, Embodiment, ownership and disownership, Consciousness and Cognition (ISSN 1053-8100) 20 (1) (2011) 82–93, https://doi.org/10.1016/j.concog.2010.09.004.
[28] R. Weir, Design of artificial arms and hands for prosthetic applications, in: Standard Handbook of Biomedical Engineering and Design, McGraw-Hill, 2004, Chap. 32.

Chapter 12

A hand-arm teleoperation system for robotic dexterous manipulation

Shuang Li[a], Qiang Li[b], and Jianwei Zhang[a]
[a] *Universität Hamburg, Hamburg, Germany,* [b] *University of Bielefeld, Bielefeld, Germany*

12.1 Introduction

The dexterous robotic hand-arm system is a promising replacement of humans for performing tedious and dangerous tasks. When autonomous manipulation of the robotic system involves complicated perception and recognition, teleoperation shows superior performance in making fast decisions and handling tricky situations. Due to the complexity and high requirements of dexterous manipulation by anthropomorphic robotic hands, teleoperation is an efficient approach to control the dexterous robotic hand-arm system.

Robotic teleoperation takes advantage of human intelligence and cognition and plays a crucial role in many applications, e.g., space exploration, earthquake rescue, medical surgery, and imitation learning. Human–robot interactive devices, such as joystick-, marker-, inertial and magnetic measurement unit (IMU)-, or electromyography (EMG)-based data gloves or wearable suits and haptic devices [1], have been typically implemented in robotic teleoperation. Regarding dexterous teleoperation, glove-based methods must be customized and easily obstruct natural joint motions, while IMU-based and EMG-based methods have less versatility and dexterity. On the other hand, IMU- [2] and EMG-based [3] devices are usually easy to use and efficient in controlling the manipulators with multiple degrees of freedom (DoFs). Zhang et al. [4] presented a teleoperation system combining an EMG armband for controlling a six-DoF multifingered hand with IMU sensors for operating a Universal Robot 10. However, the forearm's EMG signals are decoded to classify two hand motions (open and grasp), so the multifingered hand is merely capable of achieving simple grasping tasks. Thereby, these researches suggest that it is valuable to make use of IMU devices to control the robot arm.

On the other side, markerless devices have the advantages of allowing for natural, unrestricted limb/finger motions and of being less invasive [5,6]. Most markerless devices in teleoperation are low-cost RGB cameras. Leveraging

Tactile Sensing, Skill Learning, and Robotic Dexterous Manipulation
https://doi.org/10.1016/B978-0-32-390445-2.00021-0

human body tracking and hand pose estimation algorithms, markerless vision-based teleoperation [7,8] has been studied in controlling humanoid robots or dexterous robotic hands. In [9] and [10], based on the human hand pose acquired from a hand pose estimator, a retargeting method was used to map the three-dimensional position of human hand keypoints to the joint angles of the robot hand. Both methods were tested on an anthropomorphic hand in simulation. Later, using merely the observation of the bare human hand, Handa et al. [11] achieved impressive results in multiple dexterous manipulation tasks by a 23-DoF hand-arm robotic system. They collected the hand pose dataset and implemented the experiments based on a multicamera system. Nevertheless, it is difficult to reproduce the two-phased data collection procedure. Owing to the limited workspace of the camera system, they only can implement the manipulation tasks that only require a relatively small motion range. Overall, previous vision-based methods always decompose the teleoperation task into two steps: human pose perception and finger targeting. The first step performs classification tasks of human gesture recognition or regression tasks of hand pose estimation. The second step computes or sets the corresponding kinematic configuration of the robot from the perception results. In contrast to these methods, directly inferring joint angles of the robot from human hand perception is more intuitive and has no postprocessing time.

Recently, Li et al. [12] proposed the end-to-end network TeachNet to control a five-fingered robotic hand and demonstrated human gesture imitation and simplistic grasping experiments. To handle the cross-domain issues between human hands and the robot hand, TeachNet defines an extra consistency loss to exploit the geometrical resemblance between them. Although this model is efficient, it could not achieve high accuracy and thus manipulation experiments at a high dexterity level were not shown. Therefore, we will further explore an end-to-end markerless teleoperation method with higher accuracy to control the robotic hand. Consequently, compensating for the dissimilarities of appearance and anatomy between the human and robotic hands plays an essential role in the end-to-end learning method. Here, we introduce an image-to-image translation [13–15] concept for vision-based teleoperation methods. The image-to-image translation aims to map a representation of a scene into another and to discover the hidden mapping feature between two representations. In our case, we would like to find a method that can thoroughly comprehend the kinematic similarity between the human hand and the robot hand. We assume that if a model could translate the observed human hand image to the robot hand image, the model would have discerned shared embeddings representing the resemblance of pose features between two image domains. Therefore, we will investigate how to integrate the image-to-image translation network architecture into the end-to-end joint regression model.

In summary, this chapter presents a multimodal teleoperation system consisting of a novel vision-based network (Transteleop) for generating the robot joint angles and an IMU-based arm posture by observing human manipulation

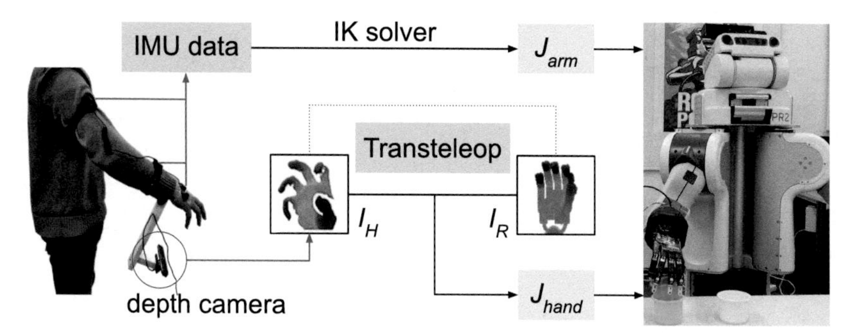

FIGURE 12.1 The hand-arm teleoperation system is built on a vision-based method, Transteleop, which generates the joint commands of a robotic hand, and an IMU-based arm control method. Transteleop gains meaningful pose information from depth images of a human hand I_H and predicts the joint angles J_{hand} of the robot hand. The wearable IMU setup provides the absolute movements of the human arm, which are used to generate the joint commands of the robot arm. The multimodal system succeeds in implementing dexterous manipulation tasks, e.g., pick-and-place, cup insertion, object pushing, and dual-arm handover.

behaviors. Transteleop takes the depth image of the human hand as input and then generates the reconstructed image of the robot hand and estimates the joint angles of the robot hand. A wearable camera holder and the IMU devices enable simultaneous control of the robot arm and hand. In real experiments, this multimodal system achieves a series of dexterous manipulation tasks that go beyond simple pick-and-place operations.

12.2 Problem formulation

The goal of this chapter is to develop a hand-arm teleoperation system for dexterous manipulation. In this system, we expect that the teleoperator performs hand movements for dexterous manipulation, and the operator is not limited in a specified place, e.g., observed by an external table-fixed camera or mocap system. We achieve this goal by two separate controllers for the anthropomorphic robotic arm and hand (see Fig. 12.1). A vision-based neural network is used to control the robot hand, and an IMU-based device is applied to control the arm. The vision part aims to train a deep learning model that is fed human hand images and predicts joint angles of the robot, while the IMU part intends to map the absolute motion of the human arm to the robot arm simultaneously.

The robot system used in this work is a PR2 robot[1] with a 19-DoF Shadow hand[2] mounted on its right arm. The right arm of PR2 only has five DoFs due to the equipped Shadow hand. The kinematic chain of the anthropomorphic Shadow Dexterous Hand [16] is shown on the right side of Fig. 12.2. The first finger (FF), the middle finger (MF), and the ring finger (RF) have four joints,

---

1 http://www.willowgarage.com/pages/pr2/overview.

2 https://www.shadowrobot.com/dexterous-hand-series/.

the distal, middle, proximal, and metacarpal joints. The little finger (LF) and the thumb (TH) are provided with an extra joint for holding the objects. There are five BioTac tactile sensors attached to each fingertip, which replaces the last phalanx and the last joint's controllability. Therefore, all five fingers make 17 DoFs, so the two in the wrist joint bring the total number of DoFs to 19.

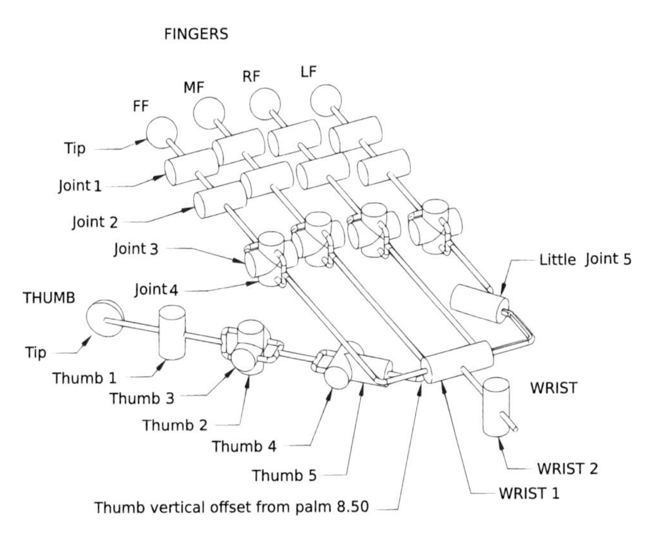

FIGURE 12.2 The kinematic chain of the Shadow hand. In this work, joint 1 of each finger is stiff. FF represents the first finger, LF represents the little finger, MF represents the middle finger, and RF represents the ring finger.

12.3 Vision-based teleoperation for dexterous hand

In this section, we describe the principles and architecture of the proposed vision-based pose estimation method (Transteleop). Transteleop combines the paired image-to-image translation method into the pose regression problem in order to capture the valuable latent hand pose information. It is trained by supervised multioutput learning on a self-built pair-wise human–robot dataset. We also explain the efficient generation method of the self-built dataset.

12.3.1 Transteleop

Image-to-image translation has been implemented by different generative models, e.g., restricted Boltzmann machines [17], autoencoder models [18,19], generative adversarial networks (GANs) [20], and their variants. Moreover, previous studies in robotics have indicated that the image-to-image translation method is capable of learning shared features between human–robot mapping pairings [21,22]. This section explores how to use this method to discover the kinematic resemblance between the human hand and the robot hand. Assume that there is an image I_R of a robotic hand from a fixed viewpoint and an image I_H of a human hand in random global orientation, and the robotic hand in the image acts the same as the human hand. Although the bone length, the global pose, and the joint range of these paired hands are distinct, the pose feature H_{share} such as the skeletal shape and the whole silhouette is rather similar. To attain an instructive feature representation H_{share}, we adopt a generative structure that maps from image I_H to image I_R and retrieves the pose J_R from the bottleneck layer of this structure as H_{share}. We formulate this regression problem as follows:

$$f_{trans} : I_H \in \mathbb{R}^2 \rightarrow H_{share} \rightarrow I_R \in \mathbb{R}^2, \tag{12.1}$$

$$f_{joint} : H_{share} \rightarrow J_R. \tag{12.2}$$

As is well known, conditional GANs have significantly improved image reconstruction quality because the discriminator pursues high reality and low blurriness of reconstructed images. However, from the manipulation and tracking control perceptive, the neural model should concentrate more on the pose feature of the input instead of the reality. And as tested in Section 12.5 (Transteleop evaluation), the conditional GAN-based model is quite useful for rendering the scenario in a high-quality way but estimates hand posture less accurately than the autoencoder model. Accordingly, we take an autoencoder structure for our proposed pipeline. As shown in Fig. 12.3, Transteleop consists of three main modules: the encoder–decoder module, the joint module, and the preprocessing module.

In the encoder–decoder module, the encoder takes a depth image of a human hand I_H and discovers the latent feature embedding H_{share} between the human hand and the robot hand. Six convolutional layers containing four downsampling layers and two residual blocks with the same output dimension are used in the encoder. Thus, given an input image of size 96×96, the encoder computes an abstract $(6 \times 6 \times 512)$-dimensional feature representation H_{human}. Instead of convolutional layers, the encoder and the decoder are linked through one fully connected (FC) layer because the pixel areas in I_H and I_R in our dataset are not matched. The human hands vary in various global poses, but the wrist pose of the robot hand is fixed (see Section 12.4 on dataset generation). This design results from the fact that an FC layer allows each unit in the decoder to reason on the entire image content. In contrast, a convolutional layer cannot directly connect all locations within a feature map. Through the embedding submodule, we extract the useful 8192-dimensional feature embedding H_{share} from H_{human}. The decoder reconstructs a depth image of the robot hand $\hat{I}_R$ from a fixed viewpoint from the latent pose feature H_{share}. One FC connects the feature from H_{share} to the $(6 \times 6 \times 512)$-dimensional robot feature vector H_{robot}. Four upconvolutional layers with learned filters and one convolutional layer for image generation follow.

Specifically, the generated image $\hat{I}_R$ should care more about the accuracy of local features such as the position of fingertips instead of global features such as

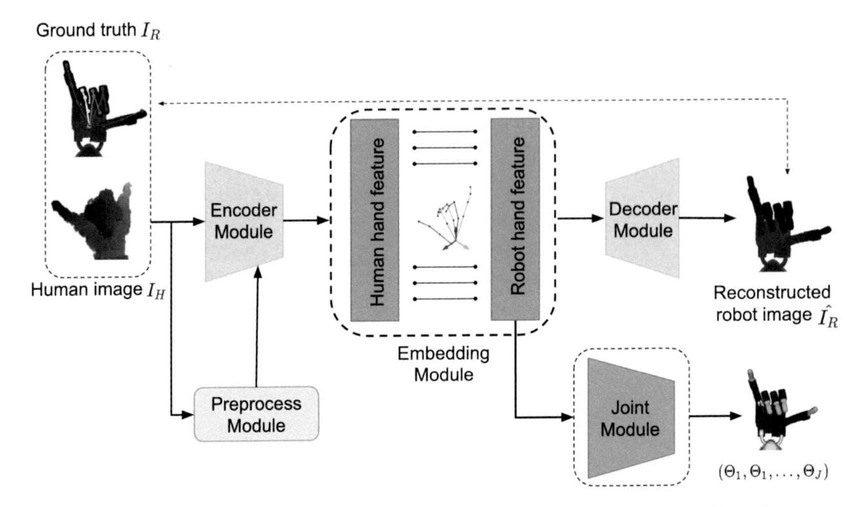

FIGURE 12.3 Transteleop architecture. Center: The encoder–decoder structure is an image-to-image translation branch, which feeds depth images of a human hand I_H and produces reconstructed depth images of the robot hand I_R. The joint module takes the pose embedding from the encoder–decoder structure and predicts the robot's joint angles J_{hand}. The preprocessing module is a spatial transformer network that explicitly permits the spatial manipulation of input images.

image style because the pixels around the joint keypoints possess more information about the hand pose. Regarding the Shadow hand, as depicted in Fig. 12.5, each finger has three keypoints. Therefore, we devised a keypoint-based reconstruction loss to capture the human hand's overall structure and highlight the pixels around the 15 keypoints of the hand. The scaling factor of each pixel error is determined by how close this pixel is to all keypoints, shown in Fig. 12.4. We regard the eight neighboring pixels of each keypoint as necessary as these keypoints themselves. The reconstruction loss L_{recon} is an L2 loss that prefers to minimize the mean pixel-wise error but does not encourage less blurring, defined as

$$\mathcal{L}_{recon} = \frac{1}{N} \sum_{i=1}^{N} \alpha_i \cdot (I_{R,i} - \hat{I}_{R,i})^2, \tag{12.3}$$

where N is the number of pixels and $\alpha_i \in [0, 1]$ is the scaling factor of the ith pixel.

The joint module deduces 19-dimensional joint angles J_{hand} from the latent feature embedding H_{robot} of the decoder. We choose H_{robot} rather than H_{share} because H_{robot} contains richer features depicting the pose feature of the robot hand. Three FC layers are employed in the joint module. The joint module is supervised with a mean squared error (MSE) loss $\mathcal{L}_{joint}$,

$$\mathcal{L}_{joint} = \frac{1}{M} \| J_{hand} - J \|_2^2, \tag{12.4}$$

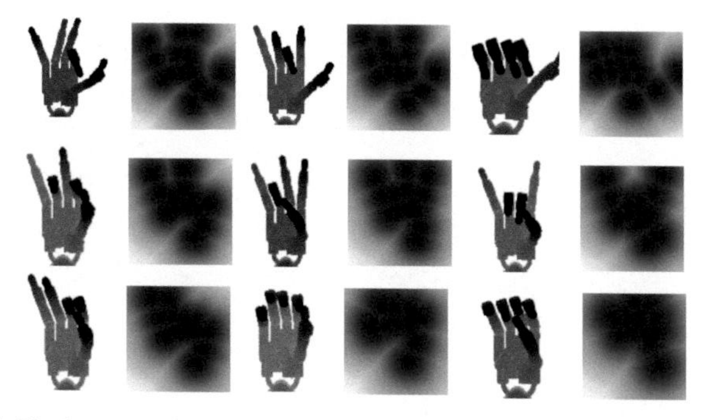

FIGURE 12.4 The heatmap of scaling matrix α. The darkness of the pixel color illustrates how important the pixel is.

where M is the number of joint angles and J denotes the ground truth joint angles.

Overall, the complete training objective is

$$\mathcal{L}_{hand} = \lambda_{recon} \cdot \mathcal{L}_{recon} + \lambda_{joint} \cdot \mathcal{L}_{joint}, \tag{12.5}$$

where λ_{recon} and λ_{joint} are the loss weights.

The preprocessing module provides spatial transformation capabilities of input images by a spatial transformer network (STN) [23]. The STN module is inserted into the beginning of Transteleop without any extra training supervision to the optimization process. This design hopes to empower Transtelep to be invariant to the orientation diversity of the human hand poses.

12.3.2 Pair-wise robot–human hand dataset generation

Transteleop demands a human–robot pairings dataset containing pair-wise human depth images and robot hand depth images, with corresponding robot joint angles and poses. Recording depth images of robot images and kinematic retargeting from the human pose to the robot pose are the main two dataset generation issues. As a matter of fact, collecting real-world robot images is time consuming and leads to noisy data. We instead use an existing human hand dataset with labeled human hand depth images, get the corresponding joint angles of the robot by an optimized retargeting method, and then manipulate the robot and capture the depth images of its current state in simulation. The pipeline of our dataset generation is shown in Fig. 12.5.

The off-the-shelf human hand dataset we chose is the BigHand2.2M Dataset [24]. The BigHand2.2M dataset comprises 2.2 million depth images of human hands with 21 annotated 3D joint locations related to the camera coordinate. The dataset was collected from 10 subjects (seven male, three female). The human hand model from the BigHand2.2M dataset has 21 joints with 31 DoFs, while the Shadow hand we use has 24 joints and moves with 19 DoFs. Even though

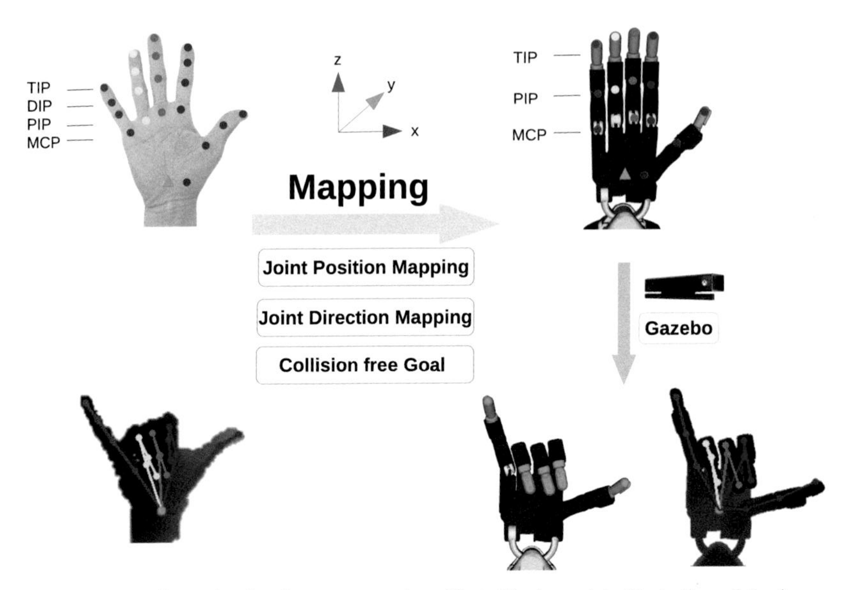

FIGURE 12.5 Illustration for dataset generation. (Top) The keypoints illustration of the human hand model from the BigHand2.2M dataset and the Shadow hand with BioTac sensors. TIP, DIP, PIP, and MCP mean fingertip joint, distal joint, proximal joint, and metacarpal joint, respectively. (Bottom left) A depth image example from the BigHand2.2M dataset. (Bottom right) The corresponding RGB and depth images of Shadow gestures were obtained from Gazebo. The colored circles denote the joint keypoint positions on the hand, and the green triangles denote the local hand frames.

the Shadow hand is humaniform, the kinematic dissimilarities between the robot hand and the human hands, such as the limited joint angles of the robot, the consistent length of the robot fingers, and the independent control of each finger on the robot, make it impossible to directly assign the joint positions of the human hand to the robot hand. In order to imitate the human hand pose, an optimized retargeting method is developed to integrate position mapping and orientation mapping of each joint and to avoid self-collisions of the robot properly. The retargeting method relies on the inverse kinematic solver bio_ik [25]. We set two wrist joints of the Shadow at 0 rad to simplify the divergence between the Shadow hand and the human hand.

We build the local human hand frame at the human wrist keypoint, and the robot hand frame is located at 34 mm above the z-axis of the robot wrist joint. Here, 34 mm is the height from the robot wrist joint to the thumb 5 joint (see Fig. 12.2). Then, we use the TIP, PIP, and MCP positions in each finger of the human hand to map the TIP, PIP, and MCP keypoints of the Shadow hand. But the mapping weights ω_{pf} for TIP joints is set bigger than the weights ω_{pp} for the PIP joints. To enforce the robot hand pose akin to the human hand posture, a direction mapping with weight ω_d is applied to five proximal phalanges and distal phalanges of the thumb. In this dataset, ω_{pf}, ω_{pp}, and ω_d are set as 1, 0.2,

(a) Example 1

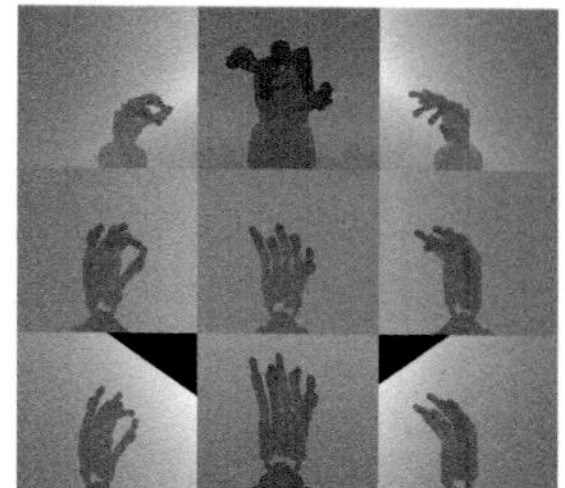

(b) Example 2

FIGURE 12.6 Two examples of the Shadow depth images from nine viewpoints corresponding to the corresponding human gestures in our dataset.

and 0.2. Additionally, when the inverse kinematic solver produces self-collision solutions a cost function F_{cost} in Eq. (12.6) is activated to provide approximate solutions by impelling the collided links outward within a collision free radius:

$$F_{cost} = \max(0, R\_col - \|P_i - P_j\|^2), \tag{12.6}$$

where P_i and P_j denote the positions of link i and link j, respectively, and R_col is the minimum collision-free radius between two links.

Consequently, the robot moves to the calculated configuration in Gazebo.[3] Taking advantage of the simulation environment, we set nine depth cameras with different observing positions to increase the diversity of the robot data. Thus nine visual samples of the robot are collected for each pose simultaneously. In Fig. 12.6 we present two examples of the robot depth images from nine viewpoints corresponding to one human hand pose. Note that in order to avoid many depth values at infinity, we set a wall behind the Shadow hand in Gazebo. The background's darkness, especially some corners of the depth images, is different because in some viewpoints, the camera is able to look at infinity, and the depth values go to infinity. With this pipeline, a human–robot pair-wise dataset that contains 400K pairs of synthetic robot depth images and human hand depth images, with corresponding robot joint angles, was collected. Fig. 12.7(a) shows nine pair-wise depth images in our dataset.

12.4 Hand-arm teleoperation system

The previous two sections study how to teleoperate an anthropomorphic hand by vision control. In this section, we investigate how to extend the hand teleoperation system to a hand-arm system. A hand-arm system implies that the teleoperator works in an unlimited workspace. Paradoxically, the field of view of the depth camera is usually restricted. The depth sensor used in our experiments is the Intel RealSense SR300, whose horizontal and vertical depth fields

[3] http://gazebosim.org/.

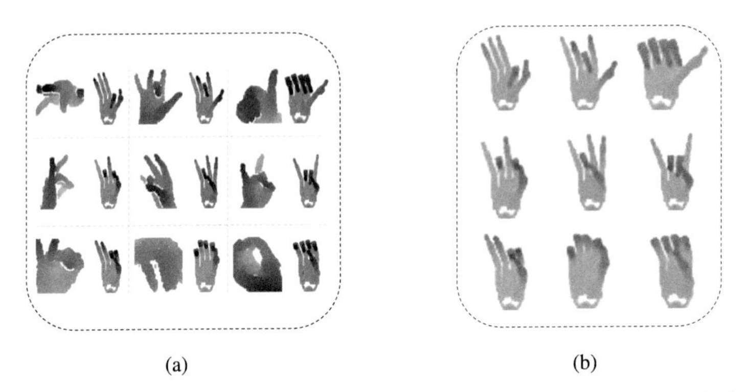

FIGURE 12.7 (a) Examples of pair-wise human–robot hand datasets. (b) Reconstructed robot images generated from Transteleop, whose corresponding robot images are the examples shown on the left.

of view are 73 degrees and 59 degrees, respectively. In case the operators implement teleoperation tasks that require a wide workspace, the human hand will disappear from the camera's view range. We solve this problem by a cheap 3D printed camera holder, which can be mounted on the forearm of the teleoperator, as shown in Fig. 12.8. During the teleportation experiments, the camera will move along with the forearm. The weight and the length of the camera holder are 248 g and 35 cm.

FIGURE 12.8 The camera holder is used to mount the camera on the human arm.

Due to the uncertainty of the camera position, we use the Perception Neuron (PN) device[4] to control the 5-DoF robot arm and two wrists of the robot hand, thus extending this teleoperation system to the hand-arm system. PN is an IMU-based motion capture device prevalent in body tracking, animation, and VR game interaction. Three IMU elements are sufficient to capture the palm, upper arm, and forearm motion regarding the robotic arm control. To register the human hand movement with the PR2 robot, we match the human spine frame to the torso link of the PR2 robot, as shown in Fig. 12.9. The spine link is defined in PN software, and the robot torso link is set in the robot URDF (Unified Robot Description Format) file. We assume the transformation ${}^{spine}T_h$ of the right human palm with respect to the human spine link is the same as the transformation ${}^{torso}T_r$ of the wrist of the robot arm in relation to the torso link of the robot. As

[4] https://neuronmocap.com.

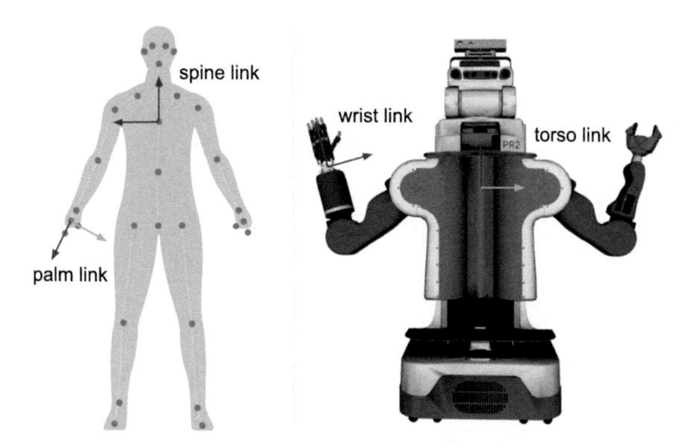

FIGURE 12.9 Illustration of the registration between the right human hand and the PR2 right robot arm. The purple dots on the human body diagram represent all possible body parts that can be tracked by Perception Neuron.

the lengths of the human arm and the robot arm are different, we update ${}^{torso}T_r$ by link length of the robot upper arm and forearm.

Given the transformation ${}^{torso}T_r$, we compute the joint angles of the robot arm by feeding this pose to the BioIK solver [25]. After this, we set the angular velocity $\mathcal{V}_t$ of each joint by calculating and scaling the feedforward joint difference between the desired joint angles of the current frame J_t^{ik} and of the previous frame J_{t-1}^{ik} and the feedback joint difference between the desired joint angle of the current frame J_t^{ik} and of the current robot joint state J_t^{robot}:

$$\mathcal{V}_{n,t} = \delta_1 \cdot (J_{n,t}^{ik} - J_{n,t-1}^{ik}) + \delta_2 \cdot (J_{n,t}^{ik} - J_{n,t}^{robot}), \qquad (12.7)$$

where n is the nth joint of the arm and δ_1 and δ_2 account for the scaling factor of each velocity term.

12.5 Transteleop evaluation

12.5.1 Network implementation details

In the preprocessing phase, we perform a morphological operation on the raw depth images and then extract a fixed-size cube around the hand. Then the image is resized to 96×96 and is normalized to $[-1, 1]$. During training, random in-plain rotation and random Gaussian noise are executed for data augmentation. Due to the invariance of the outputs J_R to the input images' spatial transformation, we do not need to modify the ground truth. Even though we collect robot images from nine viewpoints, we only use the robot images captured in straight-ahead viewpoint for training because it reduces image reconstruction difficulty and promotes convergence. To optimize our networks, we use mini-batch stochastic gradient descent and apply the Adam optimizer with a learning rate of $l_r = 0.002$ and momentum parameters $\beta_1 = 0.5,\ \beta_2 = 0.999$. We add a batch normalization (BN) layer and a rectified linear unit (ReLU) after each convolution layer. ReLU has also been employed as an activation function after all FC layers except for the last FC layer. We use $\lambda_{recon} = 1$, $\lambda_{joint} = 10$. At inference time, we only run the encoder module and the joint module for joint angle regression. The average inference time is 0.3 s, tested on a computer with Intel Core i7-5820K CPU 3.30 GHz.

12.5.2 Transteleop evaluation

We conducted the following evaluations on our paired dataset: (1) To explore the best visual representations models, we compared the Transtelop method with two network structures GANteleop, which adds a PatchGAN discriminator and an adversarial loss based on the "pix2pix" framework [13], and TeachNet [12]. (2) To validate the translation step's significance, we ran an ablation analysis by removing the decoder module, feeding human hand images. This baseline is referred to as Humanonly. (3) To examine whether the STN module learns invariance to the hand orientation and rotation, we also trained Transtelop without using STN baseline, which has the same structure as Transtelop but without the preprocessing module. (4) To show the regression results from the robot's own domain, we trained a model that removes the decoder module in Transteleop and only feeds the images of the robot hand. This baseline is referred to as Robotonly. All images fed into Robotonly are taken from a fixed third-person viewpoint. We evaluated the regression performance of Transteleop and five baselines on our test dataset using standard metrics commonly used in hand pose estimation: (1) the fraction of frames whose maximum joint angle errors are below a threshold; (2) the fraction of frames whose maximum joint distance errors are below a threshold; and (3) the average angle error over all angles.

In Figs. 12.10 and 12.11, the Robotonly model significantly outperforms other baselines over all evaluation metrics because of the matched domain and the identical viewpoint. Two Transtelop methods get higher accuracy than the Humanonly model, especially when the angle thresholds are lower than 0.2 rad. We infer that the additional reconstruction loss assists in gaining more indicative pose features. The examples of generated robot images by Transteleop are visualized in Fig. 12.7(b). Even if the reconstructed images are blurry, they are still akin to the ground truth in Fig. 12.7(a), which explicitly proves that the pose feature z_R in the embedding module could contain enough pose information. Moreover, the performance of GANteleop is worse than Transteleop because the discriminator in GANteleop focuses on pursuing realistic images and weakens the supervision of $\mathcal{L}_{joint}$. Furthermore, all methods show at least 10% improvement of the accuracy than TeachNet in the high-precision condition. It indicates that the deeper network layers play a more critical role than the supervision of the consistency loss in TeachNet. Comparing Transtelop and Transtelop without using STN, there is no significant improvement due to the STN module.

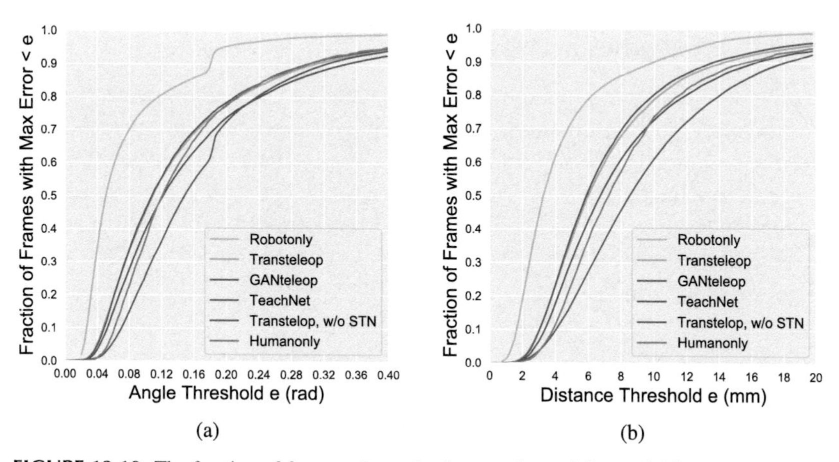

FIGURE 12.10 The fraction of frames whose absolute maximum joint angle/distance error is below a threshold between the Transteleop approach and different baselines on our test dataset. These results show that Transteleop has the best accuracy of all evaluation metrics except for the Robotonly model.

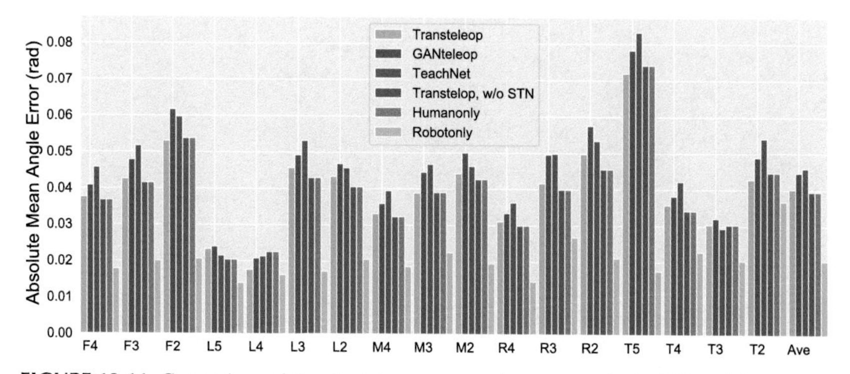

FIGURE 12.11 Comparison of the absolute average angle errors on the individual joint between the Transteleop approach and different baselines on our test dataset. F means the first finger, L means the little finger, M means the middle finger, R means the ring finger, and T means the thumb. Numbers indicate the joints of the fingers.

This suggests that the spatial transform brings a slight appearance normalizing effect to this task but does not significantly promote the hand pose transform to a canonical pose.

As illustrated in Fig. 12.11, the absolute average error of all methods is lower than 0.09 rad. The highest error happens on thumb joint 5 because there is a big discrepancy between the human thumb and the Shadow thumb. For each finger, joint 2 and joint 3 always have higher errors than other joints. The reason is that these two joints are involved in the most complex finger motions.

12.5.3 Hand pose analysis

In order to investigate which types of human hand images are more challenging to extract the pose information by Transteleop, we sort out the images in the test dataset based on their overall loss L_{stud}. Fig. 12.12(a) reveals that the complexity of the finger poses and the orientation of the human hand jointly account for the bad accuracy of hand pose estimation. Intriguingly, Fig. 12.12(b) implies that the complexity of the finger poses affects accuracy more severely than the orientation of the human hand.

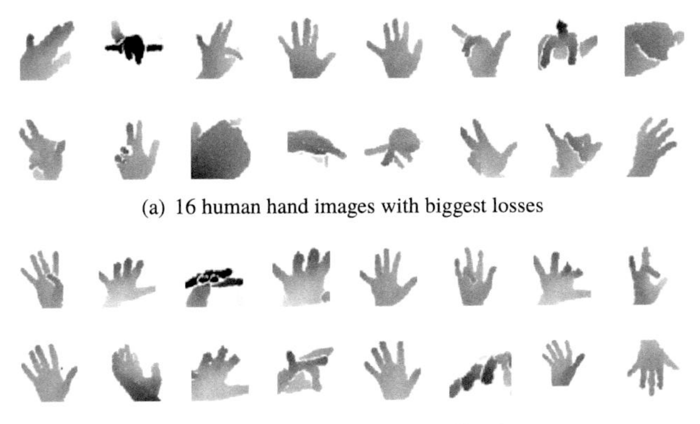

(a) 16 human hand images with biggest losses

(b) 16 human hand images with smallest losses

FIGURE 12.12 Visualization of the human hand images in the test dataset based on the loss L_{stud} sorting. From left to right, upper row to lower row, the losses of the images gradually become smaller in Fig. 12.12(b), and versa vice in Fig. 12.12(a).

12.6 Manipulation experiments

The multimodel teleoperation system was systematically evaluated across four types of physical tasks that analyze precision and power grasps, prehensile and nonprehensile manipulation, and dual-arm handover tasks. For the control of the robot arm, we set $\delta_1 = 0.7$, $\delta_2 = 0.1$. The frequency of the arm's velocity control is 20 Hz. The starting poses of the human arm were always consistent with the starting pose of the robot arm. Meanwhile, the arms of the robot always started and ended at almost similar poses over every task. The frequency of the robot hand's trajectory control is set to 10 Hz. One female and two male subjects conducted the robotic experiments, and each task was randomly performed by one of them.

To get familiar with the tasks, the operators carried out a warm-up training for each task with five nonconsecutive attempts before the real testing trials. For easier tasks, such as pick-and-place, the operators usually could accomplish the tasks after three trials. Each task was conducted five times by one demonstrator. A detailed description of the experiments is given below.

(1) Pick-and-place. In this task, we prepared two testing scenarios: pick a Pringles can and place it in a red bowl on the same table; and pick a cube on the table and place it on top of a brick on a box (see Fig. 12.13). The height of the box is 325 mm. The first scenario requires the power grasp skills of the robot, and the second scenario needs the precision grasp skills of the robot and a wide-enough workspace for the teleoperator.

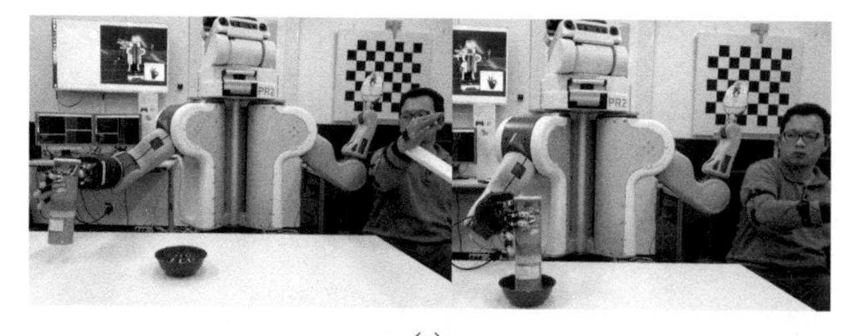

(a)

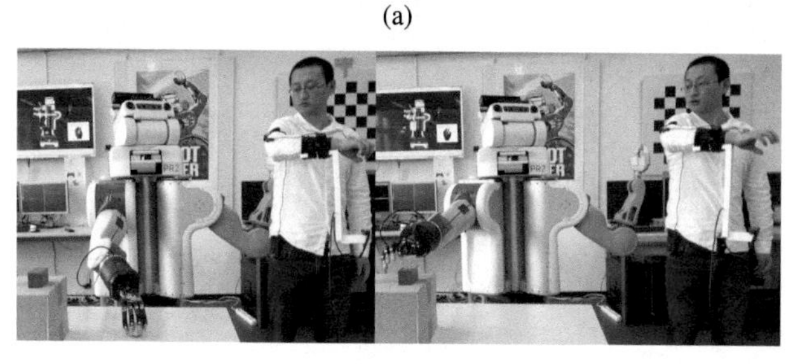

(b)

FIGURE 12.13 (a) The robot picks a Pringles can and places it in a bowl. The Pringles and the bowl are set on the same table. (b) The robot picks a wooden cube on the table and places it on a rectangular brick on a box.

(2) Cup insertion. Three concentric cups are to be inserted into each other in order of size from smallest to largest, as visualized in Fig. 12.14. The cups are randomly put on the table. This task examines the abilities of precision grasp and release.

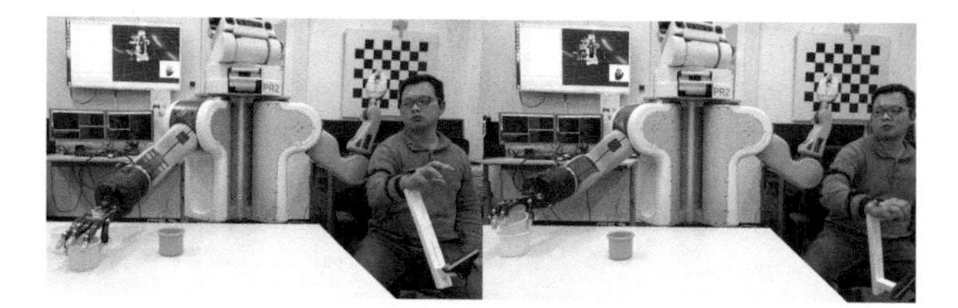

FIGURE 12.14 The robot inserts three cups into each other.

(3) Object pushing. We set random initial poses of a brick and then push the brick into a designated position. The goal pose is labeled as the red rectangular

in Fig. 12.15. The size of the rectangular is the same as the size of the brick. This task contains the challenges of pushing, sliding, and precision grasping.

FIGURE 12.15 The robot pushes a rectangular brick to a specified pose. The red rectangular on the table represents the target pose of the brick.

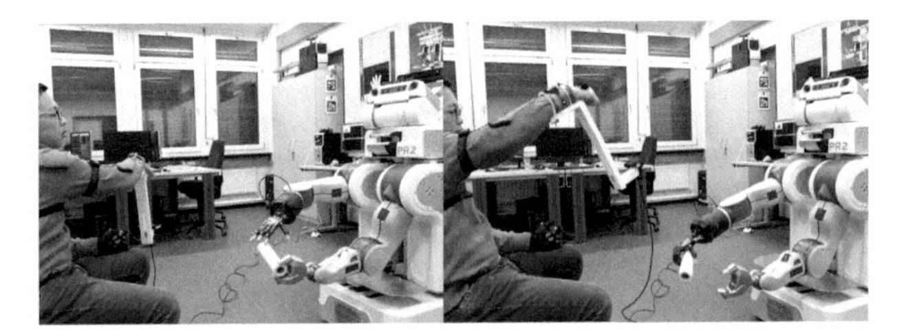

FIGURE 12.16 The PR2 robot hands a roll of paper from its left gripper over to its right hand.

(4) Dual-arm handover. The left robot arm hands a roll of paper over to its right hand. The operator also exploits the PN setup to control the left arm and left gripper of PR2. This task tests the coordination ability of the whole teleoperation system. Owing to our system's mobility, the human can sit face to face instead of parallel to the robot for better visualization. Fig. 12.16 qualitatively demonstrates this experimental scenario. Notably, in the handover task, the teleoperators needed five complete warm-up trials to adapt to the opposite operating direction of the robot.

Table 12.1 numerically shows the average completion time a teleoperator took to finish a task and the success rate. The completion time was calculated when the robot started to move until it returned to the starting pose. The high success rate and short completion time of two pick-and-place tasks and the cup insertion task indicate our system has the ability of precision and power grasps. Compared to the two pick-and-place tasks shown in Fig. 12.13, the blue brick in pick2 is much smaller than the bowl, so that the robot needed a longer time to find a correct place to land the cube. During the pushing task, the robot could quickly push the brick close to the target position using multiple fingers. Nevertheless, the operator took a long time to fine-tune the brick's orientation to make the pushing error lower than 5 mm. The handover task achieved a relatively low success rate, mainly because of the left gripper's imprecise control, so the robot accidentally lost the object. The left gripper is controlled by the IMU glove from the PN suit, but the calibration of the neurons on the gloves is unexpectedly bad. These results reflect the fact that the visual-based method is more suitable for multi-finger control than the IMU-based method.

TABLE 12.1 Average completion time and success rate of each task.

	pick1	pick2	cup	pushing	handover
Average time (s)	18.5	37.2	25.5	62.0	36.33
Average success rate	100%	100%	100%	80%	60%

12.7 Conclusion and discussion

This chapter builds a hand-arm teleportation system by combining a vision-based control method for the robotic hand and an IMU-based arm control method. Motivated by the image-to-image translation method, the vision-based model Transteleop discovers the identical kinematic features between the robot hand and the human hand and predicts the robot joint angles. The joint commands of the robot arm are calculated based on the IMU device worn on the human right arm. A self-designed camera holder bridges the hand and the arm control and enables the teleoperator to work in a flexible workspace. Ultimately, comparative evaluations on the test dataset and thorough robotic experiments verify the feasibility and reliability of our method.

Admittedly, there are some limitations to this system. The 248-g and 35-cm 3D printed camera holder plus the 108-g depth sensor has to be worn on the human's forearm, bringing much extra physical burden for the teleoperator. It would be interesting to integrate the active vision system into this teleoperation setup. The camera is mounted on the end-effector of a manipulator, and this manipulator tracks the human hand automatically. One of the primary concerns of vision-based teleoperation is the absence of haptic feedback. And the objects used in the experiments are not modeled into the planning scene of the robot, so there is no collision avoidance between these objects and the robot. In other words, the control loop is only supervised by the user. Since the visual or auditory channels could be a low-cost alternative to haptic feedback [26], to visualize the magnitude of pressure or force on each fingertip derived from the Biotac sensors during the manipulation process could be a clue to haptic feedback. And slip detection would reduce the control burden on the user and avoid unintentional collisions of the robot. The other limitation is that some high-precision tasks (such as screw tightening and in-hand brick gaiting) are still intractable to the current system. The reason is that the hand poses in the Bighand2.0 dataset are not designed for dexterous manipulation tasks. Collecting a hand dataset that concentrates more on the versatile manipulation hand poses is an effective solution.

References

[1] J.A. Fishel, T. Oliver, M. Eichermueller, G. Barbieri, E. Fowler, T. Hartikainen, L. Moss, R. Walker, Tactile telerobots for dull, dirty, dangerous, and inaccessible tasks, in: 2020 IEEE International Conference on Robotics and Automation (ICRA), IEEE, 2020, pp. 11305–11310.

[2] B. Fang, F. Sun, H. Liu, C. Liu, 3D human gesture capturing and recognition by the IMMU-based data glove, Neurocomputing 277 (2018) 198–207.

[3] C. Meeker, M. Haas-Heger, M. Ciocarlie, A continuous teleoperation subspace with empirical and algorithmic mapping algorithms for nonanthropomorphic hands, IEEE Transactions on Automation Science and Engineering (2020).

[4] H. Zhang, Z. Zhao, Y. Yu, K. Gui, X. Sheng, X. Zhu, A feasibility study on an intuitive teleoperation system combining imu with semg sensors, in: International Conference on Intelligent Robotics and Applications, Springer, 2018, pp. 465–474.

[5] Ultraleap, https://www.ultraleap.com/.

[6] J. Kofman, X. Wu, T.J. Luu, S. Verma, Teleoperation of a robot manipulator using a vision-based human-robot interface, IEEE Transactions on Industrial Electronics 52 (5) (2005) 1206–1219.

[7] D. Michel, A. Qammaz, A.A. Argyros, Markerless 3D human pose estimation and tracking based on RGBD cameras: an experimental evaluation, in: International Conference on PErvasive Technologies Related to Assistive Environments, 2017.

[8] G. Du, P. Zhang, X. Lu, Z. Li, Robot manipulator using a vision-based human–manipulator interface, Journal of Theoretical & Applied Information Technology 50 (1) (2013).

[9] D. Antotsiou, G. Garcia-Hernando, T.-K. Kim, Task-oriented hand motion retargeting for dexterous manipulation imitation, in: Proceedings of the European Conference on Computer Vision (ECCV) Workshops, 2018.

[10] F. Gomez-Donoso, S. Orts-Escolano, M. Cazorla, Accurate and efficient 3D hand pose regression for robot hand teleoperation using a monocular RGB camera, Expert Systems with Applications 136 (2019) 327–337.

[11] A. Handa, K. Van Wyk, W. Yang, J. Liang, Y.-W. Chao, Q. Wan, S. Birchfield, N. Ratliff, D. Fox, Dexpilot: vision-based teleoperation of dexterous robotic hand-arm system, in: IEEE International Conference on Robotics and Automation (ICRA), 2020, pp. 9164–9170.

[12] S. Li, X. Ma, H. Liang, M. Görner, P. Ruppel, B. Fang, F. Sun, J. Zhang, Vision-based teleoperation of shadow dexterous hand using end-to-end deep neural network, in: IEEE International Conference on Robotics and Automation (ICRA), 2019, pp. 416–422.

[13] P. Isola, J.-Y. Zhu, T. Zhou, A.A. Efros, Image-to-image translation with conditional adversarial networks, in: IEEE Conference on Computer Vision and Pattern Recognition (CVPR), 2017, pp. 1125–1134.

[14] J.-Y. Zhu, R. Zhang, D. Pathak, T. Darrell, A.A. Efros, O. Wang, E. Shechtman, Toward multimodal image-to-image translation, in: Advances in Neural Information Processing Systems (NeurIPS), 2017, pp. 465–476.

[15] D. Pathak, P. Krahenbuhl, J. Donahue, T. Darrell, A.A. Efros, Context encoders: feature learning by inpainting, in: IEEE Conference on Computer Vision and Pattern Recognition (CVPR), 2016, pp. 2536–2544.

[16] Shadow dexterous hand, http://www.shadowrobot.com.

[17] M.A. Carreira-Perpinan, G. Hinton, On contrastive divergence learning, in: International Workshop on Artificial Intelligence and Statistics, PMLR, 2005, pp. 33–40.

[18] D.P. Kingma, M. Welling, Auto-encoding variational Bayes, in: International Conference on Learning Representations (ICLR), 2014.
[19] S. Li, H. Wang, D. Lee, Hand pose estimation for hand-object interaction cases using augmented autoencoder, in: IEEE International Conference on Robotics and Automation (ICRA), 2020, pp. 993–999.
[20] I. Goodfellow, J. Pouget-Abadie, M. Mirza, B. Xu, D. Warde-Farley, S. Ozair, A. Courville, Y. Bengio, Generative adversarial nets, in: Advances in Neural Information Processing Systems (NeurIPS), 2014.
[21] L. Smith, N. Dhawan, M. Zhang, P. Abbeel, S. Levine, Avid: learning multi-stage tasks via pixel-level translation of human videos, in: Robotics: Science and Systems (RSS), 2020.
[22] P. Sharma, D. Pathak, A. Gupta, Third-person visual imitation learning via decoupled hierarchical controller, in: Advances in Neural Information Processing Systems (NeurIPS), 2019, pp. 2593–2603.
[23] M. Jaderberg, K. Simonyan, A. Zisserman, et al., Spatial transformer networks, in: Advances in Neural Information Processing Systems (NeurIPS), 2015, pp. 2017–2025.
[24] S. Yuan, Q. Ye, B. Stenger, S. Jain, T.-K. Kim, Bighand2. 2m benchmark: hand pose dataset and state of the art analysis, in: IEEE Conference on Computer Vision and Pattern Recognition (CVPR), 2017, pp. 4866–4874.
[25] P. Ruppel, N. Hendrich, S. Starke, J. Zhang, Cost functions to specify full-body motion and multi-goal manipulation tasks, in: IEEE International Conference on Robotics and Automation (ICRA), 2018, pp. 3152–3159.
[26] P. Richard, G. Burdea, D. Gomez, P. Coiffet, A comparison of haptic, visual and auditive force feedback for deformable virtual objects, in: Proceedings of the Internation Conference on Automation Technology (ICAT), vol. 49, 1994, p. 62.

Chapter 13

Neural network-enhanced optimal motion planning for robot manipulation under remote center of motion

Hang Su[a] and Chenguang Yang[b]
[a]*Dipartimento di Elettronica, Informazione e Bioingegneria, Politecnico di Milano, Milano, Italy,*
[b]*Bristol Robotics Laboratory, University of the West of England, Bristol, United Kingdom*

13.1 Introduction

It has been known for a long time that open surgery is the standard form to perform surgeries. In recent years, minimally invasive surgery (MIS) was developed and applied as a complementary surgery form to overcome some disadvantages of open surgery [1–5]. Compared with open surgery, MIS has huge advantages, such as smaller wounds, less bleeding, and, therefore, a reduced need for blood transfusions. In addition, the operation complication and postoperative pain from the wounds is significantly reduced, and postoperative recovery is also faster than in open surgery [6,7]. Except for some scenarios, such as hemorrhage occasions, which require open surgery, the application of MIS has attracted more and more attention [8]. Laparoscopy is an important MIS method. It is usually operated inside the pelvic cavities or abdominal wall for diagnosis or surgery [9,10]. Nowadays, many medical surgeries, for example, digestive system surgeries and gynecological operations, are carried out in the form of laparoscopy [8,11].

However, different from general surgery, MIS is more challenging to perform because the field of vision is limited. Generally, the minimally invasive surgeon only has two-dimensional (2D) vision of the operation space. Meanwhile, the surgeon must be careful when moving the instruments as the incisions are very small [12]. In summary, compared with open surgery, MIS has the following disadvantages: a quite long and steep learning curve, limited vision and depth sense, a constrained range of movement and flexibility, and the absence of haptic feedback. If surgeons do not receive proper training, it will lead to bad outcomes [7,12].

With the advancement of medical robotics and the increasingly stringent operating requirements in recent years, it has become a popular trend to replace

Tactile Sensing, Skill Learning, and Robotic Dexterous Manipulation
https://doi.org/10.1016/B978-0-32-390445-2.00022-2

traditional open surgery with robot-performed MIS. The progress in control techniques has shown the opportunity to overcome the shortcomings of MIS partly. As a result, the related research field has flourished. Firstly, many medical robots can provide 3D vision of the surgical space with adjustable magnification. Additionally, they have excellent flexibility that can correct tremors and improve the movements of the surgeon by motion scaling. Lastly, some medical robots who have enough joints and wristed instruments can realize human-like operations [12].

In robot-assisted MIS (RAMIS), both the cameras and tools required for surgical operations must be implanted into the body by robots to supply the surgical vision and perform the operation. During the surgical process, the surgical tools enter the patient through the narrow wound and follow the wound constraint during the entire operation, i.e., they cannot exert additional forces on the surgical incision. Hence, the end-effector of the robot can only move along the axis and spin around the entry point. Therefore, the incision provides a motion restriction which is known as the remote center of motion (RCM) [13].

In RAMIS, the RCM constraint is a prerequisite to execute surgery properly. However, such a critical constraint also brings a huge challenge for robot control. Generally, RCM constraints can be realized by mechanical methods or software methods [14,15]. The mechanical method is also called the passive RCM constraint, which mechanically constrains the pivot using circular tracking arcs, dual-parallelograms, or synchronous spherical linkages [14,16–18]. However, those methods require external mechanical constraint devices, which makes the mechanical method expensive. Meanwhile, the mechanical structure unavoidably wears out with time. Benefiting from advances in control methods [19–23], the implementation of RCM constraints by software methods in kinematic control has become feasible. The software method is also called the programmable or active RCM, which usually exploits on the redundant industrial robot and is realized by coordination of multiple joints [24]. Different from the passive constraint method, active RCM presents more flexibility and is less expensive. Though the active method has many advantages, it will pose other problems such as space occupation and less maneuverability. Hence, studying different software solutions and their RCM processing effects already becomes the main topic of robot operation. Various methods and controllers have been proposed and studied to realize the stable software constraint, including the RCM-constrained Jacobian [25], the kinematic controllers based on dyadic quaternions [26], the task space augmentation methods [27,28], and decoupled controllers [15,29,30]. In addition, optimization methods that treat RCM constraints as equational constraints also have been proposed [31].

In the RCM-constrained Jacobian matrix, the end point's Jacobian matrix is calculated by the Jacobian matrices of the manipulator and surgical tool; then, it will be inserted into a kinematic feedback controller. Hence, the robot operated under the controller will follow the RCM constraint rules [25]. However, for the controller-based dual-quaternion, the RCM constraint rule is directly programmed and maintained by the motion controller [26]. And in the task space augmentation method, the RCM constraints and the task space coordinates were combined to realize the goals of respecting the RCM constraints and the reference trajectory [27]. In addition, the decoupled control of the null space and the task space [15,27] has been proved to be an effective method, in which the null space and task space are treated separately and then combined with special weights. In addition, a pseudoinverse Jacobian matrix is introduced in this method to solve some special cases where a Jacobi matrix is not available in robot kinematics [15].

Model predictive control (MPC) is a model-based control method which needs the accurate modeling of the controlled object. Although it has been widely used in various industrial applications, it is still challenging to apply it in surgical robots, because accurate modeling of the robot is really challenging. Hence, only a few old studies of laparoscopic applications have been conducted [32–35]. However, in the case where the robot has a well-defined equation of motion, MPC has obvious merits. Firstly, the MPC method is easier to implement than the decoupled way. In the MPC method, only the task space needs to be considered. Secondly, benefiting from its modeling and optimization properties, the MPC method can predict the future trajectory and the inputs will be set to the optimal value at the start [36]. Lastly, by using the MPC strategy, constraints can be easily introduced into the inputs and the states of the system, such as imposing a speed constraint to avoid the joint speed reaches high values [37–39]. For these reasons, the MPC method was selected to realize the robot's RCM constraints and trajectory tracking in the MIS.

In this chapter, we present the method that uses the MPC method to realize the RCM constraint, and show its usefulness in a simulation environment based on the KUKA LWR4+. The goals in this application are controlling the robot to follow the designed trajectory and keeping the surgical tool through the RCM points simultaneously. This means that the controller should control both the orientation and the position of the surgical tool. Because the surgical tool is installed on the robot end-effector, the orientation and the position of the robot wrist must also be regulated. Firstly, we consider the RCM constraint and trajectory tracking problems of the surgical tool in 2D space. To simply analyze progress, the surgical tool was modeled as a rigid rod with virtual translation velocity and virtual rotation velocity. Therefore, the RCM constraint and trajectory tracking can be converted into the control problem of the rigid rod. For such a dynamic system with a certain model, the MPC is easy to employ to establish the controller. In fact, the medical robot is working in the 3D space, so the kinematic controller should have the ability to implement the RCM constraint and trajectory tracking in 3D space. To simplify the problem, the surgical tool was projected to the y-z plane and the x-y plane, respectively. Then, the 3D problem can be treated as two 2D problems to use the abovementioned 2D MPC solution. The true location and orientation of the surgical tool's end may be determined by mapping it from 2D to 3D. For this rigid rod model, the end point of the surgical tool has a certain relation to the robot wrist joint under the RCM constraint. Once the movement of the wrist joint is determined, the required posture of the robot can be calculated by the inverse kinematic and realized. This method transforms the trajectory tracking and RCM constraint problem into a control problem for surgical tools, so that the solution can be easily replicated to other manipulators. In addition, the RBFNN-based approximation method is applied to control the external disturbances and kinematic uncertainties. Finally, by comparing the decoupling method with the MPC control method without RBFNN, the method's efficacy and accuracy are proved in a simulated environment using the KUKA LWR4+ robot. It should be mentioned that the proposed method could also work with other general serial robot manipulators.

13.2 Problem statement

13.2.1 Kinematics modeling

As a single anthropomorphic manipulator, KUKA LWR 4+ is a lightweight redundant manipulator which has seven degrees of freedom. The ability to complete redundant operations is given because of the humanoid structure of the single manipulator, which gives the end-effector of the manipulator the ability to accomplish some complex tasks and enhance the collaboration performance between human and robot in the process of surgery. According to the Denavit–Hartenberg (D-H) convention, the initial position and the link frames of the manipulator are chosen. The D-H parameters represent the joint angle θ_i, the twist angle between joint axes α_i, the length of the link d_i, and the displacement between joint axes a_i. The D-H parameters of seven joints are shown as Table 13.1 [40–42].

TABLE 13.1 D-H parameters of the KUKA LWR4+ robot.

Joint i	α_i (rad)	a_i	d_i (m)	θ_i
1	$\pi/0.2$	0	d_1	θ_1
2	$pi/2$	0	0	θ_2
3	$pi/2$	0	d_2	θ_3
4	$\pi/2$	0	0	θ_4
5	$\pi/2$	0	d_3	θ_5
6	$\pi/2$	0	0	θ_6
7	0	0	0	θ_7

The rotation transformation homogeneous matrix ${}^{i-1}T_i$ is used to describe the connecting relationship between two adjacent links. Furthermore, multiplying the T matrices of the joints from the base to the end-effector we can calculate the pose of the tool [43]:

$$^{B}T_E = ^{B}T_1^1 T_2^2 T_3^3 T_4^5 T_6^6 T_E, \tag{13.1}$$

where B represents the base, E is used to describe the end-effector, and T is a 4×4 matrix.

13.2.2 RCM constraint

Considering the actual process in MIS, the motion of the manipulator is not as flexible as other platforms. All the necessary tools must be implanted into the patient via the incision point, and it should be ensured that the wound edge will not be touched and damaged at any time. Thus, the motion track of the manipulator has to be away from the center point, which means the manipulator can only move along the surgical orientation or rotate around the center point. The surgical tools used in MIS are usually composed of a slender shaft and a cutting or gripping head. Hence, it is possible to use a fixed-length probe to simulate the movement and path planning of a surgical tool under RCM constraints. Likewise, here we take a segment line with fixed length l to simulate the surgical tool, which can move under the RCM constraint. Finally, the simulation is built in a 2D space, as shown in Fig. 13.1.

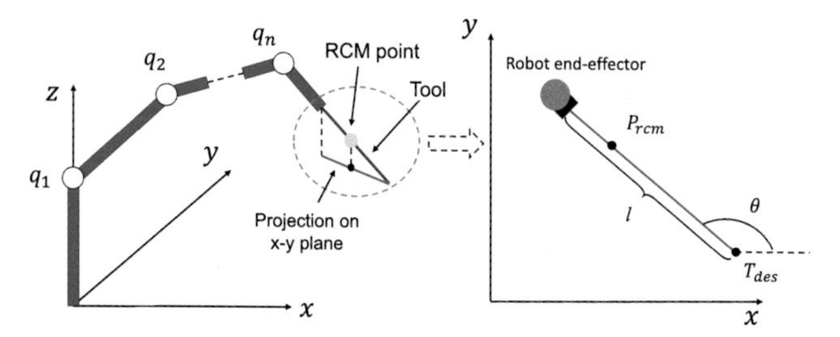

FIGURE 13.1 The surgical tool in 3D space and the projection of the surgical tool in 2D, taking the x-y plane as an example. The surgical tool is attached to the robot end-effector. The tool tip will follow the desired trajectory T_{des} meanwhile passing the RCM point P_{rcm}.

In the above model, T_{des} represents the desired trajectory and P_{rcm} refers to the RCM constraint point. Under the premise of satisfying the RCM constraint, the surgical tool needs to follow the desired trajectory T_{des} meanwhile crossing the RCM point. Because the surgical tool is modeled as a rigid rod, we can ensure the unique pose of the surgical tool tip under the limitation of two conditions mentioned before, which means the pose of the surgical tip is certain by T_{des} and P_{rcm}. In our experiment, we construct a vector that points from the tool tip to the center of the RCM constraint, referring to the intended position of the surgical instrument under the RCM constraint:

$$tool = P_{rcm} - T_{des}. \tag{13.2}$$

Then the following equation can calculate the angel of the vector:

$$\bar{\theta} = acos\left(\frac{x_{rcm} - x_{actual}}{|tool|}\right), \tag{13.3}$$

where $|tool| = \sqrt{(x_{rcm} - x_{actual})^2 + (y_{rcm} - y_{actual})^2}$. It is noted that the chose angle is related to the x-axis.

Using the tool vector, the wrist joint pose in geometric space can be presented by the desired trajectory T_{des} and the RCM point P_{rcm}. The specific form of the expression is as follows:

$$\theta_{robot} = \bar{\theta}, \tag{13.4}$$

$$P_{robot} = T_{des} - l \cdot \frac{tool}{|tool|}. \tag{13.5}$$

Therefore, we establish a clear correspondence between the robot wrist joint pose and the desired trajectory under the premise of satisfying the RCM constraint. Furthermore, the main control targets are the position and orientation of the manipulator.

13.2.2.1 2D RCM constraint

In reality, the control strategy is related to the time parameters. It is presented by a sequence of control points spread across the desired trajectory [44–48]. In the process of motion tracking, the wrist joint and surgical tool need to keep satisfying the RCM constraint mentioned above. However, another parameter that should be considered is velocity. As shown in Fig. 13.2, x_{actual}, y_{actual}, and θ_{actual} refer to characterizing the surgical tool's posture and v_x and v_y were used to present the tool tip virtual linear velocities while ω is the virtual angular velocity. Similarly, (x_{rcm}, y_{rcm}) represent the coordinates of the RCM constraint. Finally, defining the vector of the desired trajectory as (x_d, y_d, θ_d) represents the desired pose trajectory of the manipulator. If the manipulator has a good performance in pose tracking, the goal of following the desired trajectory under the RCM constraint will also be accomplished. Furthermore, we establish a combined kinematic model containing the manipulator and the surgical tool in virtual space to simplify the model.

At the initial time, the tool tip has the same location as the starting point of the desired trajectory. The surgical tool's posture at time t can be calculated by substituting the time and velocity into the following kinematic model:

$$\begin{aligned} x(t) &= x(t-1) + v_x T, \\ y(t) &= y(t-1) + v_y T, \\ \theta(t) &= \theta(t-1) + \omega T. \end{aligned} \tag{13.6}$$

Here T represents the time period between t and $t-1$. After separating the state and output, we can recast the kinematic model (6) in a more general matrix

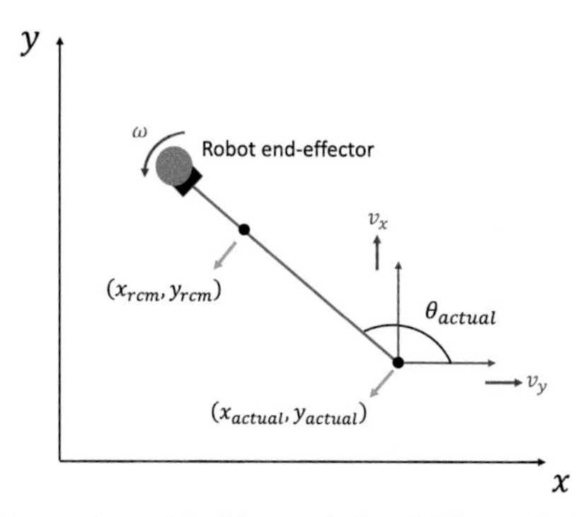

FIGURE 13.2 The 2D kinematic model of the surgical tool. The surgical tool frame is built on the end of the surgical tool and aligns with the robot base frame (blue). The surgical tool is assumed to move along the x- and y-axes with the virtual velocities v_x and v_y. Meanwhile, the surgical tool also can rotate around the tool tip with ω.

form:

$$\begin{bmatrix} x(t) \\ y(t) \\ \theta(t) \end{bmatrix} = \begin{bmatrix} 1 & 0 & 0 \\ 0 & 1 & 0 \\ 0 & 0 & 1 \end{bmatrix} \begin{bmatrix} x(t-1) \\ y(t-1) \\ \theta(t-1) \end{bmatrix} + \begin{bmatrix} T & 0 & 0 \\ 0 & T & 0 \\ 0 & 0 & T \end{bmatrix} \begin{bmatrix} v_x \\ v_y \\ \omega \end{bmatrix}. \tag{13.7}$$

Then, it can be stated more clearly as

$$X(t) = A \cdot X(t-1) + B \cdot u(t). \tag{13.8}$$

Here $u(t)$ represents the surgical tool's velocity vector. Meanwhile, velocity also is a critical control parameter in the kinematic model. In addition, the new actual position can be calculated by the following equation:

$$\begin{bmatrix} x_{actual} \\ y_{actual} \\ \theta_{actual} \end{bmatrix} = \begin{bmatrix} 1 & 0 & 0 \\ 0 & 1 & 0 \\ 0 & 0 & 1 \end{bmatrix} \begin{bmatrix} x(t) \\ y(t) \\ \theta(t) \end{bmatrix}. \tag{13.9}$$

This can simply be presented as

$$Y(t) = C \cdot X(t). \tag{13.10}$$

Once the surgical tool's position has been ensured based on the kinematic model, the MPC controller will be designed to produce the control law $u(t)$ to follow the desired trajectory.

13.2.2.2 3D RCM constraint

The RCM constraint problem in 3D is similar to the RCM constraint problem in 2D, so it is feasible to extend the RCM constraint expressions from the 2D plane

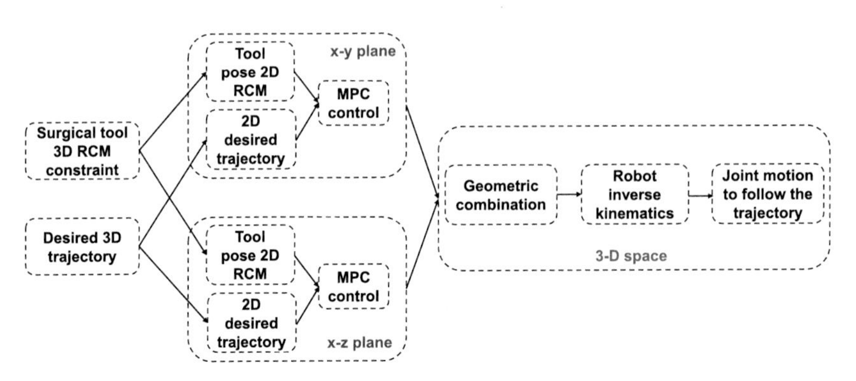

FIGURE 13.3 Conceptual scheme of the projection decomposition method.

to 3D space. However, a simpler way to deal with this problem is to decompose the RCM constraint. Here, we propose a dimensionality reduction method based on coordinate decomposition. It reduces the dimensionality of the 3D problem by projecting the surgical tools and RCM constraint relations in 3D space to mutually orthogonal planes x-y and x-z. Therefore, the 3D RCM constraint problem is smoothly converted into two 2D RCM constraint problems. In Cartesian space, it is possible to determine the posture of a segment in three dimensions by its projection on two orthogonal planes. In this section, we analyze the RCM constraint problem according to the x-y and x-z projection planes. Likewise, the surgical tool tip's desired trajectory in the x-y and x-z planes can also be obtained by using the projection decomposition method. Meanwhile, passing this method, we can apply the 2D RCM constraint in these two separated planes.

In the x-y plane, utilizing the method of geometric projection laws, we can easily get the vectors including the tip coordinates x, y and its orientation θ_{xy}. Similarly, we also can obtain the corresponding tool tip parameters x, z, and θ_{xz} from the x-z plane. Particularly, the x-coordinate obtained from two planes has the same value. Hence, the wrist joint's coordinates that meet the RCM constraint can be written as

$$\begin{aligned} x_{robot} &= x_{tip} + l \cdot \cos(\theta_{xy}) \cdot \cos(\theta_{xz}), \\ y_{robot} &= y_{tip} + l \cdot \sin(\theta_{xy}) \cdot \cos(\theta_{xz}), \\ z_{robot} &= z_{tip} + l \cdot \cos(\theta_{xy}) \cdot \sin(\theta_{xz}). \end{aligned} \tag{13.11}$$

Here l represents the length of the surgical tool, x_{tip}, y_{tip}, and z_{tip} are the positions of the surgical tool, and x_{robot}, y_{robot}, and z_{robot} refer to the coordinates of another end of the surgical tool. The manipulator can be well controlled to have a good performance in desired trajectory tracking under the RCM constraint in 3D space with the trajectory of the end joint. The projection decomposition method's elaborated framework is shown in Fig. 13.3.

13.3 Control system design

13.3.1 Controller design method

In this chapter, MPC is characterized by a horizon M. The model predicts the potential output at time $i+1$ based on the M anticipated frames at time i. To construct the MPC controller, we use the discrete kinematic model described in Section 13.2.1:

$$\begin{aligned} x_i(k+1) &= A \cdot x_i(k) + B_i \cdot u_i(k) + d_i(k), \\ y(k) &= C_i \cdot x_i(k), \end{aligned} \tag{13.12}$$

where $x_i \in R^n, u_i \in R^r, y \in R^q, d_i \in R^n$, and n, r, and q present state dimension, the input dimension, and the output dimension, respectively. It must be pointed out that, during operations, the organ motion is rhythmic owing to the breathing rate, which disrupts trajectory tracking mission with RCM restrictions. These external disturbance-induced dynamics are unpredictable and nonlinear, and are defined as $d_i \in R^n$ [49].

The MPC for the discrete model in (13.8) and (13.10) is as follows:

$$Y(k) = F_y \cdot x(k) + G_y \cdot U(k), \tag{13.13}$$

where

$$Y(k) = \begin{bmatrix} y(k+1) \\ \vdots \\ y(k+p) \end{bmatrix}_{qP\times 1}, \tag{13.14}$$

$$U_i(k) = \begin{bmatrix} u_i(k) \\ \vdots \\ u_i(k+M-1) \end{bmatrix}_{Mr\times 1}, \tag{13.15}$$

$$F_y = \begin{bmatrix} CA_i \\ \vdots \\ CA_i^p \end{bmatrix}_{pq\times n}, \tag{13.16}$$

$$G_y = \begin{bmatrix} CB_i & 0 & 0 \\ \vdots & \vdots & 0 \\ CA_i^{M-1}B_i & \cdots & CB_i \\ \vdots & \vdots & \vdots \\ CA_i^{P-1}B_i & \cdots & C\sum_{i=0}^{P-M} A_i^i B_i \end{bmatrix}_{pq\times Mr}. \tag{13.17}$$

The prediction steps are presented as p, while the control steps are M.

To determine the optimal trajectory, we need to minimize the following cost function after obtaining the various outputs $Y(k)$:

$$\min_{U_i(k)} = \| W(k) - Y(k) \|^2_{Qy} + \| U(k) \|^2_{Ry} . \quad (13.18)$$

Here, $W(k)$ is a $t \times qp$ matrix having the tip's referred posture at moment k and t represents the number of time intervals in the simulation experiment. The optimal solution is given by

$$U(k) = (G_y^T Q_y G_y + R_y)^{-1} G_y^T Q_y (W(k) - F_y x(k)). \quad (13.19)$$

So we can define

$$u(k) = d^T U(k), \quad (13.20)$$

where

$$d = \begin{bmatrix} 1 & 0 & 0 & 0 & 0 & 0 & 0 & 0 & 0 & 0 & 0 & 0 \\ 0 & 1 & 0 & 0 & 0 & 0 & 0 & 0 & 0 & 0 & 0 & 0 \\ 0 & 0 & 1 & 0 & 0 & 0 & 0 & 0 & 0 & 0 & 0 & 0 \end{bmatrix} . \quad (13.21)$$

Finally, we could get the updated x and y values in (13.12). During the simulation, the process is repeated.

13.3.2 RBFNN-based approximation

As previously stated, organ motion is noise for trajectory tracking missions with RCM restrictions and is defined as $d_i \in R^n$. The RBFNN-based approximation technique is presented to compensate for the disturbance to obtain an accurate trajectory by regulating the tip despite the external disturbance. We define the following smooth function $f(Z) : R^q \to R$, and then the RBFNN is used to evaluate the external disturbance:

$$f_{nn}(Z_{in}) = W^T S(Z_{in}), \quad (13.22)$$

where $Z_{in} \in \Omega \subset R^q$ represents the input, $W = [w_1, w_2, \cdots, w_m] \in R^m$ is the NN weight, and $m > 0$ is the NN node number of the hidden layer, $S(Z_{in}) = [S_1(Z_{in}), S_2(Z_{in}), \cdots, S_i(Z_{in})]^T$, and $S_i(Z_{in})$ is an activation function that has been selected as a Gaussian function:

$$S_i(Z_{in}) = \exp\left[\frac{-(Z_{in} - u_i^T)(Z_{in} - u_i)}{\eta_i^2}\right], \quad i = 1, 2, \cdots, m, \quad (13.23)$$

where $u_i = \left[u_{i1}, u_{i2}, \cdots, u_{iq}\right]^T \in R^q$ and η_i are the receptive field's center and the variance, respectively.

According to the activation function definition, $S(Z_{in})$ is bounded and described as

$$S(Z_{in}) \leqslant \xi, \tag{13.24}$$

where ξ is a positive constant scalar.

With a sufficiently large node m, we have

$$f_{nn}(Z_{in}) = W^{*T} S(Z_{in}) + \varepsilon, \tag{13.25}$$

where W^* is the optimal weight in a compact set $\Omega_{Z_{in}} \subset R^q$. The approximation error should satisfy $\|\varepsilon\| \leqslant \rho$, where ρ is a small unknown constant scalar.

The optimal weight vector in $\Omega_{Z_{in}} \subset R^q$ can be represented as

$$W^* = \arg\min_{Z_{in} \in R} \left\{ \sup \left| f_{nn}(Z_{in}) - W^T S(Z_{in}) \right| \right\}. \tag{13.26}$$

To approximate the disturbance, the following decoupled compensator is chosen. The input is X_d, $\dot{X}_d$, X, and $\dot{X}$, the output is the angular velocity of each axis with compensation, and an adaptive RBFNN term $u_{rbfnn}(k)$

$$u_{rbfnn}(k) = \Theta^T S\left(X_d, \dot{X}_d, X, \dot{X}\right) \tag{13.27}$$

is adopted to compensate the disturbance. The final control term is as follows:

$$u_d(k) = u(k) + u_{rbfnn}(k). \tag{13.28}$$

13.3.3 Control framework

To solve the RCM constraints in 3D space, we performed projection decomposition to split the issue into two 2D RCM constraint problems. The wrist joint coordinate and velocity can be obtained by implementing MPC constraints in two orthometric planes. Because the control process is operated in the 2D space, we must use geometric relations to integrate the wrist joint position and velocity into 3D space. Once the trajectory of the wrist joint is determined, both the angle and the angular speed of all the joints in the robot can be computed by inverse kinematics, so that we can complete the trajectory tracking task. To correct for the organ motion disturbance, the RBFNN-based approximation technique is used. Fig. 13.4 depicts the complete control structure.

13.4 Simulation results

In this section, we perform two kinds of comparison experiments. Firstly, the proposed technique is compared to conventional decoupling methods that do not use RBFNN-based approximation [15]. The experiment result shows that the proposed method has a huge advantage in improving the performance of

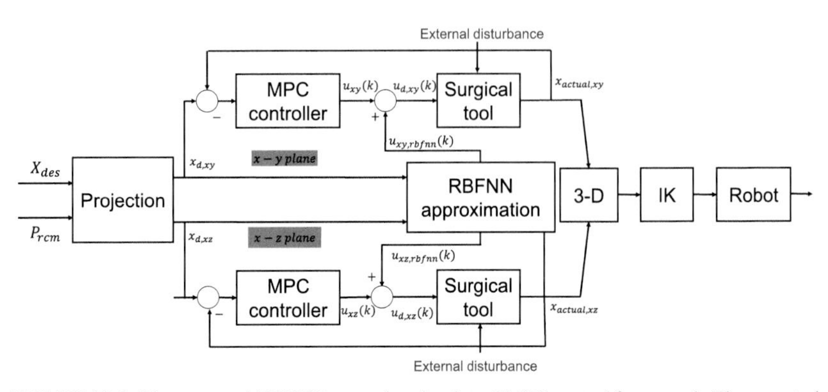

FIGURE 13.4 The proposed RBFNN approximation-based MPC control framework. The expected location and angular trajectory in the x-y plane is $x_{d,xy}$, (x_d, y_d, θ_d) is the desired position and angle trajectory in the x-y plane, and $x_{actual,xy}$ is the actual position and angle trajectory of the surgical tool in the x-y plane; $u_{xy}(k)$ is the output of the x-y plane MPC controller, $u_{xy,rbfnn}(k)$ is the compensation term, and $u_{d,xy}(k) = u_{xy}(k) + u_{xy,rbfnn}(k)$. For clarity, the meanings of these values on the x-z plane are removed here.

trajectory tracking and RCM constraints. Moreover, the RBFNN approximation-based compensation method is compared with the presented method, which is based on the MPC but without interference compensation consisting of RBFNN-based approximation. The experimental result illustrates that the proposed method is more stable under external disturbances caused by organ motion. All experiments were implemented and verified in the MATLAB® environment. Also, we adopt a KUKA LWR4+ robot in the experiment. The robot is applied with clear initial settings and specific structure details. Based on the explicit virtual robot model, we built an RCM constraint simulation scenario as shown in Fig. 13.5. In this scenario, the red circle represents the RCM constraint and the blue rod is used to simulate the surgical tool. We keep the RCM in an absolutely anchored position while controlling the tip to move according to the desired trajectory. For the experimental process, we examine the Zig-Zag and sine trajectories, which are shown in Fig. 13.6. We focused more on the projection of the spatial trajectory in the horizontal plane during the experiment, so the Z-direction's position is specified as constant. In addition, during surgery, the body is assumed to remain stationary, but the motion of the organs accompanied by rhythmic breathing will usually present periodic perturbations. This introduces challenges for trajectory tracking and the RCM constraint. Hence, to simulate such interference, we set the periodic sinusoidal disturbances. And the following evaluation metrics are presented to facilitate the analysis of simulation results:

$$
\begin{aligned}
e_{traj_x} &= x_{actual} - T_{des,x},\\
e_{traj_y} &= y_{actual} - T_{des,y},\\
e_{traj_z} &= z_{actual} - T_{des,z},
\end{aligned}
\tag{13.29}
$$

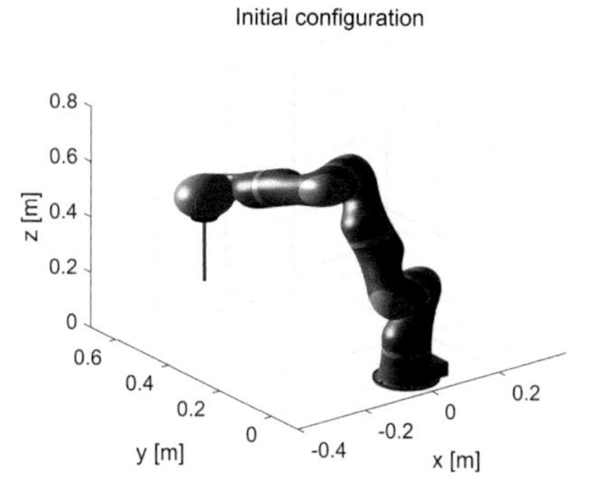

FIGURE 13.5 Simulation setup in MATLAB.

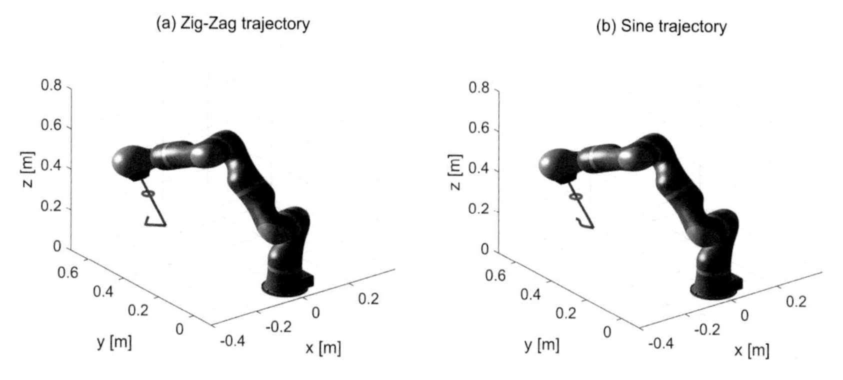

FIGURE 13.6 Simulation trajectories.

$$e_{traj} = \sqrt{e^2_{traj_x} + e^2_{traj_y} + e^2_{traj_z}}, \tag{13.30}$$

$$e_{rcm} = \frac{\|(P_{rcm} - X_{actual}) \times tool\|}{\|P_{rcm} - X_{actual}\|}. \tag{13.31}$$

TABLE 13.2 Root mean square error (RMSE) of the trajectory tracking and RCM constraint in Experiments one and two. Unit [mm].

		Zig-Zag trajectory		Sine trajectory	
		$RMSE_{traj}$	$RMSE_{rcm}$	$RMSE_{traj}$	$RMSE_{rcm}$
Experiment one	Decoupled	1.0741	4.1321	0.8271	2.8418
	MPC	0.7098	2.1×10^{-12}	0.5485	2.1×10^{-12}
Experiment two	Without RBFNN	0.2937	0.2763	0.2901	0.2833
	RBFNN	0.0640	0.0620	0.0641	0.0453

The Zig-Zag trajectory and the Sine trajectory of the first comparative experiment are demonstrated in Fig. 13.7 (1) and Fig. 13.7 (2), respectively. From the experiment results, our MPC-based method outperforms the decoupled method

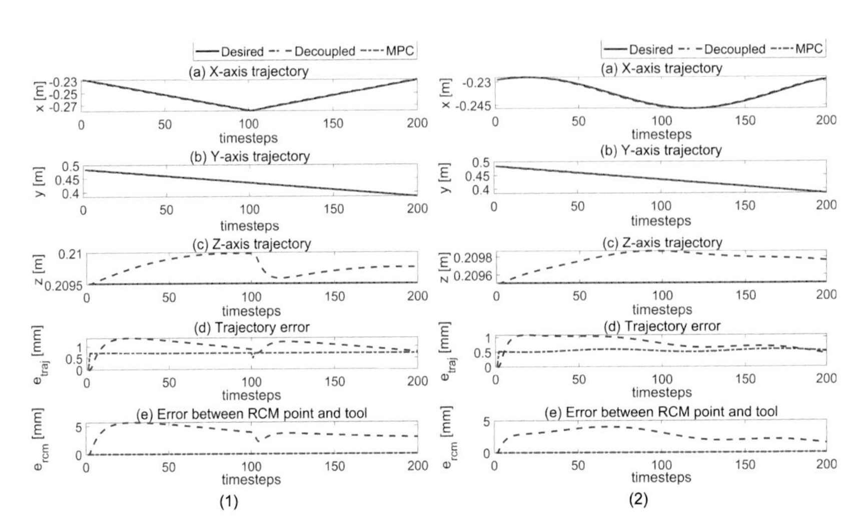

FIGURE 13.7 Experiment one. (1) Zig-Zag trajectory. (2) Sine trajectory.

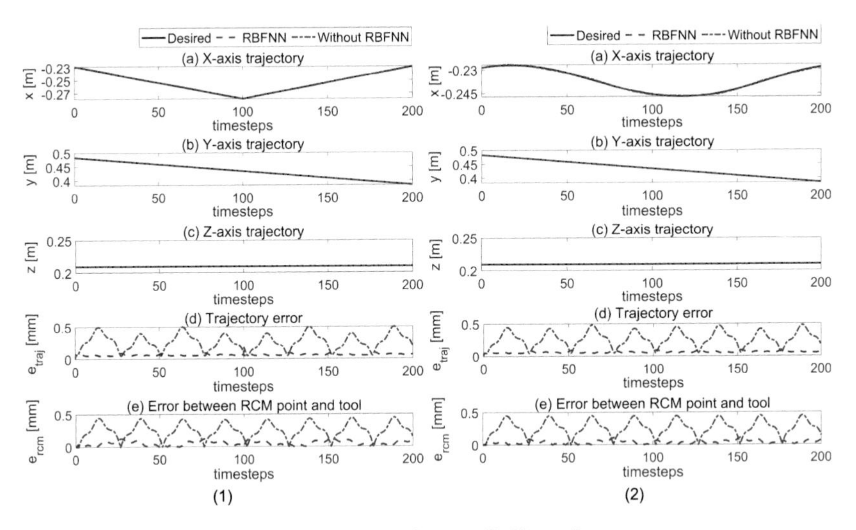

FIGURE 13.8 Experiment two. (1) Zig-Zag trajectory. (2) Sine trajectory.

with respect to the error in both trajectory and RCM. As far as the trajectory error is concerned, our method limits the error to 1 mm, while the decoupling method has an error greater than 1 mm. A similar advantage of error performance is more evident in the RCM case, where, in the proposed method, the error is suppressed around 0, while in the comparison method, the error even reaches 5 mm.

The Zig-Zag trajectory and the Sine trajectory of the second comparative experiment are demonstrated in the left and right of Fig. 13.8, respectively. According to the experimental results, we can know that without compensation, the method is greatly affected by the interference from the rhythmical motions of organs. On the contrary, through the RBFNN approximation-based disturbance compensation, the method can effectively reduce the RCM error to zero. This shows that the RBFNN approximation-based disturbance compensation mechanism has significant advantages for suppressing external disturbances. The contributions of RBFNN approximation-based disturbance compensation mechanism benefit from accurate estimation and compensation to the disturbances.

We emphasized the benefits of the suggested method based on the discussion and analysis of the experimental and demonstration results. The main contribution of the proposed method is the improvement of performance on the trajectory tracking and RCM accuracy, as it is shown in Table 13.2, when compared to a decoupled control-based method, as well as its performance in the minimization of external disturbances caused by the rhythmical movements of the human organs.

13.5 Conclusion

In this part, the advantages and potential of the MPC-based method in the field of RAMIS have been illustrated. The main works and contributions of the proposed method are listed as follows in order of importance.

(1) We map the 3D trajectory tracking problem considering RCM constraints to the 2D trajectory tracking problem, which effectively reduces the control complexity and improves the achievability.

(2) We assign a kinetic model to the control problem of the surgical tool, which is qualified in representing the motion characteristics of the real plant and facilitating the design of the control law.

(3) We designed an RBFNN-based approximation mechanism, which effectively suppresses the unavoidable periodic perturbations caused by the human body.

Extensive validation experiments were performed to prove the advantages of the proposed method. Through the experimental results, the proposed method outperformed the previous proposed decoupled method in terms of trajectory tracking accuracy and RCM maintenance. In addition, the proposed method with RBFNN approximation-based mechanism has adequate quality in estimating and suppressing external disturbances such as organ perturbations. In future works, we will focus on addressing the deployment of the proposed approach on a practical robotics platform. To do so, we will further consider other details of the physical robots, such as mechanical constraints, specific deployment scenarios, etc., which are not thoroughly considered in the current work. Also, we will improve our approach to meet the requirements of controlling physical plants, such as the KUKA LWR4+. Possible measures include adjusting the parameters of the MPC, altering the values of the D-H parameters, etc. Moreover, through our evaluation of this work, we believe that it has sufficient generalizability and the related results can be used for other robots and scenarios [50,51].

References

[1] C. Yang, D. Huang, W. He, L. Cheng, Neural control of robot manipulators with trajectory tracking constraints and input saturation, IEEE Transactions on Neural Networks and Learning Systems (2021).
[2] H. Huang, C. Yang, C.P. Chen, Optimal robot-environment interaction under broad fuzzy neural adaptive control, IEEE Transactions on Cybernetics (2020).
[3] U.E. Ogenyi, J. Liu, C. Yang, Z. Ju, H. Liu, Physical human-robot collaboration: robotic systems, learning methods, collaborative strategies, sensors, and actuators, IEEE Transactions on Cybernetics (2021).
[4] D. Huang, C. Yang, Y. Pan, L. Cheng, Composite learning enhanced neural control for robot manipulator with output error constraints, IEEE Transactions on Industrial Informatics 17 (1) (2019) 209–218.
[5] H. Huang, T. Zhang, C. Yang, C.P. Chen, Motor learning and generalization using broad learning adaptive neural control, IEEE Transactions on Industrial Electronics 67 (10) (2019) 8608–8617.
[6] J. Lei, J. Huang, X. Yang, Y. Zhang, K. Yao, Minimally invasive surgery versus open hepatectomy for hepatolithiasis: a systematic review and meta analysis, International Journal of Surgery 51 (2018) 191–198.
[7] B. Jaffray, Minimally invasive surgery, Archives of Disease in Childhood 90 (5) (2005) 537–542.
[8] A. Buia, F. Stockhausen, E. Hanisch, Laparoscopic surgery: a qualified systematic review, World Journal of Methodology 5 (4) (2015) 238.
[9] G.S. Litynski, Erich Mühe and the rejection of laparoscopic cholecystectomy (1985): a surgeon ahead of his time, JSLS: Journal of the Society of Laparoendoscopic Surgeons 2 (4) (1998) 341.
[10] T. Schlich, C.L. Tang, Patient choice and the history of minimally invasive surgery, The Lancet 10052 (388) (2016) 1369–1370.
[11] W. Reynolds Jr, The first laparoscopic cholecystectomy, JSLS: Journal of the Society of Laparoendoscopic Surgeons 5 (1) (2001) 89.
[12] H.R. Patel, A. Linares, J.V. Joseph, Robotic and laparoscopic surgery: cost and training, Surgical Oncology 18 (3) (2009) 242–246.
[13] N. Aghakhani, M. Geravand, N. Shahriari, M. Vendittelli, G. Oriolo, Task control with remote center of motion constraint for minimally invasive robotic surgery, in: 2013 IEEE International Conference on Robotics and Automation, IEEE, 2013, pp. 5807–5812.
[14] R.H. Taylor, A. Menciassi, G. Fichtinger, P. Fiorini, P. Dario, Medical robotics and computer-integrated surgery, in: Springer Handbook of Robotics, 2016, pp. 1657–1684.
[15] H. Su, C. Yang, G. Ferrigno, E. De Momi, Improved human–robot collaborative control of redundant robot for teleoperated minimally invasive surgery, IEEE Robotics and Automation Letters 4 (2) (2019) 1447–1453.
[16] G.S. Guthart, J.K. Salisbury, The intuitive/sup tm/telesurgery system: overview and application, in: Proceedings 2000 ICRA. Millennium Conference. IEEE International Conference on Robotics and Automation. Symposia Proceedings (Cat. No. 00CH37065), vol. 1, IEEE, 2000, pp. 618–621.
[17] R.H. Taylor, J. Funda, B. Eldridge, S. Gomory, K. Gruben, D. LaRose, M. Talamini, L. Kavoussi, J. Anderson, A telerobotic assistant for laparoscopic surgery, IEEE Engineering in Medicine and Biology Magazine 14 (3) (1995) 279–288.
[18] E. Kobayashi, K. Masamune, I. Sakuma, T. Dohi, D. Hashimoto, A new safe laparoscopic manipulator system with a five-bar linkage mechanism and an optical zoom, Computer Aided Surgery 4 (4) (1999) 182–192.

[19] J. Chen, H. Qiao, Motor-cortex-like recurrent neural network and multi-tasks learning for the control of musculoskeletal systems, IEEE Transactions on Cognitive and Developmental Systems (2020).

[20] S. Liu, M. Sun, L. Feng, H. Qiao, S. Chen, Y. Liu, Social neighborhood graph and multigraph fusion ranking for multifeature image retrieval, IEEE Transactions on Neural Networks and Learning Systems (2021).

[21] J. Chen, H. Qiao, Muscle-synergies-based neuromuscular control for motion learning and generalization of a musculoskeletal system, IEEE Transactions on Systems, Man, and Cybernetics: Systems (2021).

[22] X. Huang, W. Wu, H. Qiao, Computational modeling of emotion-motivated decisions for continuous control of mobile robots, IEEE Transactions on Cognitive and Developmental Systems (2021).

[23] S. Zhong, J. Chen, X. Niu, H. Fu, H. Qiao, Reducing redundancy of musculoskeletal robot with convex hull vertexes selection, IEEE Transactions on Cognitive and Developmental Systems 12 (3) (2019) 601–617.

[24] M. Ceccarelli, International Symposium on History of Machines and Mechanisms, Springer, 2004.

[25] H. Sadeghian, F. Zokaei, S.H. Jazi, Constrained kinematic control in minimally invasive robotic surgery subject to remote center of motion constraint, Journal of Intelligent & Robotic Systems 95 (3) (2019) 901–913.

[26] M.M. Marinho, M.C. Bernardes, A.P. Bó, A programmable remote center-of-motion controller for minimally invasive surgery using the dual quaternion framework, in: 5th IEEE RAS/EMBS International Conference on Biomedical Robotics and Biomechatronics, IEEE, 2014, pp. 339–344.

[27] J. Sandoval, G. Poisson, P. Vieyres, A new kinematic formulation of the rcm constraint for redundant torque-controlled robots, in: 2017 IEEE/RSJ International Conference on Intelligent Robots and Systems (IROS), IEEE, 2017, pp. 4576–4581.

[28] X. Wei, N. Chen, Composite hierarchical anti-disturbance control for nonlinear systems with dobc and fuzzy control, International Journal of Robust and Nonlinear Control 24 (2) (2014) 362–373.

[29] W. He, T. Meng, X. He, S.S. Ge, Unified iterative learning control for flexible structures with input constraints, Automatica 96 (2018) 326–336.

[30] W. He, T. Meng, Adaptive control of a flexible string system with input hysteresis, IEEE Transactions on Control Systems Technology 26 (2) (2017) 693–700.

[31] M. Li, A. Kapoor, R.H. Taylor, A constrained optimization approach to virtual fixtures, in: 2005 IEEE/RSJ International Conference on Intelligent Robots and Systems, IEEE, 2005, pp. 1408–1413.

[32] J. Gangloff, R. Ginhoux, M. de Mathelin, L. Soler, J. Marescaux, Model predictive control for compensation of cyclic organ motions in teleoperated laparoscopic surgery, IEEE Transactions on Control Systems Technology 14 (2) (2006) 235–246.

[33] D. Wang, Neural network-based adaptive dynamic surface control of uncertain nonlinear pure-feedback systems, International Journal of Robust and Nonlinear Control 21 (5) (2011) 527–541.

[34] Z. Li, J. Deng, R. Lu, Y. Xu, J. Bai, C.-Y. Su, Trajectory-tracking control of mobile robot systems incorporating neural-dynamic optimized model predictive approach, IEEE Transactions on Systems, Man, and Cybernetics: Systems 46 (6) (2015) 740–749.

[35] Z. Li, K. Zhao, L. Zhang, X. Wu, T. Zhang, Q. Li, X. Li, C.-Y. Su, Human-in-the-loop control of a wearable lower limb exoskeleton for stable dynamic walking, IEEE/ASME Transactions on Mechatronics (2021).

[36] J.A. Rossiter, Model-Based Predictive Control: a Practical Approach, CRC Press, 2003.

[37] L. Wang, Model Predictive Control System Design and Implementation Using MATLAB®, Springer Science & Business Media, 2009.

[38] M. Katliar, F.M. Drop, H. Teufell, M. Diehl, H.H. Bülthoff, Real-time nonlinear model predictive control of a motion simulator based on a 8-dof serial robot, in: 2018 European Control Conference (ECC), IEEE, 2018, pp. 1529–1535.
[39] D.Q. Mayne, E.C. Kerrigan, E. Van Wyk, P. Falugi, Tube-based robust nonlinear model predictive control, International Journal of Robust and Nonlinear Control 21 (11) (2011) 1341–1353.
[40] U.Z.A. Hamid, H. Zamzuri, P. Raksincharoensak, M.A.A. Rahman, Analysis of vehicle collision avoidance using model predictive control with threat assessment, in: 23rd ITS World Congress, 2016.
[41] H. Su, W. Qi, Y. Hu, H.R. Karimi, G. Ferrigno, E. De Momi, An incremental learning framework for human-like redundancy optimization of anthropomorphic manipulators, IEEE Transactions on Industrial Informatics (2022).
[42] H. Su, W. Qi, C. Yang, J. Sandoval, G. Ferrigno, E. De Momi, Deep neural network approach in robot tool dynamics identification for bilateral teleoperation, IEEE Robotics and Automation Letters 5 (2) (2020) 2943–2949.
[43] H. Su, A. Mariani, S.E. Ovur, A. Menciassi, G. Ferrigno, E. De Momi, Toward teaching by demonstration for robot-assisted minimally invasive surgery, IEEE Transactions on Automation Science and Engineering (2021).
[44] C. Yang, J. Luo, C. Liu, M. Li, S.-L. Dai, Haptics electromyography perception and learning enhanced intelligence for teleoperated robot, IEEE Transactions on Automation Science and Engineering 16 (4) (2018) 1512–1521.
[45] C. Yang, C. Chen, N. Wang, Z. Ju, J. Fu, M. Wang, Biologically inspired motion modeling and neural control for robot learning from demonstrations, IEEE Transactions on Cognitive and Developmental Systems 11 (2) (2018) 281–291.
[46] C. Yang, G. Peng, Y. Li, R. Cui, L. Cheng, Z. Li, Neural networks enhanced adaptive admittance control of optimized robot–environment interaction, IEEE Transactions on Cybernetics 49 (7) (2018) 2568–2579.
[47] C. Yang, Y. Jiang, J. Na, Z. Li, L. Cheng, C.-Y. Su, Finite-time convergence adaptive fuzzy control for dual-arm robot with unknown kinematics and dynamics, IEEE Transactions on Fuzzy Systems 27 (3) (2018) 574–588.
[48] C. Zeng, C. Yang, H. Cheng, Y. Li, S.-L. Dai, Simultaneously encoding movement and semg-based stiffness for robotic skill learning, IEEE Transactions on Industrial Informatics 17 (2) (2020) 1244–1252.
[49] R. George, S.S. Vedam, T. Chung, V. Ramakrishnan, P.J. Keall, The application of the sinusoidal model to lung cancer patient respiratory motion, Medical Physics 32 (9) (2005) 2850–2861.
[50] Y. Chen, L. Wang, K. Galloway, I. Godage, N. Simaan, E. Barth, Modal-based kinematics and contact detection of soft robots, Soft Robotics (2021).
[51] S. Yu, T.-H. Huang, X. Yang, C. Jiao, J. Yang, Y. Chen, J. Yi, H. Su, Quasi-direct drive actuation for a lightweight hip exoskeleton with high backdrivability and high bandwidth, IEEE/ASME Transactions on Mechatronics 25 (4) (2020) 1794–1802.

Chapter 14

Towards dexterous in-hand manipulation of unknown objects

A visuotactile feedback-based method

Qiang Li, Robert Haschke, and Helge Ritter
Center for Cognitive Interaction Technology (CITEC), Bielefeld University, Bielefeld, Germany

14.1 Introduction

A major challenge to exploit the potential of multifingered robot hands for carrying out manual actions that so far are restricted to human dexterity is to enable large-range in-hand manipulation. Even if the geometry of the object, the kinematics of the hand, and the friction conditions at all contact points are accurately known, the planning and execution of controlled regrasp sequences constitutes a difficult task. Since in most practical cases such information is at best available only partly, some works [7,49,29] have focused on the development of solutions that remain feasible even in the absence of detailed information about the object, the contact friction, and the hand kinematics.

Many of these approaches are based on ideas of robust control [5], i.e., methods that are little affected by deviations from their underlying model assumptions. In this work, we present a novel approach that differs from these lines of research in that it integrates active exploration into the determination of a regrasp sequence: regrasping is split into successive repositionings of a free finger that identifies a suitable next contact point by small exploratory movements across the object's surface in the vicinity of its current contact while simultaneously monitoring a quality measure that combines grasp stability and manipulability in such a way that it can be evaluated under very weak information assumptions.

In addition, readily available information from tactile fingertips and finger joints is used for a feedback controller to compensate deviations between the actual and the modeled object motion such that a very simple model with practically no assumptions beyond local surface smoothness is sufficient. As a result, the approach is capable to generate regrasp motion sequences towards a desired object pose. We demonstrate the feasibility of the approach within a physics-

Tactile Sensing, Skill Learning, and Robotic Dexterous Manipulation
https://doi.org/10.1016/B978-0-32-390445-2.00023-4

based simulation employing a 22-DoF model of the anthropomorphic Shadow Robot Hand manipulating a spherical object (without using this geometry information in the control method). A preliminary experiment also proves the feasibility of our method on a real robot platform. To this end, we use the robot arm with the tactile sensor array mounted as the big fingertip demonstrating the unknown object's surface tactile servoing exploration and local manipulation experiment evaluation, which are at the core of the proposed manipulation framework.

The chapter is arranged as follows. In Section 14.2 we summarize the state of the art and discuss how our approach is positioned w.r.t. existing methods. In Section 14.3, we introduce our object manipulation strategy composed of a local manipulation controller and a global finger gait planner. Subsequently, in Section 14.4, we introduce the control scheme for online exploration of the object surface aiming for an optimization of finger manipulability and grasp stability. Both parts of the manipulation framework are evaluated and discussed in Section 14.5 employing the physics-based simulation. Subsequently, in Section 14.6, we demonstrate the feasibility of the in-hand manipulation on a real robot platform. Finally, Section 14.7 summarizes our work and provides an outlook.

14.2 State of the art

Roughly, there are four different lines of research coping with the problem of in-hand manipulation. The first line of research follows an analytic modeling and control approach, requiring rather strong assumptions and detailed knowledge about the situation: the hand kinematics, object properties like shape, mass, and mass distribution, the contact locations and friction coefficients, and the local surface geometry of both the object and fingertips are assumed to be known. Based on this knowledge object grasping and manipulation can be formulated as an offline optimization problem precomputing optimal joint-level finger trajectories, which are subsequently executed on the robot in feedforward fashion.

For example, [3,6] designed a torque-based control scheme to realize finger motions on an object surface assuming complete knowledge about geometric and frictional properties of the finger–object interaction in case of spinning, rolling, or sliding contacts. Considering unknown or inaccurate dynamics models of both the object and finger segments, [36] proposed to use an adaptive control strategy identifying the unknown mass and moment parameters during execution of the manipulation action. They proved the feasibility of their adaptive control strategy in simulation. [54] proposed a control law only depending on current joint angle measurements to realize object manipulation.

The second line of research is based on the idea of extensive planning, using a simulation of object–hand interaction [32] to optimize the grasping and manipulation processes in physics-based simulation prior to execution of the robot. Grasp poses optimized w.r.t. certain quality criteria [31] become arranged in a pose graph [37] to plan manipulation sequences using state-of-the-art motion planning methods like RRT [56] or PRM [41], tackling, e.g., the problem of screwing a light bulb [55].

A major drawback of this approach is that hand pose trajectories, which are planned in a preceding offline process, are applied on a real platform in a forward fashion, not accounting for inevitable errors occurring during execution. Furthermore, the planning stage strongly depends on knowledge about the geometry of the object and fingertips. Nevertheless humans can easily manipulate objects without this detailed knowledge. We suppose that the incredible dexterity of human manipulation originates from tight control loops employing tactile sensor feedback. Consequently we propose to employ tactile feedback to estimate contact positions and forces and introduce a manipulation strategy solely based on this feedback.

The third research line is to learn the skill of in-hand manipulation directly. Different approaches were studied already in robot learning domain. One majority of research is reinforcement learning (RL). Normally this is a procedure of end-to-end learning. Via the self-exploration interaction with the environment, a robotic hand can learn a mapping/controller from different perception modalities to its joints' command for a given task. For example, in [13] a mapping was learned from image pixels to joints' commands in physics simulation. In [53], such a mapping was learned from tactile sensing to finger motion. Along the RL branch, the representative in-hand manipulation work was done by openAI because of its well-known work using the Shadow hand to rate a cube [1]. The whole training procedure was done in the physics simulation and then transferred to a real robotic hand. In order to improve the control robustness of the learned controller, authors considered not only the object's appearance but also the friction as the input of the network. Different subbranches, e.g., considering computational cost and the robot's accuracy, were also explored in the RL-based approach. In [57], a scalable solution was proposed for the learning algorithm. In [50], the learning strategy was presented and the required minimum visual features during in-hand manipulation with a low-cost robotic hand were studied. Another way obtaining the skill of in-hand manipulation is to learn from demonstration. Along this line, the learning was done at two levels, namely local manipulation and global controllers. The global controllers were used to imitate a human hand's high-level behavior, and they can be the gaiting coordinator [16] or the local manipulation's bootstrap. In [16], a skill for fingertips rotating an object was learned. The robotic hand imitated the gaiting pattern of the human fingers employing a central pattern generator (CPG) as the representation model. In [4], an RL method was developed to fine-tune parameters of CPGs for skills improvement. The authors of [15] worked towards the relocation task of an object using whole hand manipulation. Authors used experts' experience to bootstrap local controllers which were explored further with RL approaches.

The fourth line of research is interactive control approach. Unlike the first line, this approach does not assume the known object, but exploits the sensorized hand to model it by actively controllable exploration. Along this branch, robust reactive and exploration controllers are required to bootstrap the interaction between the robotic hand and the unknown object [33,20], which are the focus of this chapter. Along this line, in [12] a controller was devised that is capable to spin a pen of known shape at an impressive speed. In [38], hybrid force/position controllers were proposed to realize unknown object grasping by sliding the contact points on the object surface to optimize grasp stability. A relevant research area considers the problem of monitoring grasp quality during manipulation. The monitored quality provides a potential field in which the robotic hand can explore and interact with the object. Given a set of contact points, the size and shape of the grasp wrench space are evaluated, i.e., the set of wrenches applicable to the object through these contacts without slippage. The problem of finding optimal contact points on the surface of the object is typically decoupled from this [51]. The simplest approach to evaluate the grasp quality is to compute the magnitude of the smallest external wrench which the grasp can resist without slippage. This corresponds to the radius of largest sphere which can be inscribed into the wrench space [9,34]. Extending this worst-case measure to a task-aware stability measure, in [28,11] it was proposed to use the volume of the largest task-specific wrench-ellipsoid instead. Adjusting the shape and orientation of the ellipsoid, one can emphasize specific wrench sets, which are more important for a given task. Finally, in [10], [2], and [42] a real-time capable implementation was proposed to find optimal grasp forces considering exact friction cone constraints.

14.3 Reactive object manipulation framework

Reactive feedback-based strategies for object manipulation appear most suitable as a starting point for our approach, because feedforward execution of manipulation trajectories cannot account for real-life deviations from the planned trajectory: The initial object pose might be estimated incorrectly, fingers might unpredictably slide or roll or even lose contact at all.

Conceptually, we divide the object manipulation process into two stages: a local manipulation controller and a global finger gait planner. While the local controller moves the object by a small amount [21], the global planner supervises this motion and determines an appropriate sequence of finger gaits for regrasping in order to eventually continue the object motion when the local motion approaches the joint limits or the boundary of the object's configuration space.

In order to find a suitable new finger posture to continue the manipulation process, we devise a control strategy for the second stage, sliding an active finger over the object surface and finding a new grasp point while locally optimizing a given quality measure. Although the proposed gradient-based method cannot find the global optimum, it turns out to be highly sufficient for efficient regrasping.

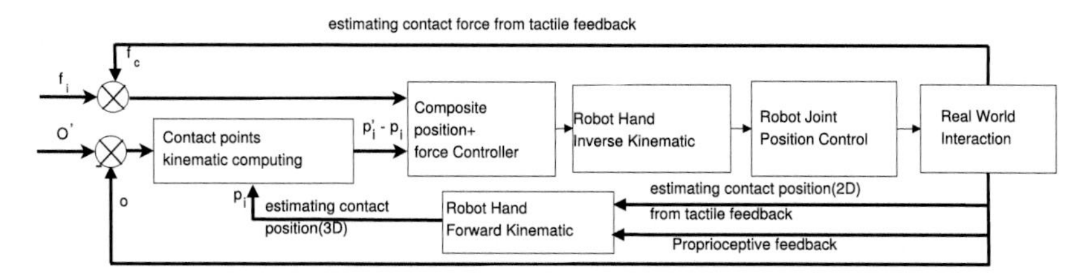

FIGURE 14.1 Low-level, local manipulation control scheme.

Aiming at real-world scenarios and manipulating also unknown objects, we avoid as much information about the objects as possible. Especially, we assume that the global object shape, mass, and mass distribution, as well as local contact properties like surface geometry and friction properties are not known to the robot. Furthermore, we do not explicitly model sliding or rolling. Rather, we assume that we can estimate the current object pose, contact locations, and normal force magnitudes employing vision, joint angle sensors, and modern tactile sensors [45].

In the following three subsections we introduce the first stage, realizing small-scale object motions. The corresponding controller can be subdivided into two branches: a contact position and a contact force controller, which are combined to yield final desired fingertip motions. Subsequently these Cartesian fingertip motions are transformed into an appropriate joint-space motion employing standard inverse kinematics approaches. Fig. 14.1 summarizes this local manipulation control scheme.

14.3.1 Local manipulation controller – position part

In order to realize a small object motion $M = O^{-1} \cdot O'$ from the current object pose O to the targeted pose O', we need to determine appropriate finger joint motions. To avoid the need for a detailed object model, we make the essential assumption that current contact positions $\vec{p}_i^o$ do not move relative to the object within a control cycle. Of course, this is only an approximation. However, the sensor feedback available in the next control cycle will allow us to recognize and correct for undesired contact motion, e.g., due to slipping or rolling. From the current contact positions w.r.t. the object frame by $\vec{p}_i^o$, we can easily compute the contact positions $\vec{p}_i'$ (w.r.t. the palm) targeted in the current control cycle as follows:

$$\vec{p}_i' = O' \cdot \vec{p}_i^o = O \cdot M \cdot O^{-1} \vec{p}_i. \tag{14.1}$$

From this we can compute the required positional changes $\Delta \vec{p}_i = \vec{p}_i' - \vec{p}_i$ for all contact points. Because the local object and fingertip geometries as well as grasp stability measures are not explicitly taken into account, the actual grasp configuration might have changed after application of the computed hand pose.

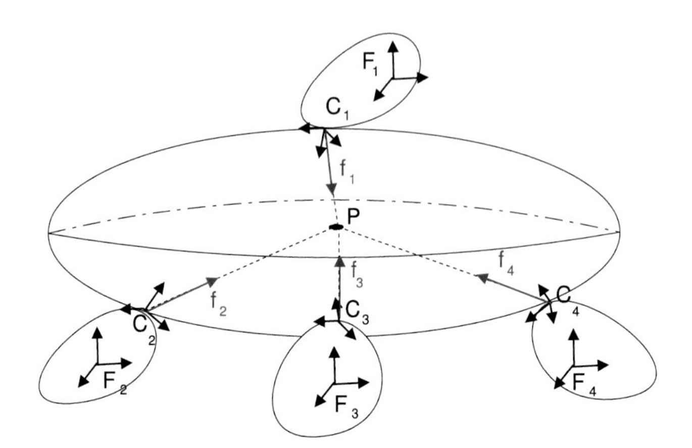

FIGURE 14.2 Planner employs centroid $\bar{\vec{p}}$ of contact points [25].

This corresponds to sliding or rolling contacts or even to a loss of a contact. We tolerate such changes and use fast feedback to correct for these unmodeled deviations.

14.3.2 Local manipulation controller – force part

A mere kinematic consideration of the manipulation task is not sufficient. In order to maintain a stable grasp and to not break the object, we have to control contact forces as well.

Conventional contact force planners strive for a globally optimal contact force distribution ensuring grasp stability, i.e., all contact forces staying within corresponding friction cones, the totally applied force exactly resisting external forces (i.e., gravity), and limiting local contact forces. This general solution is meaningful only if the contact force is controllable. However, we assume that there is no 3D contact force feedback (obtained directly or indirectly), but only the force magnitude is available from tactile sensors. Following concepts from [52] the central idea is to plan the force direction such that the resultant moment will be zero and to plan the force magnitudes along these directions such that the resultant force applied to the object becomes zero.

Obviously the resultant moment is zero if the contact force directions of all fingers intersect in one point. As illustrated in Fig. 14.2, we chose the intersection point as the centroid $\bar{\vec{p}}$ of all contact points $\vec{p}_i$, i.e., $\bar{\vec{p}} = \frac{1}{N} \sum_{i=1}^{N} \vec{p}_i$. Accordingly, normalized contact force direction vectors can be computed as follows:

$$\hat{\vec{f}_i} = \frac{\bar{\vec{p}} - \vec{p}_i}{\|\bar{\vec{p}} - \vec{p}_i\|}, \tag{14.2}$$

where $\vec{f_i}$ denotes the contact force of the ith finger. The force magnitudes f_i are constrained by

$$\sum_{i=1}^{N} \vec{f_i} = \sum_{i=1}^{N} f_i \cdot \hat{\vec{f_i}} = 0\,, \tag{14.3}$$

which defines a system of linear equations. Denoting the matrix of normalized force directions with $\hat{F} \in \mathbb{R}^{3\times N}$ and the vector of desired force magnitudes with $F \in \mathbb{R}^N$, we can summarize the latter equation in matrix form:

$$\hat{F} \cdot F = 0. \tag{14.4}$$

We find a positive solution ($f_i > 0$) to this equation using singular value decomposition (SVD). As the problem is underdetermined, there exists a nonzero null space of $\hat{F}$. The desired solution can be expressed as a linear combination of the null space basis. The superposition coefficients should be selected appropriately to guarantee the contact force to fall within the range of tactile sensor measurements and prevent crushing the objects.

14.3.3 Local manipulation controller – composite part

Eventually, both contributions – from the position controller and the force controller – need to be combined into a common control signal determining the desired motion of each fingertip. Conventional solutions for a simultaneous position and force control are the hybrid position/force control and the indirect force control. The hybrid position/force controller decouples the control problem based on the task constraint defined by the contact frame: Force is controlled along the surface normal of the contact, while position is controlled in the tangent plane.

However, in our scenario, both components cannot be decoupled, because there is a motion component along the force direction and vice versa. That is why we have chosen the indirect force control scheme and propose a composite position+force controller, whose schema diagram is shown in Fig. 14.3. The control signals $\vec{u}_1$ and $\vec{u}_2$, denoting desired fingertip motions originating from the position resp. force controller, are additively superimposed to yield the final motion $\vec{u}$ of a single fingertip sent to the inverse kinematics module. The position controller is realized as a proportional controller (P-controller) parametrized by diagonal gain matrix k_P^p, while the force controller is a proportional integral controller (PI-controller) parametrized by diagonal gain matrices k_P^f and k_I^f. The PI-controller is used in order to guarantee a higher priority level for force control. TM is the transformation matrix of the contact reference frame, which is used to transform the scalar normal force error $\Delta\vec{f}$ into a vector-valued force parallel to the contact normal direction (z-axis). Forces tangential to the contact point are left unmodeled. The stiffness coefficient k_{stiff} finally transfers the force deviation into a displacement error. The whole controller can be summarized as

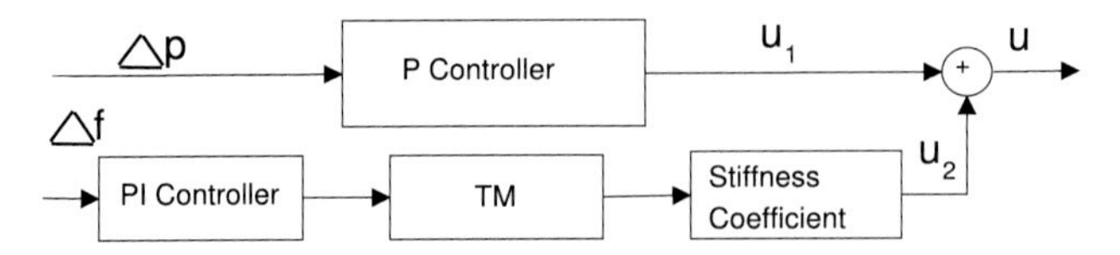

FIGURE 14.3 Composite position+force control scheme [25].

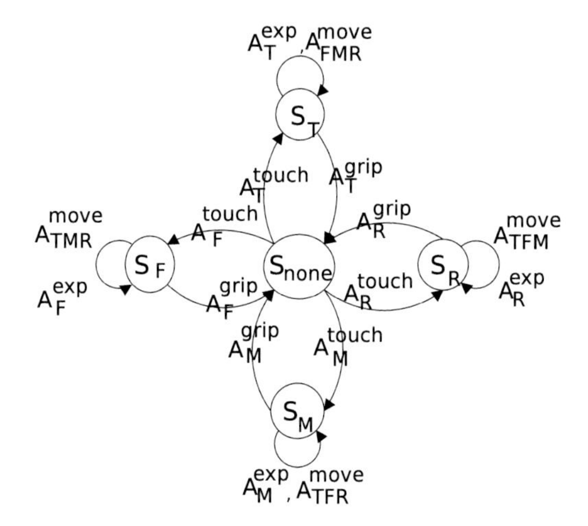

FIGURE 14.4 High-level, global regrasp planner.

follows:

$$\vec{u} = k_P^p \cdot \Delta\vec{p} + k_{\text{stiff}} \cdot TM \cdot \left(0, 0, k_P^f \cdot \Delta\vec{f} + k_I^f \cdot \int \Delta\vec{f}\right)'. \tag{14.5}$$

14.3.4 Regrasp planner

The workspace limits of the multifingered hand restrict the object motion that can be achieved by the local manipulation controller. If the target pose is too far away, the hand must regrasp the object in order to reach the target pose in two or more steps of the local manipulation controller.

Finding a suitable regrasp is the task of the regrasp planner. It uses a strategy that is loosely inspired by human manipulation skills: the movement of the local manipulation controller uses three "passive" fingers to realize the object motion, while the fourth finger takes the role of "actively" exploring the object to identify an optimal next contact point from which it can – together with two of the previously "passive" fingers – form a new "passive" group such that the local manipulation process can seamlessly be continued.

For simplification we employ a static finger gait as described by the state machine shown in Fig. 14.4. All fingers successively explore the object surface for a better contact position, while the remaining fingers move or stabilize the object. To this end, we distinguish the following states:

S_{none}:	TFMR are all grip fingers,	
S_R:	TFM grip fingers	and R exploration finger,
S_M:	TFR grip fingers	and M exploration finger,
S_F:	TMR grip fingers	and F exploration finger,
S_T:	FMR grip fingers	and T exploration finger.

Fingers are abbreviated as follows: T: thumb, F: forefinger, M: middle finger, R: ring finger. State transitions are accompanied by a common sequence of object manipulation and exploration motions as follows: After holding the object with all fingers in the initial state S_{none}, an individual finger becomes the actively exploring finger, contacting the object with very small force. In this state, the three remaining, passively holding fingers move the object. After this local manipulation is performed, the active finger slides over the object surface to find a suitable contact point for further manipulation. Finally, the active finger reestablishes object contact with a contact force determined by the force planner (Section 14.3.2), thus returning to the state S_{none}. Please note that the contact force planner considers only the passively grasping fingers and assumes zero contact force for the actively exploring finger. State transitions are triggered by an external process cycling through the states S_{none}–S_R–S_{none}–S_M–S_{none}–...

14.4 Finding optimal regrasp points

For rotary object manipulation studied in our previous work [24,23] we could employ precomputed contact points for regrasping. However, for more general object shapes we require a more elaborate search process to find suitable new contact points, which allow to continue the object manipulation motion. One major contribution of the present work is a feedback controller to accomplish this task in an online fashion, which is introduced in the following.

We formalize the contact point selection as an optimization problem by considering two quality criteria, the grasp stability and the manipulability, as an objective function to be maximized in an exploratory search process.

14.4.1 Grasp stability and manipulability

Classical grasp planning theory aims for force-closure grasps which can resist external disturbance wrenches from arbitrary directions without slipping or rolling. Grasp quality measures rely on an analysis of the grasp wrench space, i.e., the set of wrenches applicable to an object through a set of normalized contact forces, which is closely related to the grasp matrix G [39]. The most popular methods employ (1) the minimal singular value of G or (2) the determinant of G. While the former method yields a worst-case criterion measuring the distance to an unstable grasp configuration along the worst wrench direction, the latter evaluates the volume of the wrench space, which averages over all possible wrench directions and is invariant under a change of the torque reference system. We employ the latter criterion:

$$\phi_{\text{stability}} = \sqrt{\det(G_{\text{passive}} G^t_{\text{passive}})}. \tag{14.6}$$

Because the exploration phase aims for a recomposition of the group of fingers passively holding the object in the next exploration phase, grasp stability is evaluated considering only contacts of those fingers (which are known beforehand due to the fixed finger gait). This is denoted by the subscript "passive" to G in Eq. (14.6).

The manipulability [35] measures the distance of the current hand pose to a singular configuration and thus expresses the capability of the current pose to actively move all fingertips into an arbitrary Cartesian direction. The manipulability is calculated from the Jacobian matrix of the robot hand, which is a block-diagonal matrix formed from the finger Jacobians J_i, assuming uncoupled finger motion. Because during exploration only the Jacobian and thus the manipulability of the active finger changes, we can reduce calculations to the appropriate submatrix J_i:

$$\phi_{\text{manipulability}} = \sqrt{\det(J_i J_i^t)}. \tag{14.7}$$

Both criteria are complementary: while the grasp stability criterion only considers the contact configuration, the manipulability focuses on the finger motion range, avoiding singular configurations. Hence, both criteria need to be combined effectively. However, a simple linear superposition of both measures is not promising, because often both criteria are conflicting with each other: an increase of stability generates a decrease of manipulability and vice versa, such that a linear superposition finds the least compromise only (see Section 14.5 for an example). Additionally, it would be difficult to find a suitable weighting of both components, because they are not normalized.

Hence, we employ a hierarchical combination of both criteria, which are alternatively chosen for optimization in an adaptive fashion. Because we aim for in-hand object manipulation, we propose to normally use manipulability as the primary criterion which is optimized as long as the grasp stability criterion fulfills a given minimal threshold. Hence, if the grasp stability is larger than the threshold, we generate active finger motion following the estimated gradient of the manipulability index, thus maximizing this criterion. Otherwise, if grasp stability is below the threshold, finger motion will be generated along the gradient of grasp stability in order to increase stability to the necessary level first.

14.4.2 Object surface exploration controller

Given a selected objective function ϕ, the task of the exploratory motion controller is to generate a sliding motion over the (unknown) object surface which

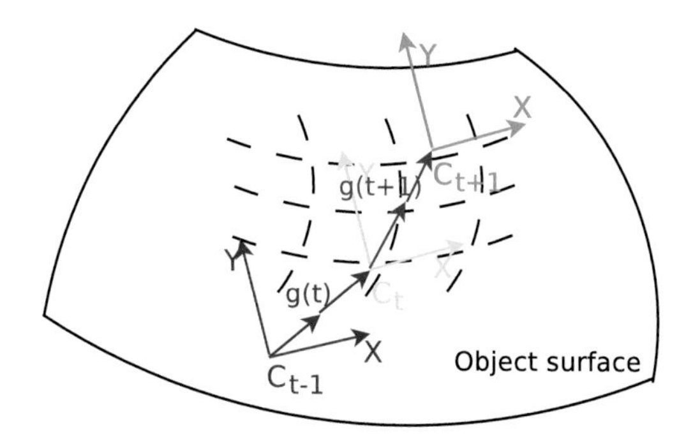

FIGURE 14.5 Basic control scheme for surface exploration.

finds and follows the objective's gradient $\nabla\phi$. Because both quality criteria are nontrivial, nonlinear functions of the contact point motion $\dot{\vec{c}}$, we do not try to find a closed-form solution of the gradient, but aim for its online estimation.

The basic concept of the control scheme is illustrated in Fig. 14.5. For a number n of control cycles, the exploratory motion follows the estimated gradient $\tilde{\nabla}\phi$, which is projected onto the tangential plane at the current contact point:

$$\dot{\vec{c}}^r = \eta \cdot T_c^r \cdot P(\vec{c}) \cdot (T_c^r)^{-1} \cdot \tilde{\nabla}^r \phi\,, \tag{14.8}$$

where η is the gain factor, T_c^r is the homogeneous transformation from the contact reference frame (script c) to the global reference frame (script r), and $P(\vec{c}) = \mathrm{diag}(1, 1, 0, 1)$ is a projector mapping onto the tangent plane at the contact point $\vec{c}$. This targeted contact velocity is fed into a hybrid position/force controller (cf. [21]), which tries to realize this motion in the tangent plane while simultaneously maintaining a (small) normal contact force onto the object. After n control cycles we reestimate the gradient $\nabla_{\vec{c}}\phi$ from the slope of the objective function during motion $\Delta\vec{c}$ from $\vec{c}$ to $\vec{c}'$. This procedure is iterated until convergence or until a predefined number of $N_{\max}$ control cycles have been executed.

The complete control algorithm is summarized in Algorithm 14.1. In order to obtain a smooth estimation of the gradient, in line 8 we apply a sliding average using a smoothing coefficient of $\lambda = 0.9$. Please note that the estimated gradient $\tilde{\nabla}\phi$ is represented w.r.t. the time-constant global reference frame. The contact reference frames, which are used to compute $\nabla\phi$, vary over time and thus are not suitable for a time-consistent representation.

Further note that the targeted motion is restricted to the tangent plane of the contact point in order to maintain contact. Of course, due to approximation errors and foremost due to the unknown object shape, the real contact motion might also have a normal component, thus changing the contact force or losing contact at all. However, the hybrid position/force controller accounts for these deviations from the planned motion by maintaining a given contact force. Obviously this control algorithm only assumes that the object surface changes

Algorithm 14.1 Object surface exploration, maximizing ϕ.

1: $i = 0$ {initialize cycle count}
2: $\tilde{\nabla}\phi^r \propto \mathcal{N}(0, \sigma = 0.15)$ {randomly initialize gradient}
3: **while** $++i \leq N_{\max}$ **and** $\| P \cdot (T_c^r)^{-1} \cdot \tilde{\nabla}\phi^r \| \nleq \varepsilon$ **do**
4: **if** $i \mod n = 0$ **then** {update $\tilde{\nabla}\phi^r$ every n cycles}
5: $\Delta\vec{c}_c = P(\vec{c}) \cdot T_r^c \cdot (\vec{c}'_r - \vec{c}_r)$
6: $\nabla\phi^c = [\frac{\phi' - \phi}{\Delta\vec{c}_x^c}, \frac{\phi' - \phi}{\Delta\vec{c}_y^c}, 0]^t$
7: limit norm of $\nabla\phi^c$ to $\nabla_{\max}$
8: $\tilde{\nabla}\phi^r \leftarrow \lambda \cdot \tilde{\nabla}\phi^r + (1 - \lambda) \cdot T_c^r \cdot \nabla\phi^c$
9: $\phi' \leftarrow \phi, \quad \vec{c}' \leftarrow \vec{c}$
10: **end if**
11: $\dot{\vec{c}}^r \leftarrow \eta \cdot T_c^r \cdot P(\vec{c}) \cdot (T_c^r)^{-1} \cdot \tilde{\nabla}\phi^r$
12: **end while**

smoothly along every contact trajectory, such that gradient estimates can be computed and small motion deviations along the surface normal can be compensated.

14.5 Evaluation in physics-based simulation

The object manipulation algorithm is first validated in a physics-based simulation experiment. We employ the Vortex physics engine [17] to obtain real-time contact information (contact position and normal force magnitude) and the object's pose (position and orientation). The former information will also be accessible in the real world, exploiting modern tactile fingertip sensors providing a moderate spatial resolution. Exploiting the known sensor shape and kinematic model of the hand, we can calculate contact positions relative to the hand and correlate them to a visually tracked, coarse object model.

In order to model noisy real-world sensors, artificial white noise is superimposed on the feedback provided by the physics engine. Particularly, the standard deviation of added Gaussian noise for the contact positions is 0.3 cm, and it is 0.1 for the contact forces. Hence, the positional noise resembles the spatial resolution of tactile sensors in the Shadow Robot Hand. Note that we do not model calibration errors, which usually result in systematic deviations.

We test one exemplary object – a distorted and discretized sphere of ca. 5 cm diameter – which has to be rotated in-place by a 22-DoF Shadow hand model (see Fig. 14.6). Note that the shape of the object and other parameters like friction properties are not available to the manipulation strategy. Also note that the object's surface is not smooth, but rather composed from smaller planar patches with edges in between. For initial grasping, we assume that the object is located in a suitable pose relative to the hand and desired grasp points are known. The simulation resembles our real robot setup to facilitate future transfer

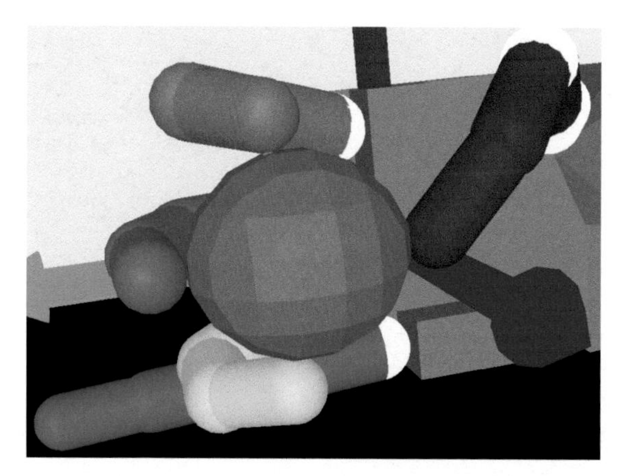

FIGURE 14.6 Simulation scenario: The Shadow hand model shall rotate an unknown object using the four fingers thumb (red), forefinger (gold), middle finger (green), and ring finger (yellow). The green y-axis is parallel to the thumb and middle finger contact normals.

to the real world, once the required tactile feedback is robustly available from fingertip sensors. The whole manipulation process comprises three phases:

1. Grasp the object while it is fixed in the world, which is necessary to avoid that the object is kicked off.
2. Unfreeze the object and stabilize the grasp employing active force control in order to prepare manipulation.
3. Actually manipulate the object.

We focus the evaluation on the last item only. The realization of a stable grasp, which is the objective of the first two stages, is reported in our previous work [40].

14.5.1 Local object manipulation

First we discuss the results of realizing three local manipulation actions employing the motion controller presented in Section 14.3:

1. Translating the object by 5 mm parallel to the y-axis, which is also parallel to the thumb and middle finger contact normals, thus demonstrating the effectiveness of the composite position/force controller (Fig. 14.7).
2. Rotating the object around the y-axis back and forth by 0.2 rad (Fig. 14.8).
3. Tracking a figure 8 sized 5×10 mm in the x-y plane (Fig. 14.9). The center of the figure is defined as the object position when the unfreezing object command is given. We need to point out that the center of number 8 is not the same between the no noise and noise situations. In the noise case, the object is kicked off a small deviation from the predefined point at the time instant of giving the unfreezing command, which is the reason for the initial deviation.

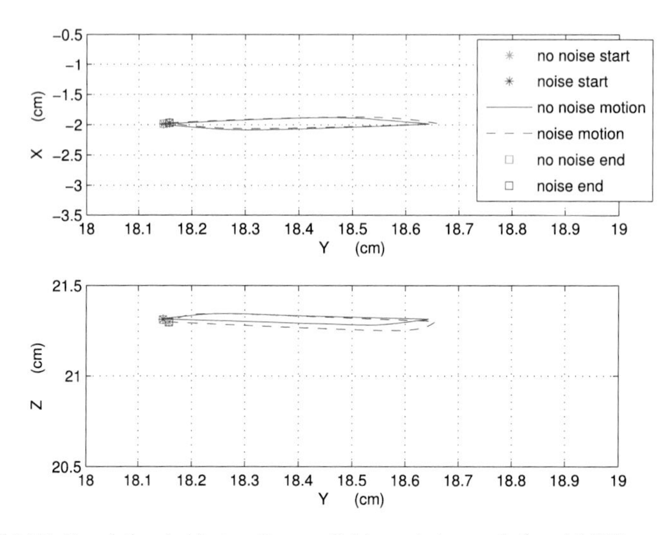

FIGURE 14.7 Translational object motion parallel to contact normals (y-axis) [25].

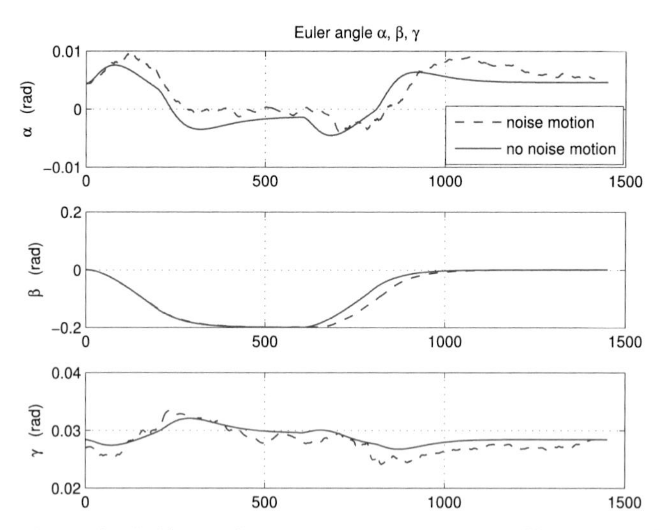

FIGURE 14.8 Rotational object motion around the y-axis. Note the different scales of the outer and middle figures [25].

All three manipulation actions were performed in two different settings: with and without adding artificial measurement noise to the sensor values obtained from the physics simulation engine. All simulation results show that the proposed composite position+force controller can successfully realize manipulation actions of unknown objects, although fingertip rolling or sliding is not explicitly modeled.

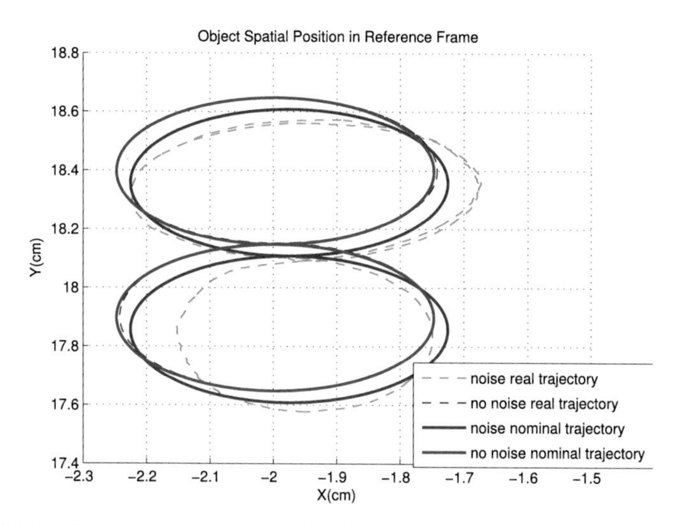

FIGURE 14.9 The object tracking a figure 8 in the x-y plane [25].

However, the experiments also show that the tracking performance significantly decreases if noisy sensor feedback is used – especially in the last experiment. The peak tracking error increases to 20% in this case. Naturally, the controller gains in Eq. (14.5) directly affect the tracking accuracy. Although a large P-gain k_P^p can improve tracking accuracy, this will also lead to oscillations in force control. Consequently, an adaptive control scheme which automatically regulates these control parameters could be employed to further improve tracking accuracy. A video illustrating the discussed manipulation actions can be found in [22].

14.5.2 Large-scale object manipulation

A second set of experiments was performed to demonstrate the interplay of small-scale object manipulation, globally controlled finger gaiting, and finger repositioning in order to realize large-scale object motions. To this end, the following sequence of actions is performed iteratively:

1. Choose an active finger (according to the state machine shown in Fig. 14.4), which releases the grip and contacts the object with a small force only (transition to state S_X).
2. Rotate the object by the small amount of 5 degrees using the three remaining, passive fingers.
3. Explore the object surface with the active finger according to Algorithm 14.1.
4. Reestablish grip with all fingers (transition to state S_{none}) and continue the manipulation process with the next active finger.

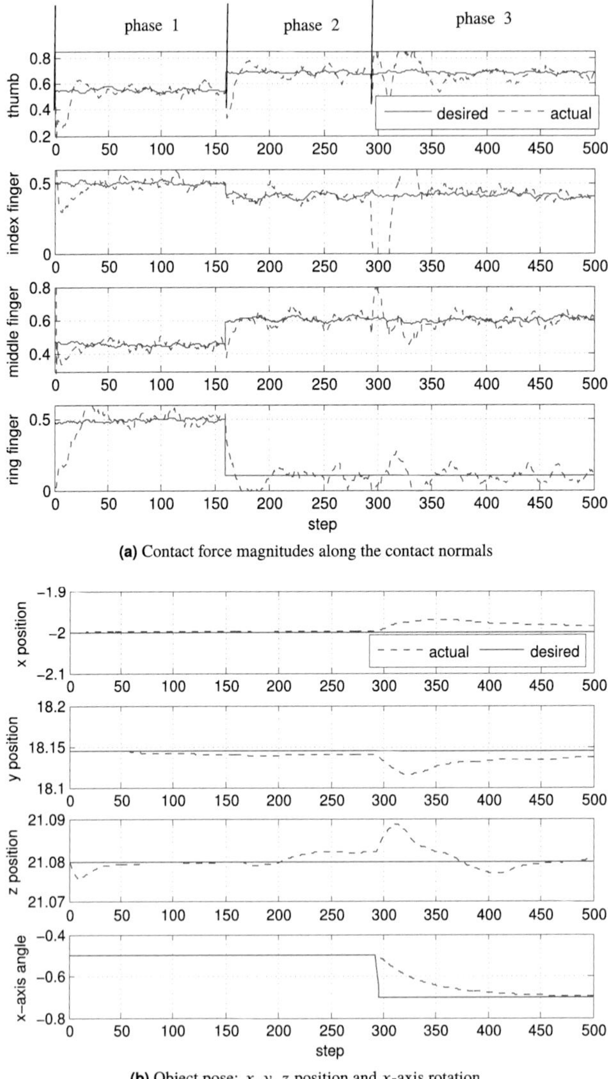

(a) Contact force magnitudes along the contact normals

(b) Object pose: x, y, z position and x-axis rotation

FIGURE 14.10 Force and motion trajectories while rotating the object.

14.5.2.0.1 Relaxing grip and rotating the object

The trajectories shown in Fig. 14.10 illustrate the actions A_R^{touch} and A_{TFM}^{move}, when the ring finger relaxes its grip (phase 2) and the remaining fingers rotate the object around the x-axis (phase 3). The first phase in this figure corresponds to the state S_{none}, where all four fingers hold the object employing the active force controller described in Section 14.3.2.

In Fig. 14.10a, the contact force evolution during all three phases is shown, while Fig. 14.10b shows the motion trajectories of the object position (x, y, z) and its rotation angle around the x-axis. Target trajectories are visualized as red solid lines, while actual trajectories are depicted as blue dotted lines. Note that the noisy force trajectories are mainly due to the artificially superimposed sensor noise, which shall model real-world conditions.

At the transition from phase 1 to phase 2, the desired contact force for the ring finger (last subgraph of Fig. 14.10a) is lowered from 0.5 to 0.1. The other fingers slightly adapt their contact forces to account for the omission of the fourth contacting finger. Finally, in phase 3, the object rotation is realized while maintaining the force profile, which is best seen in the last subgraph of Fig. 14.10b. The simulation results show that the object position error is less than 0.5 mm and the orientation error is less than 0.01 rad.

14.5.2.0.2 Object surface exploration

After realizing a small-scale object motion, the fingers need to be repositioned on the object's surface in order to allow for further manipulation movements. To this end the surface exploration controller presented in Section 14.4 is used. Fig. 14.11 shows the evolution of the quality measures during an exploratory motion of the ring finger (action A_R^{exp}) under two different conditions: The red solid lines result from the motion when only the manipulability index is optimized, while the blue dotted lines result from the motion when both quality criteria are optimized simultaneously using a linear superposition with weights $\frac{1}{2}$ each. Both motions start from the same initial configuration.

As can be seen from the graphs, both criteria are conflicting with each other, resulting in only minor changes to the overall quality measure in the superimposed case. However, when the optimization focuses onto a single criterion, the proposed control strategy can successfully estimate the gradient direction after a few iteration steps and subsequently follow this gradient for maximization. The decrease of the manipulability during the first 75 time steps is due to a poor random initialization of the gradient direction (line 2 of Algorithm 14.1), which is only slowly overcome due to the sliding average (line 8). As can be seen from the stability graphs, the grasp is stable during the course of manipulation, although stability may decrease. Repeating the exploratory motion 50 times using different initial gradient directions always leads to a successful maximization of the objective function.

14.5.2.0.3 Continuous object manipulation

Finally, Fig. 14.12 shows the evolution of the contact points w.r.t. the object frame (a) and the corresponding quality criteria (b) during a continuous rotation of the object following the finger gait pattern as defined by the state machine in Fig. 14.4: exploring with the ring, middle, and index fingers and the thumb, in this order.

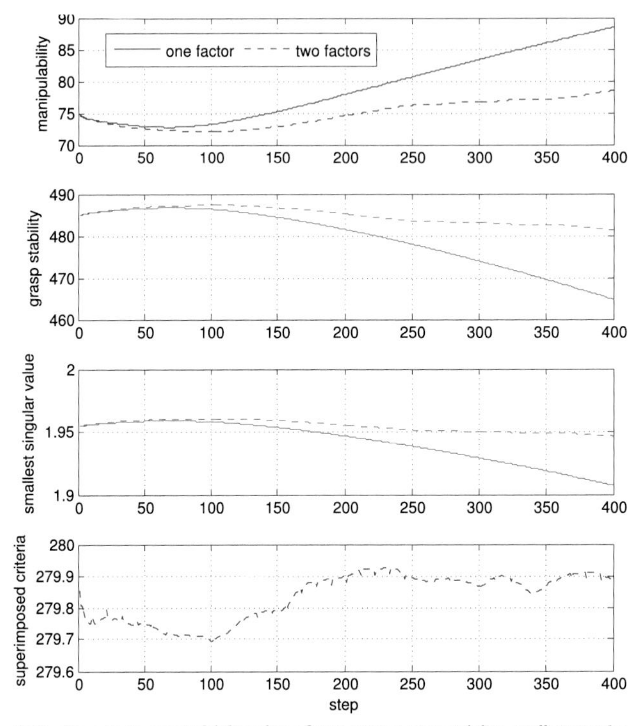

(a) Quality criteria: manipulability of ring finger, average grasp stability, smallest singular value of G, superimposed criteria

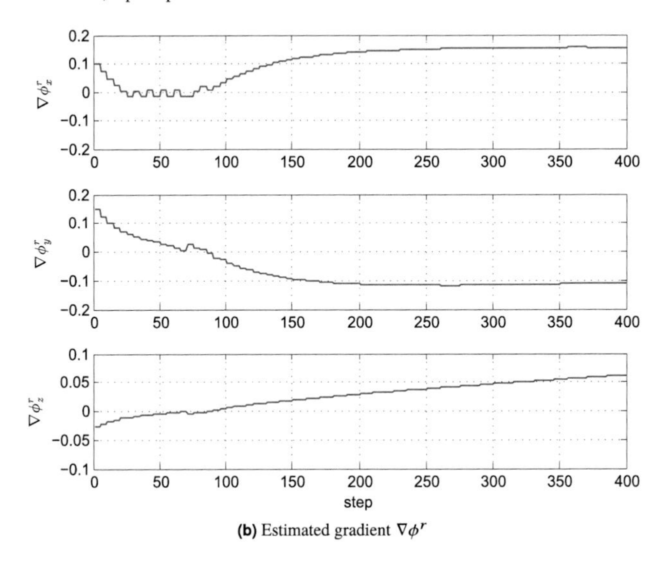

(b) Estimated gradient $\nabla\phi^r$

FIGURE 14.11 Ring finger exploring the object surface.

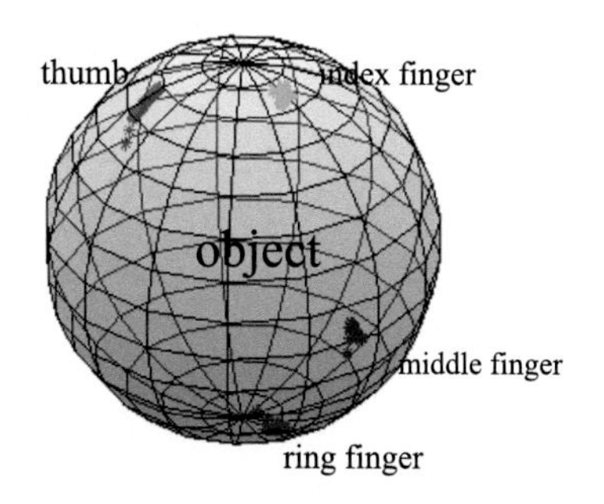

(a) Contact point cloud generated by object exploration

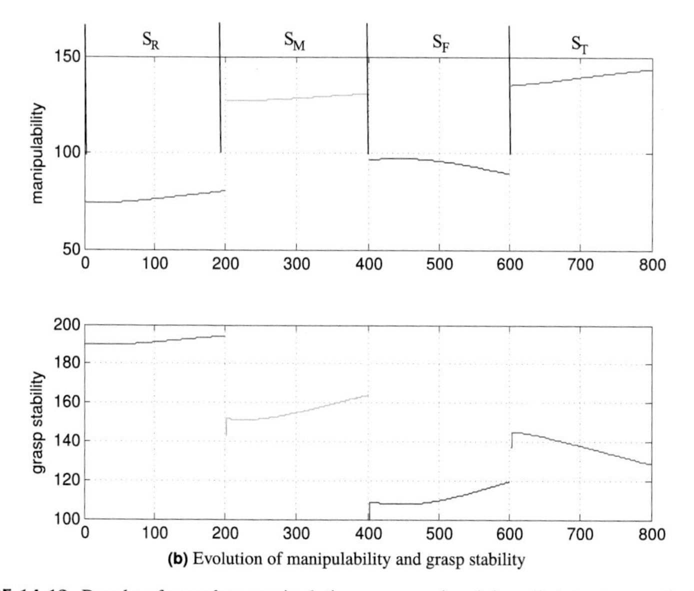

(b) Evolution of manipulability and grasp stability

FIGURE 14.12 Results of complete manipulation sequence involving all states, i.e., exploring with all fingers in turn in one manipulation period.

As can be seen from the evolution of the objective functions, the algorithm always maximizes the manipulability of the exploring finger – except in state S_F, where the grasp stability is chosen as the objective function. In this state, the grasp stability – evaluated for the group of passive fingers in the final state – drops to a low value, because the thumb will become active in the final phase and is thus excluded from the holding task. The complete manipulation sequence is illustrated in [26].

Please note that the exploration process also reveals valuable shape information of the object as illustrated by the contact point cloud in Fig. 14.12a. If object manipulation is continued, more and more contact points are sampled and can serve as a basis to reconstruct the object shape [30].

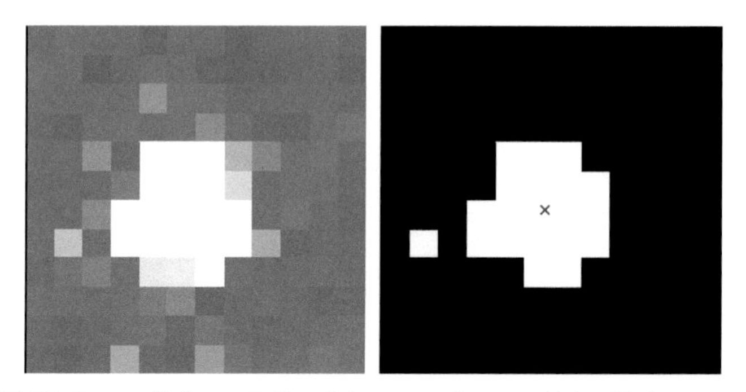

FIGURE 14.13 Raw tactile image (left) and the extracted contact blob with the center of gravity denoting the contact location (right).

14.6 Evaluation in a real robot experiment

The key components of our proposed manipulation framework are the two capabilities of (1) a single finger to realize the hybrid position+force controller sliding across the unknown surface of the object and (2) two fingers locally manipulating the unknown object, while parsimoniously making use of tactile sensor information. Although appropriately miniaturized tactile sensors for our Shadow Robot Hands are not yet available, we attempt to prove the feasibility of our framework in the real robot experiments using the larger tactile sensor array [48] mounted on the Kuka lightweight robot arm [14] as the big fingertip.

The tactile sensor module comprises an array of 16×16 tactels providing normal contact force information with 12-bit resolution. The distance of neighbor tactels is 5 mm and the sampling frequency is as high as 1.9 kHz. In our case, we want to control the contact position/force on the sensor. To this end we employ blob extraction methods from computer vision (ICL, [8]) to estimate the contact region and subsequently compute the contact position as the center of gravity of all force measurements within the blob (cf. Fig. 14.13). The contact force is defined as the average value of all tactel measurements. Regardless of the coarse sensor spacing, we achieve a spatial resolution of 0.5–1.0 mm due to this ensemble averaging. More detailed computing and calibration about the estimated contact position/force can be found in [27]. Together with the high frame rate, the sensor is thus suitable for real-time robot control.

14.6.1 Unknown object surface exploration by one finger

In the experiment, the Kuka robot arm is operated in joint-space compliance mode at 250 Hz. For all joints, the stiffness parameters are 200 Nm/rad and the damping parameters are 0.2 Nm·s/rad. The single-finger experimental setup is shown in Fig. 14.14. An error-reducing motion of the sensor array can be easily calculated in its local reference frame: Contact position errors are corrected by an appropriate tangential motion of the sensor, while normal force errors are

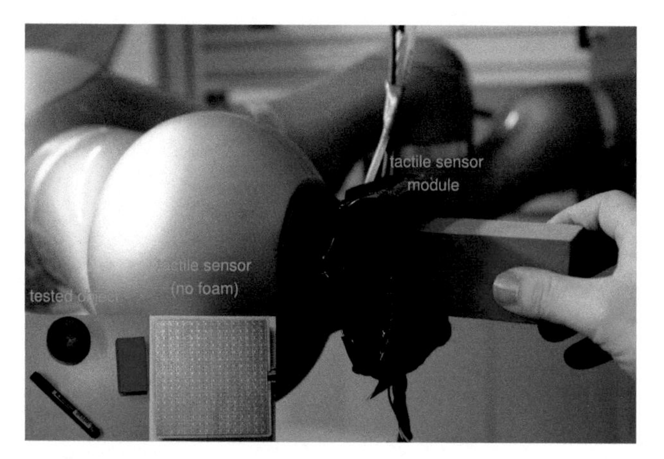

FIGURE 14.14 Experiment scenario and tested object.

corrected by a motion normal to the contact sensor plane. For both controllers (position and force) a PID-type controller is employed. Using an inverse kinematics solver the resulting Cartesian end-effector motion is transformed into the corresponding joint-space motion. The robot arm control is realized using the control basis framework (CBF, [43]) providing the inverse kinematics solver and the open Kuka control software (OpenKC, [47,46]).

14.6.1.0.1 Normal force control

In the first experiment we only control the normal force applied to the object. As the quality of force control heavily depends on the stiffness of objects (softer objects allow for a larger motion range given a fixed force range) we evaluated the control performance on various objects of different stiffnesses: a rigid pen, a toy box from rigid foam, and a soft ball. The results for maintaining a desired pressure level of $p = 1$ (2 kPa or 2 N spread over an area of 40 tactels) are shown in Fig. 14.15. As expected, stiffer objects take longer to converge to a stable tracking result (response time) and exhibit stronger force oscillations given similar deflections. However, in all cases the desired force level will eventually be well maintained, as shown in Table 14.1. The steady-state error and standard deviation are computed from the time series starting after the response time and lasting until 20 s. All values are obtained by averaging over 20 trials.

14.6.1.0.2 Contact position control

In the second experiment, we focus on contact position tracking and neglect the applied normal force. The goal is to maintain the center of gravity (COG) of the contact region at the center of the tactile sensor frame. The experimental results are shown in Fig. 14.16. As can be seen from the top subfigure, an initial position offset is corrected within less than 2 s. The steady-state error and response time

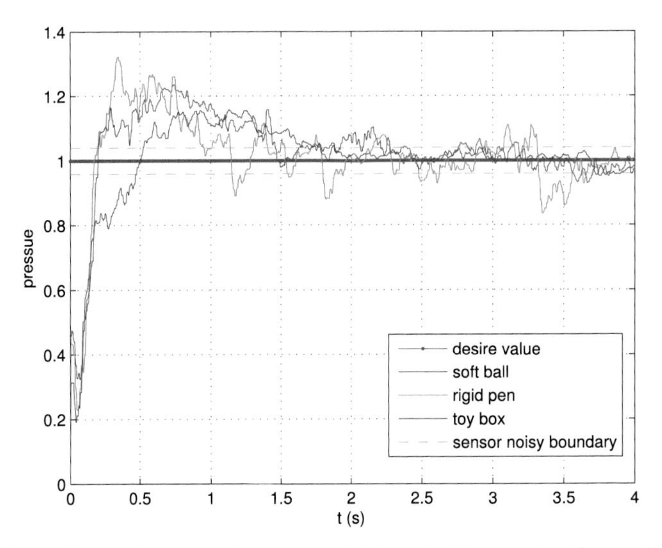

FIGURE 14.15 Normal force control on objects of different stiffnesses. Tracking results get close to the sensor's noise level of 0.04.

TABLE 14.1 Normal force control: tracking results.

Object	Steady-state error	Standard deviation	Response time
Rigid pen	0.0032	0.039	2.5 s
Toy box	0.0026	0.039	2 s
Soft ball	0.0010	0.043	2 s

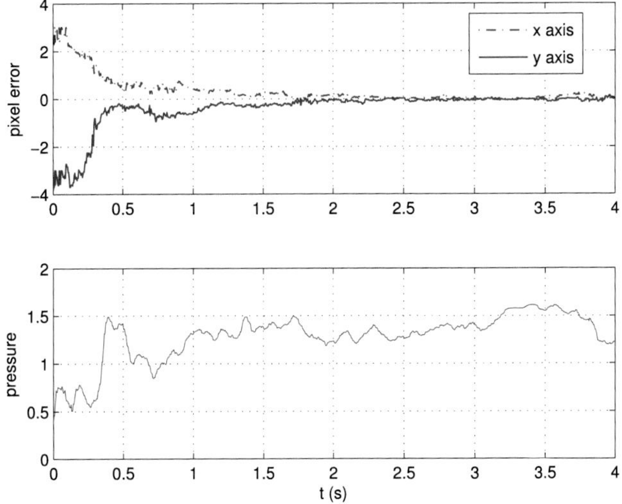

FIGURE 14.16 Contact position control. Above: Position error. Below: Normal force evolution.

TABLE 14.2 Contact position control: tracking results.

Axis	Steady-state error	Standard deviation	Response time
X	−0.0027 pixel	0.0440 pixel	2 s
Y	−0.0406 pixel	0.0509 pixel	2 s

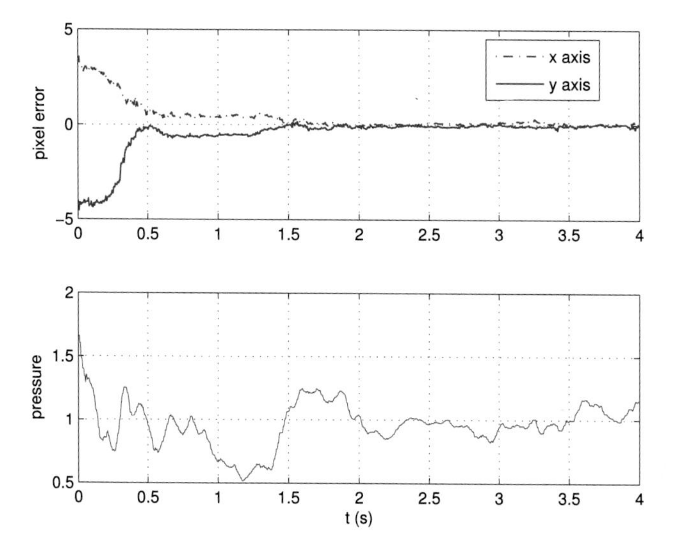

FIGURE 14.17 Hybrid position and force control.

are summarized in Table 14.2. As can be seen from the bottom subfigure, the normal force applied in this experiment evolves randomly as it is not controlled. Note that a large normal force – due to friction – will also cause large tangential forces, rendering the sliding motion more difficult.

14.6.1.0.3 Simultaneous position and force control

Finally, we combine both controllers to evaluate the more realistic scenario of contact location tracking while maintaining a given small contact force as required by the surface exploration controller. The evolution of the errors in contact position and force are shown in Fig. 14.17. We obtain similar position tracking results as before (upper subfigure) and the force tracking result is again within the noise level of the sensor. Quantitative results are summarized in Table 14.3.

Using the hybrid position+force controller as the control basis, we can realize the more complex surface exploration experiment, where the sensors roll and slide across an unknown object surface [27].

TABLE 14.3 Hybrid position/force control: tracking results.

Axis	Steady-state error	Standard deviation	Response time
X	0.0041 pixel	0.1146 pixel	1.8 s
Y	0.0082 pixel	0.1158 pixel	1.8 s
Force	−0.0014	0.1335	2 s

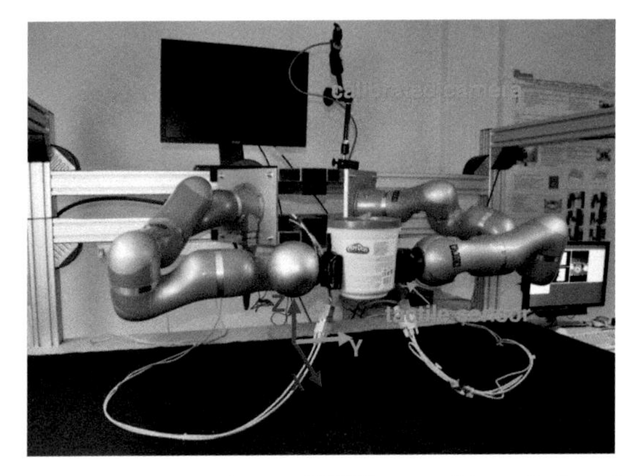

FIGURE 14.18 Experimental setup using two KUKA arms with attached tactile sensor arrays as large fingertips.

14.6.2 Unknown object local manipulation by two fingers

We evaluate the method using two KUKA LWR arms, each equipped with a tactile sensor module. Thus, both arms act as two large fingers with tactile-sensitive fingertips.

The information about the shape, size, and friction of the manipulated object is not available to the robot. The experimental setup is shown in Fig. 14.18. The camera observes the object from the top. The vision feedback frequency is 30 Hz, and we limit the tactile feedback to the same frame rate. They are both processed by a smoothing filter averaging within a window of 20 frames. Proprioceptive feedback and joint angle control rates are both fixed at 125 Hz. The detailed arm-eye and tactile calibration process can be found in [18]. We extract joint angle measurements and send joint control commands via the KUKA FRI interface [44,46]. All controller gains are manually tuned to guarantee the stability of controllers in all manipulation experiments.

The whole manipulation process comprises three phases:

1. Vision-guided grasping of the object.
2. Moving along the world's Z-, X-, and Y-axes in sequence.
3. Rotating around the Z- and X-axes.

Each phase is described in detail in the following.

14.6.2.0.1 Vision-guided grasping

The first stage is to guide the arms to contact the object, exert the planned grasp force (see Section 14.3.2), and hold the object. The force planner – originally designed for multifingered hands – has been simplified for the two-arm scenario: While the contact force vectors still point towards the centroid of contact locations (see Fig. 14.2), their magnitudes are set equally to a predefined value.

To establish object contact, we apply a simple, hard-coded opposition strategy: Starting from the estimated pose (x, y, z) of the marker attached to the object we attempt to drive both fingertips, i.e., arm end-effectors, to the virtual grasp point $(x, y, z - z_0)$ slightly below the marker frame, where the offset z_0 is a constant determined by the size of the tactile sensor module. The approaching motion of both arms is stopped as soon as contact to the object is detected by the tactile sensor.

The coarse calibration of the tactile sensors does not provide accurate enough force feedback to stably hold the object with a pure force-feedback controller. Rather, the object will slowly drift away. However, exploiting visual feedback about the object position too, the composite position/force controller (see Section 14.3.3) successfully accomplishes the grasping task.

14.6.2.0.2 Translating motion

In the first experiment, the object is moved 10 cm along the world's z-, x-, and y-axes in sequence. The resulting trajectories for force and positional errors are shown in Fig. 14.19. As can be seen from the deflections in the bottom subfigure, a new target pose was set after 5, 22, and 35 seconds. In all cases the positional error quickly decays to the noise level.

Considering the force error trajectories, we see that the motion along the y-axis generates most deviations. This is because this motion direction is parallel to the contact normal, thus heavily demanding the composite force/position controller. As soon as the positional error along the y-axis stabilizes around zero, also the force error starts to decay.

14.6.2.0.3 Rotating motion

Secondly, we will show how the object *orientation* can be controlled. In each experiment, firstly the object will be lifted 10 cm along the z-axis before being rotated around the z- and x-axes. The results of both experiments are shown in Figs. 14.20 and 14.21. Again, the positional and rotational errors quickly decay after setting a new target pose. However, rotational errors are corrected more slowly due to a more conservative choice of controller gains.

Looking at the force error trajectories, we observe that the errors do not completely decay anymore. This is due to the fact that the physically applied force direction is not normal to the sensor surface anymore. However, the sensor

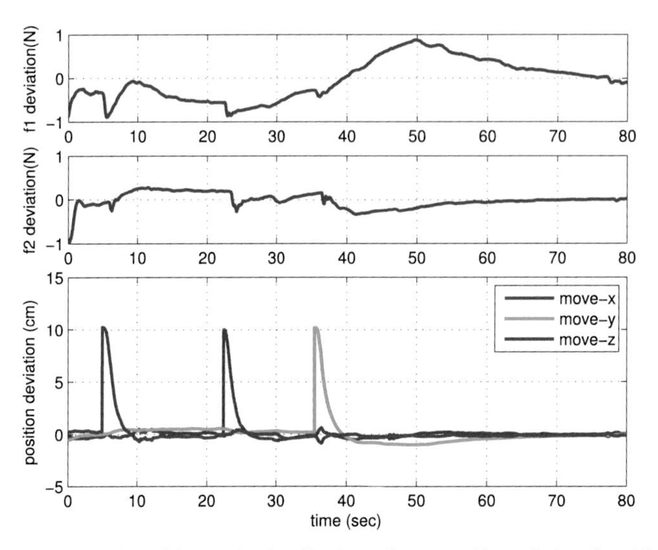

FIGURE 14.19 Evolution of force (f_1, f_2 of both tactile sensors/fingertips) and positional errors during translating motion; x, y, and z errors are mapped to red, green, blue colors, respectively.

only measures normal forces. Again, incorporating visual feedback and employing the composite position/force controller we can realize stable object rotation nevertheless.

Using our proposed manipulation strategy, many objects which have a different contact geometry with fingertip can be manipulated. For example, it is possible to generate the across-edge behavior during the course of manipulating one cubic-shaped object. All experiments are also shown in [19].

14.7 Summary and outlook

We proposed a visuotactile feedback-based controller framework for in-hand manipulation of unknown objects. This framework allows (1) to realize small-scale object motions based on parsimonious prior assumptions about the manipulated object and (2) to slide a finger across an object's surface simultaneously estimating and optimizing a given smooth objective function to realize the optimization regrasp. Combining both capabilities it was shown that large-scale object motions can be realized by alternating small object motions and finger repositioning, thus circumventing natural joint- and work-space boundaries.

Our controller framework makes as little use of prior knowledge as possible. Neither the object shape nor detailed contact properties need to be known. The controller relies on tactile feedback to obtain the current contact point, contact normal (which typically is known from the local finger geometry), and contact force magnitude. Vision feedback can be employed to track the object pose. As soon as sufficiently sensitive tactile sensors become available for multifingered

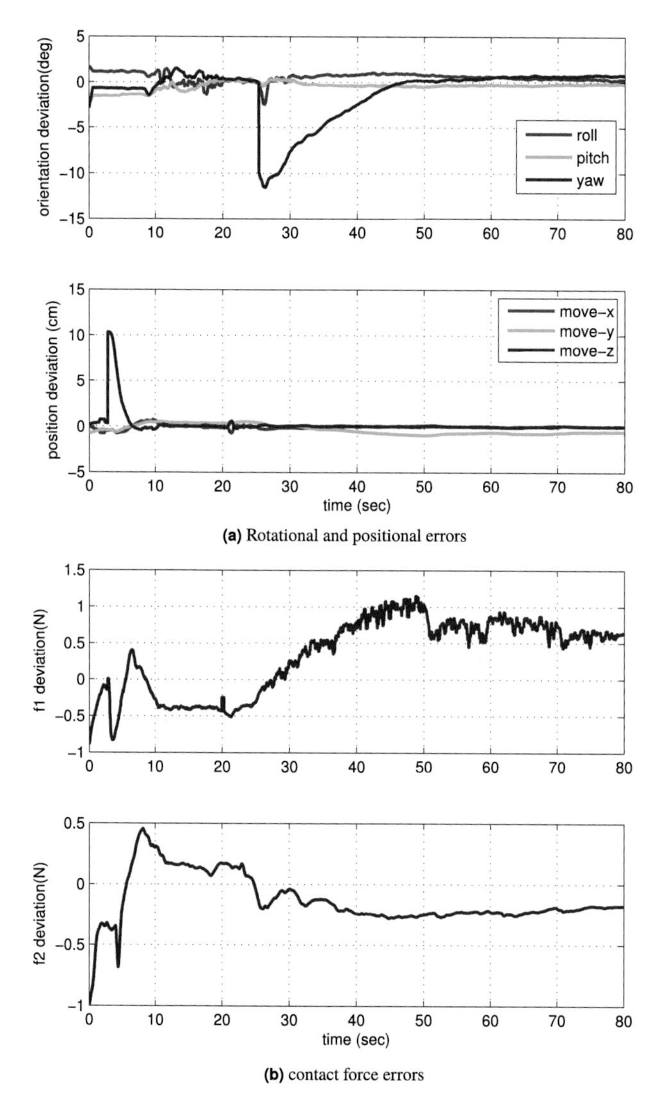

FIGURE 14.20 Error evolution rotating around world's z-axis.

robot hands, like the Shadow Robot Hand, the proposed algorithm can be easily transferred to real-world scenarios.

In the meantime we have proven the feasibility of the approach using the tactile sensor array mounted on the robot arm as the large fingertip to implement the unknown object surface local area exploration and two fingers' local manipulation. Due to its closed-loop characteristics, the approach is very robust to sensor noise and small deviations from the planned motion, which was demonstrated in both physics-based simulation and real-world experiments.

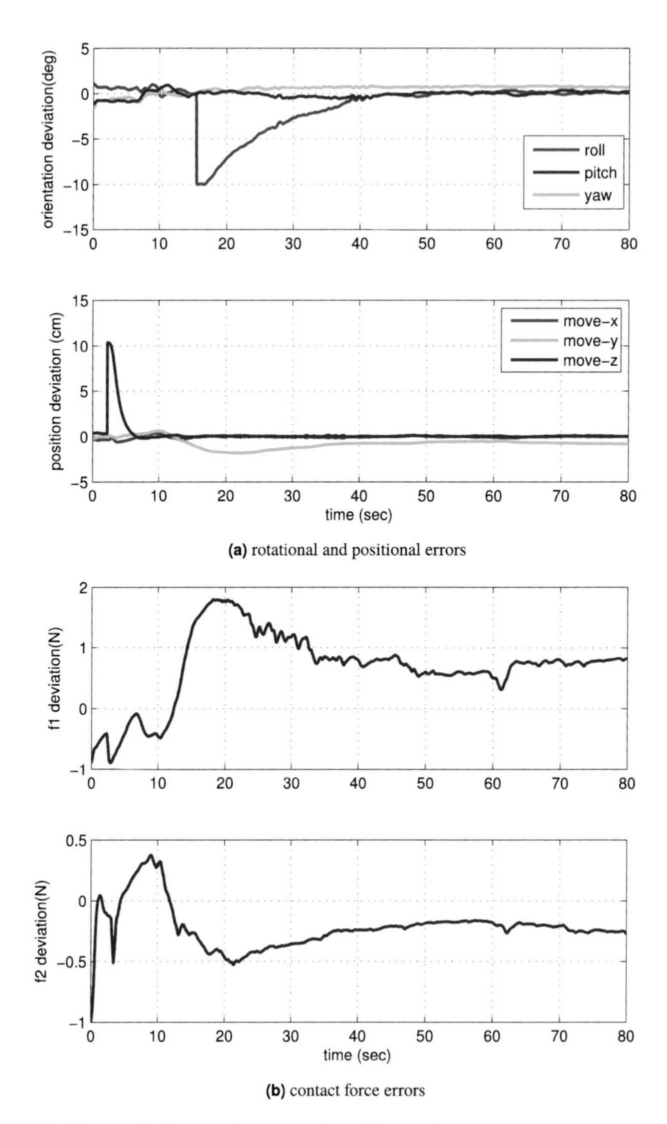

FIGURE 14.21 Error evolution rotating around world's x-axis.

In the future, we will transfer and evaluate our manipulation strategy to real anthropomorphic robot hands covered with tactile sensors as skin and deal with in-hand manipulation of more complex objects.

Acknowledgment

This work is supported in part by the "DEXMAN" project (Project number: 410916101) funded by the Deutsche Forschungsgemeinschaft (DFG, German Research Foundation) and

"Autonomous Exploration of Manual Interaction Space" supported from Honda Research Institute Europe. Qiang Li gratefully acknowledges the technical help from Christof Elbrechter, Martin Meier, Matthias Schöpfer, and Carsten Schürmann during the preparation of the robot experiments.

References

[1] OpenAI: Marcin Andrychowicz, Bowen Baker, Maciek Chociej, Rafal Jozefowicz, Bob McGrew, Jakub Pachocki, Arthur Petron, Matthias Plappert, Glenn Powell, Alex Ray, et al., Learning dexterous in-hand manipulation, The International Journal of Robotics Research 39 (1) (2020) 3–20.

[2] M. Buss, T. Schlegl, Multi-fingered regrasping using on-line grasping force optimization, in: Robotics and Automation, 1997. Proceedings, 1997 IEEE International Conference on, vol. 2, IEEE, 1997, pp. 998–1003.

[3] J. Chen, M. Zribi, Control of multifingered robot hands with rolling and sliding contacts, The International Journal of Advanced Manufacturing Technology 16 (1) (2000) 71–77.

[4] Anna Lisa Ciancio, Loredana Zollo, Eugenio Guglielmelli, Daniele Caligiore, Gianluca Baldassarre, Hierarchical reinforcement learning and central pattern generators for modeling the development of rhythmic manipulation skills, in: Development and Learning (ICDL), 2011 IEEE International Conference on, vol. 2, IEEE, 2011, pp. 1–8.

[5] Jefferson A. Coelho, Roderic A. Grupen, A control basis for learning multifingered grasps, Journal of Robotic Systems 14 (7) (1997) 545–557.

[6] A.B.A. Cole, J.E. Hauser, S.S. Sastry, Kinematics and control of multifingered hands with rolling contact, Automatic Control, IEEE Transactions on 34 (4) (1989) 398–404.

[7] Hao Dang, J. Weisz, P.K. Allen, Blind grasping: stable robotic grasping using tactile feedback and hand kinematics, in: Robotics and Automation (ICRA), 2011 IEEE International Conference on, May 2011, pp. 5917–5922.

[8] Christof Elbrechter, Image component library (ICL), http://www.iclcv.org/, 2012.

[9] C. Ferrari, J. Canny, Planning optimal grasps, in: Robotics and Automation, 1992. Proceedings, 1992 IEEE International Conference on, IEEE, 1992, pp. 2290–2295.

[10] L. Han, J.C. Trinkle, Z.X. Li, Grasp analysis as linear matrix inequality problems, Robotics and Automation, IEEE Transactions on 16 (6) (2000) 663–674.

[11] R. Haschke, J.J. Steil, I. Steuwer, H. Ritter, Task-oriented quality measures for dextrous grasping, in: Computational Intelligence in Robotics and Automation, 2005. CIRA 2005. Proceedings. 2005 IEEE International Symposium on, June 2005, pp. 689–694.

[12] T. Ishihara, A. Namiki, M. Ishikawa, M. Shimojo, Dynamic pen spinning using a high-speed multifingered hand with high-speed tactile sensor, in: Humanoid Robots, IEEE International Conference on, 2006, pp. 258–263.

[13] Kapil D. Katyal, Edward W. Staley, Matthew S. Johannes, I-Jeng Wang, Austin Reiter, Phillipe Burlina, In-hand robotic manipulation via deep reinforcement learning, in: Proceedings of the Workshop on Deep Learning for Action and Interaction, in Conjunction with Annual Conference on Neural Information Processing Systems, Barcelona, Spain, vol. 9, 2016.

[14] Kuka, Kuka roboter gmbh, http://www.kuka-robotics.com/en/products/addons/lwr/, 2012.

[15] Vikash Kumar, Abhishek Gupta, Emanuel Todorov, Sergey Levine, Learning dexterous manipulation policies from experience and imitation, arXiv preprint, arXiv:1611.05095, 2016.

[16] Yuichi Kurita, Jun Ueda, Yoshio Matsumoto, Tsukasa Ogasawara, Cpg-based manipulation: generation of rhythmic finger gaits from human observation, in: Robotics and Automation, 2004. Proceedings. ICRA'04. 2004 IEEE International Conference on, vol. 2, IEEE, 2004, pp. 1209–1214.

[17] CM LABS, Vortex physics simulation, http://www.vxsim.com/, 2012.

[18] Qiang Li, Christof Elbrechter, Robert Haschke, Helge Ritter, Integrating vision, haptics and proprioception into a feedback controller for in-hand manipulation of unknown objects, in: Intelligent Robots and Systems. IEEE International Conference on, 2013.

[19] Qiang Li, Robert Haschke, Helge Ritter, Video of in-hand manipulation with kuka lwr, https://www.youtube.com/watch?v=UgCv5ESAYfc&t=5s, 2013.
[20] Qiang Li, Robert Haschke, Helge Ritter, A visuo-tactile control framework for manipulation and exploration of unknown objects, in: 2015 IEEE-RAS 15th International Conference on Humanoid Robots (Humanoids), IEEE, 2015, pp. 610–615.
[21] Qiang Li, Robert Haschke, Helge Ritter, Bram Bolder, Simulation results for manipulation of unknown objects in hand, in: Robotics and Biomimetics, IEEE International Conference on, 2011.
[22] Qiang Li, Robert Haschke, Helge Ritter, Bram Bolder, Video of local manipulation of unknown object, https://sites.google.com/site/qiangliresearch/qiang-li-s-homepage/my-project/manual-intelligence/unknown-object-reactive-manipulation, 2011.
[23] Qiang Li, Robert Haschke, Helge Ritter, Bram Bolder, Video of rotary object continuous rotation, https://sites.google.com/site/qiangliresearch/qiang-li-s-homepage/my-project/manual-intelligence/largescaleobjectmanip, 2011.
[24] Qiang Li, Robert Haschke, Helge Ritter, Bram Bolder, Rotary surface object manipulation by multifingered robot hand, in: 7th German Conference for Robotics, 2012.
[25] Qiang Li, Robert Haschke, Helge Ritter, Bram Bolder, Towards unknown objects manipulation, IFAC Proceedings Volumes 45 (22) (2012) 289–294.
[26] Qiang Li, Robert Haschke, Helge Ritter, Bram Bolder, Video of find optimized grasping points, https://sites.google.com/site/qiangliresearch/qiang-li-s-homepage/my-project/manual-intelligence/tactile-based-grasp-point-optimization, 2012.
[27] Qiang Li, Carsten Schürmann, Robert Haschke, Helge Ritter, A control framework for tactile servoing, in: Proc. Robotics: Science and Systems, 2013.
[28] Z. Li, S.S. Sastry, Task-oriented optimal grasping by multifingered robot hands, Robotics and Automation, IEEE Journal of 4 (1) (1988) 32–44.
[29] Emanuele Luberto, Yier Wu, Gaspare Santaera, Marco Gabiccini, Antonio Bicchi, Enhancing adaptive grasping through a simple sensor-based reflex mechanism, IEEE Robotics and Automation Letters 2 (3) (2017) 1664–1671.
[30] M. Meier, M. Schöpfer, R. Haschke, H. Ritter, A probabilistic approach to tactile shape reconstruction, Robotics, IEEE Transactions on 27 (99) (2011) 1–6.
[31] A.T. Miller, P.K. Allen, Examples of 3d grasp quality computations, in: Robotics and Automation, 1999. Proceedings. 1999 IEEE International Conference on, vol. 2, IEEE, 1999, pp. 1240–1246.
[32] A.T. Miller, P.K. Allen, Graspit! A versatile simulator for robotic grasping, Robotics & Automation Magazine, IEEE 11 (4) (2004) 110–122.
[33] Andrés Montaño, Raúl Suárez, Model-free in-hand manipulation based on commanded virtual contact points, in: 2019 24th IEEE International Conference on Emerging Technologies and Factory Automation (ETFA), IEEE, 2019, pp. 586–592.
[34] Y. Nakamura, K. Nagai, T. Yoshikawa, Dynamics and stability in coordination of multiple robotic mechanisms, The International Journal of Robotics Research 8 (2) (1989) 44–61.
[35] Yoshihiko Nakamura, Advanced Robotics: Redundancy and Optimization, 1st edition, Addison-Wesley Longman Publishing Co., Inc., Boston, MA, USA, 1990.
[36] T. Naniwa, S. Arimoto, V.-P. Vega, A model-based adaptive control scheme for coordinated control of multiple manipulators, in: Intelligent Robots and Systems '94. International Conference on, vol. 1, Sep 1994, pp. 695–702.
[37] T. Phoka, A. Sudsang, Contact point clustering approach for 5-fingered regrasp planning, in: Intelligent Robots and Systems, IEEE/RSJ International Conference on, IEEE, 2009, pp. 4174–4179.
[38] R.J. Platt Jr., Learning and generalizing control-based grasping and manipulation skills, PhD thesis, Citeseer, 2006.
[39] R.M. Murray, Z.X. Li, S.S. Sastry, A Mathematical Introduction to Robotic Manipulation, CRC Press, 1994.

[40] Frank Röthling, R. Haschke, Jochen J. Steil, Helge J. Ritter, Platform portable anthropomorphic grasping with the Bielefeld 20-dof shadow and 9-dof tum hand, in: Proc. Int. Conf. on Intelligent Robots and Systems (IROS), San Diego, California, USA, Oct 2007, pp. 2951–2956.
[41] J.P. Saut, A. Sahbani, S. El-Khoury, V. Perdereau, Dexterous manipulation planning using probabilistic roadmaps in continuous grasp subspaces, in: Intelligent Robots and Systems, IEEE International Conference on, 2007, pp. 2907–2912.
[42] T. Schlegl, M. Buss, T. Omata, G. Schmidt, Fast dextrous re-grasping with optimal contact forces and contact sensor-based impedance control, in: Robotics and Automation, 2001. Proceedings 2001 ICRA. IEEE International Conference on, vol. 1, IEEE, 2001, pp. 103–108.
[43] Florian Schmidt, Control basis framework, https://github.com/fps/CBF, 2011.
[44] Günter Schreiber, Andreas Stemmer, Rainer Bischoff, The fast research interface for the kuka lightweight robot, in: IEEE Workshop on Innovative Robot Control Architectures for Demanding (Research) Applications How to Modify and Enhance Commercial Controllers (ICRA 2010), 2010.
[45] Carsten Schürmann, R. Haschke, Helge J. Ritter, Modular high speed tactile sensor system with video interface, in: Tactile Sensing in Humanoids – Tactile Sensors and Beyond @ IEEE-RAS Conference on Humanoid Robots (Humanoids), Paris, France, 2009.
[46] M. Schöpfer, F. Schmidt, M. Pardowitz, H. Ritter, Open source real-time control software for the kuka light weight robot, in: Intelligent Control and Automation (WCICA), 2010 8th World Congress on, IEEE, 2010, pp. 444–449.
[47] Matthias Schöpfer, Open kuka control, http://opensource.cit-ec.de/projects/openkc, 2011.
[48] C. Schürmann, R. Koiva, R. Haschke, H. Ritter, A modular high-speed tactile sensor for human manipulation research, in: World Haptics Conference (WHC), 2011, IEEE, June 2011, pp. 339–344.
[49] Wenceslao Shaw-Cortez, Denny Oetomo, Chris Manzie, Peter Choong, Tactile-based blind grasping: a discrete-time object manipulation controller for robotic hands, IEEE Robotics and Automation Letters 3 (2) (2018) 1064–1071.
[50] Avishai Sintov, Andrew Kimmel, Bowen Wen, Abdeslam Boularias, Kostas Bekris, Tools for data-driven modeling of within-hand manipulation with underactuated adaptive hands, in: Learning for Dynamics and Control, PMLR, 2020, pp. 771–780.
[51] R. Suárez, M. Roa, J. Cornella, Grasp quality measures, Technical Report, Universitat Politecnica de Catalunya (UPC), 2006.
[52] K. Tahara, S. Arimoto, M. Yoshida, Dynamic object manipulation using a virtual frame by a triple soft-fingered robotic hand, in: Robotics and Automation, IEEE International Conference on, 2010, pp. 4322–4327.
[53] H. van Hoof, T. Hermans, G. Neumann, J. Peters, Learning robot in-hand manipulation with tactile features, in: 2015 IEEE-RAS 15th International Conference on Humanoid Robots (Humanoids), Nov 2015, pp. 121–127.
[54] T. Wimboeck, C. Ott, G. Hirzinger, Passivity-based object-level impedance control for a multifingered hand, in: Intelligent Robots and Systems, 2006 IEEE/RSJ International Conference on, IEEE, 2006, pp. 4621–4627.
[55] Z. Xue, J.M. Zollner, R. Dillmann, Dexterous manipulation planning of objects with surface of revolution, in: Intelligent Robots and Systems, IEEE International Conference on, 2008, pp. 2703–2708.
[56] M. Yashima, Manipulation planning for object re-orientation based on randomized techniques, in: Robotics and Automation, IEEE International Conference on, vol. 2, 2004, pp. 1245–1251.
[57] Henry Zhu, Abhishek Gupta, Aravind Rajeswaran, Sergey Levine, Vikash Kumar, Dexterous manipulation with deep reinforcement learning: efficient, general, and low-cost, in: 2019 International Conference on Robotics and Automation (ICRA), IEEE, 2019, pp. 3651–3657.

Chapter 15

Robust dexterous manipulation and finger gaiting under various uncertainties

Yongxiang Fan
FANUC Advanced Research Laboratory, FANUC America Corporation, Union City, CA, United States

15.1 Introduction

In-hand manipulation has broad potential applications in industry and agriculture. Fig. 15.1 (A) shows an industrial circuit board assembly task, where a circuit board is picked and placed onto a platform before it is regrasped from a different orientation in order to perform the final insertion procedure. Fig. 15.1 (B) shows a fruit packaging task, where each fruit has to be rotated and arranged in certain orientation. By properly manipulating the object in hand, in-hand manipulation can simplify the procedure and reduce cycle time.

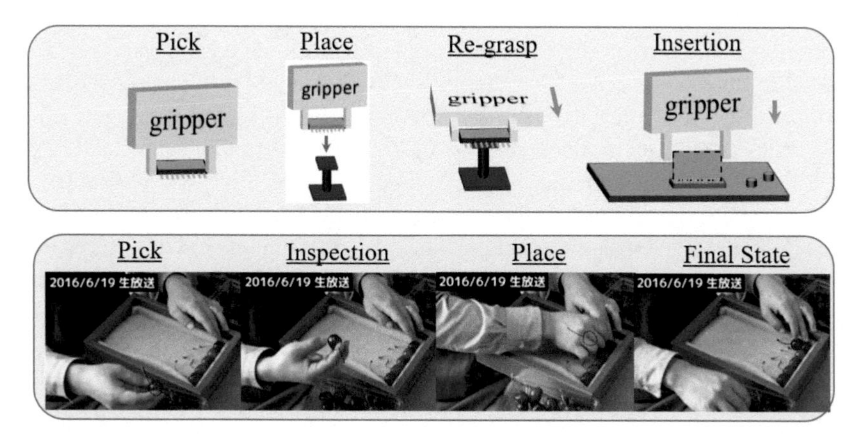

FIGURE 15.1

Achieving general in-hand manipulation is challenging. First, a robotic hand has to manipulate objects with various types of uncertainties shown in Fig. 15.2.

Type I uncertainty is caused by inaccurate 3D model perception and density distribution of the object. It is separated into object dynamics uncertainty and

Tactile Sensing, Skill Learning, and Robotic Dexterous Manipulation
https://doi.org/10.1016/B978-0-32-390445-2.00024-6

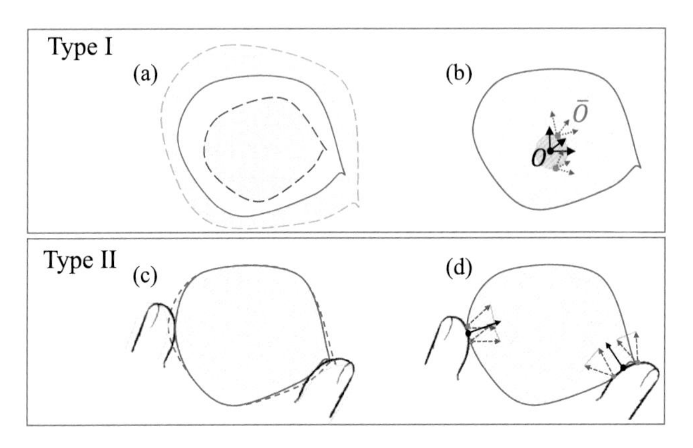

FIGURE 15.2

center of mass (COM) uncertainty, as shown in Fig. 15.2 (A) and Fig. 15.2 (B). The uncertain terms in object dynamics uncertainty include both mass and moment of inertia, which are parameters of the inertia matrix, Coriolis, and gravity of the object. The uncertain terms in COM uncertainty include COM position and principal axes of the object, which influence the object dynamics and the force transformation from contact to object. The black frame in Fig. 15.2 (B) shows the actual COM and principal axes, and red frames show possible variations. Type II uncertainty is caused by the inaccurate sensing and modeling of contacts under various surface properties. It is separated into contact dynamics uncertainty and tactile uncertainty, as shown in Fig. 15.2 (C) and Fig. 15.2 (D). The uncertain terms in contact dynamics uncertainty include stiffness, damping, coefficients of Coulomb friction, torsional friction, and rolling friction, which influence the force transformed onto the object. The uncertain terms in tactile uncertainty include contact position and orientation, which affect the contact force transformation and joint torque generation and may cause problems such as friction cone constraints violation and unexpected slippage. The black ones in Fig. 15.2 (D) show actual position/normal, while the red ones show the variations. Besides the disturbances from the contact dynamics, there might be unexpected contact and external perturbations from the environment. The controller should be designed with disturbance rejection ability.

Moreover, to perform long-range complex manipulations, the robotic hand may have to change its grasping status by relocating fingers during the manipulation, which gives the hand more dexterity and robustness. Such strategy is called finger gaits planning. However, the optimization of the finger gaiting under complicated grasp quality metrics is computationally expensive [1]. The optimization searches optimal contact points on a nonlinear object surface by maximizing object stability and hand manipulability metrics [2]. These two metrics are represented in different spaces, and associated by nonlinear forward kinematics with a high degree of freedoms (DoFs). The optimization becomes more challenging when the object 3D surface model is not available.

To solve the first challenge, a robust controller for contact uncertainties was proposed in [3]. The controller is designed for a linear time-invariant (LTI) system linearized around an equilibrium point. A force-position controller using 6D tactile sensors was implemented to realize adaptive grasping [4]. Nonlinearities were ignored due to its constant-pose grasping property. In order to consider parameter variations caused by nonlinearities, a linear parameter-varying (LPV) control with smooth scheduling was applied in [5], assuming that the nonlinearities can be approximated through linear varying parameters. To deal with dynamics uncertainties, a disturbance observer (DOB) was proposed in [6] for tracking control. The nonlinearities and parameter uncertainties are lumped into a disturbance term. It assumes full state feedback, while in dexterous hands, the velocity feedback is difficult due to the size constraints and cost issues. Feedback linearization was applied to control an unmanned aerial vehicle [7]. A linear state observer and a DOB are combined to observe the state and the lumped disturbance. Similar to [6], the structures of the parameter uncertainties are ignored, and the linear state observer assumes a perfect model for state estimation.

To have large-scale object motion, a task-specific finger gaiting policy was trained in [8] by the covariance matrix adaptation method, given the goal states of objects. However, the learned policies cannot be adapted to other objects and tasks. A high-speed hand and a high-speed vision system were applied in [9] to perform dynamic regrasping. However, the object model should be precisely known, and the presented success rate (35%) is not suitable for many applications. Impedance parameters were learned from human demonstration for robust grasping and dexterous manipulation in [10]. A tangle topology was used in [11] to reproduce object pose from learned human demo. However, the object gravity is not considered during their gait-changing process. A set of controllers is used in [12] to realize unknown object grasping by sliding on the surface to maximize grasp stability. The unknown surface of the object was explored in [13,14] by designing a global regrasping planner and searching local optimal contact points. However, predefined finger gaits are used in these approaches, and the exploration of local optima does not incorporate necessary constraints, such as joint velocity and acceleration limitations. As a result, these approaches tend to be slow in regrasping and manipulation, and the predefined finger gaits might be inapplicable to other objects and robotic hands. A sampling-based method was proposed in [15] to plan finger gaits. In [14], a contact-invariant optimization method was used to compute the states of the hand and the object, given the high-level goals. These approaches are not computationally efficient for real-time finger gaits planning. A grasp point optimization method was introduced in [16] for in-hand manipulation. To achieve continuous object motion, a global regrasp planner was defined to relocate fingers by estimating the best sliding directions. However, the performance of the planner is unknown on complex objects with uncertainties.

This chapter introduces a comprehensive architecture for dexterous in-hand manipulation. The architecture consists of a robust manipulation controller (RMC) for robust manipulation under various uncertainties and external disturbances and a finger gaits planner to relocate fingers away from joint limits in order to achieve long-range object motion.

The proposed RMC is formulated as a robust control and a contact force optimization. First, the nonlinearities are reduced by feedback linearization on a nominal model. Compared with LPV that assumes linear variations of parameters [5], the proposed method is more computationally efficient for broad-scale manipulations. Second, the robust controller is formulated as a μ-synthesis problem, and the structures of the uncertainties are considered by descriptor form, instead of treating uncertainties as a lumped disturbance [6], which results in information loss and a larger disturbance to resist. Moreover, by using the contact force optimization, the complicated contact modeling is bypassed and the contact force is regulated.

To handle the long-range object movement, we further propose a finger gaits planner to relocate the finger that deteriorates the manipulability while keep the remaining fingers moving the object. The finger gaits planner is formulated at the velocity level, instead of the position level. At each time step, the optimal joint velocities are computed to improve the hand manipulability as well as the object grasp quality, and the computed joint velocities are fed into motors by a velocity-force controller. Instead of optimizing the sliding direction from either manipulability or grasp quality as [16], we jointly optimize both qualities at high frequency without delay. The low-level velocity-force controller utilizes velocity-force control to detect the object surface and searches motions in tangent space.

The contributions of this chapter are as follows. (1) RMC is robust to large uncertainties and disturbances. It can handle 50% mass and 80% moment of inertia uncertainties, $(10, 10, 15)$-mm COM deviations, and $(0.3, 0.3, 0.3)$-rad principal axis variations, is robust to 10-fold differences in stiffness, and can withstand $(5, 5, 5)$-mm contact position and $(0.3, 0.3, 0.3)$-rad orientation variations. (2) RMC bypasses complex modeling of the contact dynamics. Instead, the robust controller is able to handle the contact dynamic uncertainties. (3) The velocity-level finger gaits planner is cast into a linear programming (LP), which is computationally efficient and can be solved in real-time (<1 ms). (4) The velocity-level finger gaits planner does not require precise 3D reconstruction for exact object surface modeling.

The remainder of this chapter is organized as follows. Section 15.2 shows the overall dual-stage optimization-based planner framework. Section 15.3 and Section 15.4 introduce the dynamics of the hand-object system and the RMC to resist uncertainties and external disturbances. Section 15.5 introduces the finger gaits planner for finger relocation. Section 15.6 shows simulation and experiment results. Section 15.7 summarizes the chapter.

15.2 Dual-stage manipulation and gaiting framework

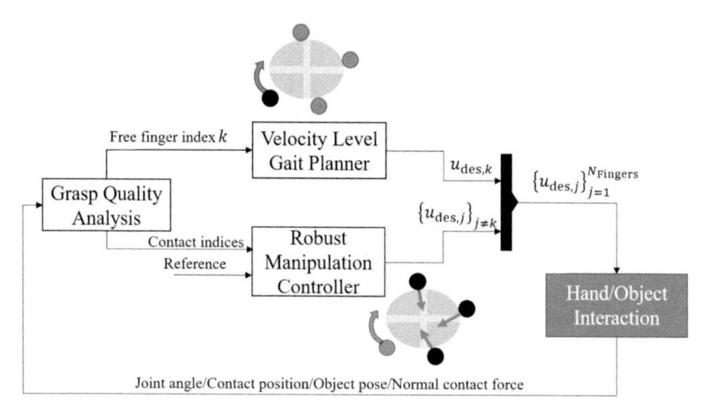

FIGURE 15.3

Fig. 15.3 shows the proposed dual-stage optimization-based planner framework. First, grasp quality analysis is conducted by combining hand manipulability and object grasp quality, and the free finger index k is chosen to change gait once the overall quality drops below a predefined threshold. A velocity-level finger gaits planner is evoked by this event, and the planner generates the control command $u_{\text{des,k}}$ to drive the selected finger towards the better quality region. The remaining fingers are controlled by an RMC to generate the desired control command $\{u_{\text{des,j}}\}_{j\neq k}$, to manipulate the object stably and track the reference motion of the object, as shown in Fig. 15.3. If the overall quality is above the threshold, all fingers will be controlled by RMC. The detailed introduction of RMC is shown in Section 15.4.

A joint-level tracking controller is used to track the desired control command $\{u_{\text{des, j}}\}_{j=1}^{N_{\text{finger}}}$, where N_{finger} denotes the number of fingers in simulation. The tracking controller uses a PID scheme and runs at a higher frequency in comparison with the finger gaits planner and the RMC (500 Hz). The control command may be of types torque, force, current, or velocity, depending on the hardware and controller design on hand. Since we do not focus on the low-level tracking control, we assume the hand joints track torque commands for the rest of this chapter unless otherwise specified.

15.3 Modeling of uncertain manipulation dynamics

15.3.1 State-space dynamics

The dynamics of the hand and object are

$$\begin{aligned} M_h(q)\ddot{q} + C_h(q,\dot{q})\dot{q} + N_h(q,\dot{q}) + J_h^T(q,x_o)f_c &= \tau, \\ M_o(x_o)\ddot{x}_o + C_o(x_o,\dot{x}_o)\dot{x}_o + N_o &= G(q,x_o)f_c, \end{aligned} \tag{15.1}$$

where $M_{h/o}$, $C_{h/o}$, and $N_{h/o}$ are inertia matrices, Coriolis matrices, and gravities for the hand/object, q, $\dot{q}$, and $\ddot{q} \in \mathbb{R}^{n_q}$ denote the joint angle, velocity, and acceleration, with n_q as the total DoFs of the hand, x_o, $\dot{x}_o$, and $\ddot{x}_o \in \mathbb{R}^{n_x}$ are a local parameterization of object position, velocity, and acceleration, where n_x is the dimension of the pose of the object, with $n_x = 6$ for 3D manipulation ($n_x = 3$ for 2D manipulation), $f_c \in \mathbb{R}^{d_c n_c}$ and $\tau \in \mathbb{R}^{n_q}$ are the contact force vector and joint torque vector, where d_c is the dimension of each contact and n_c is the contact number, $J_h \in \mathbb{R}^{(d_c n_c) \times n_q}$ is the hand Jacobian, and $G \in \mathbb{R}^{n_x \times (d_c n_c)}$ is the grasp map [2].

If the contacts are fixed with respect to both object and fingertips, then

$$J_h(q, x_o)\dot{q} = G^T(q, x_o)\dot{x}_o \tag{15.2}$$

holds. Eq. (15.2) assumes the contact forces remain in the friction cone. The object and hand dynamics in Eq. (15.1) can be connected by Eq. (15.2):

$$M(q, x_o)\ddot{x}_o + C(q, \dot{q}, x_o, \dot{x}_o)\dot{x}_o + N(q, x_o) = G J_h^{-T} \tau, \tag{15.3}$$

where

$$\begin{aligned} M &= M_o + G J_h^{-T} M_h J_h^{-1} G^T, \\ C &= C_o + G J_h^{-T} C_h J_h^{-1} G^T + G J_h^{-T} M_h \frac{d(J_h^{-1} G^T)}{dt}, \\ N &= N_o + G J_h^{-T} N_h. \end{aligned} \tag{15.4}$$

In some applications such as fruit harvesting, only the rough values of the mass m_o and the inertia $\mathcal{I}_o$ of the object can be estimated. Therefore, M_o, C_o, and N_o cannot be exactly known and would exhibit some uncertainties. Suppose that the inertia, Coriolis, and gravity can be represented as

$$M = \bar{M} + \tilde{M}_o, \quad C = \bar{C} + \tilde{C}_o, \quad N = \bar{N} + \tilde{N}_o \tag{15.5}$$

with nominal values

$$\begin{aligned} \bar{M} &= \bar{M}_o + G J_h^{-T} M_h J_h^{-1} G^T, \\ \bar{C} &= \bar{C}_o + G J_h^{-T} C_h J_h^{-1} G^T + G J_h^{-T} M_h \frac{d(J_h^{-1} G^T)}{dt}, \\ \bar{N} &= \bar{N}_o + G J_h^{-T} N_h, \end{aligned}$$

where $\bar{M}_o$, $\bar{C}_o$, and $\bar{N}_o$ are nominal object inertia, Coriolis, and gravity and $\tilde{M}_o$, $\tilde{C}_o$, and $\tilde{N}_o$ are corresponding uncertainties. The torque command τ can be related to the object-centered force F:

$$\tau = J_h^T (G^\dagger F + N_G \lambda), \tag{15.6}$$

where N_G is the matrix composed by the basis of the null space of G and λ is a free variable to control the magnitude and direction of the contact force.

The state space equation can be derived by plugging Eq. (15.5) and Eq. (15.6) into Eq. (15.3):

$$\left(\underbrace{\begin{bmatrix} \mathbb{I} & \mathbb{O} \\ \mathbb{O} & \bar{M} \end{bmatrix}}_{\bar{M}_{\text{aug}}} + \underbrace{\begin{bmatrix} \mathbb{O} & \mathbb{O} \\ \mathbb{O} & \tilde{M}_o \end{bmatrix}}_{\tilde{M}_{\text{aug}}}\right)\underbrace{\begin{bmatrix} \dot{x}_o \\ \ddot{x}_o \end{bmatrix}}_{\dot{x}} + \left(\underbrace{\begin{bmatrix} \mathbb{O} \\ \bar{N} \end{bmatrix}}_{\bar{N}_{\text{aug}}} + \underbrace{\begin{bmatrix} \mathbb{O} \\ \tilde{N}_o \end{bmatrix}}_{\tilde{N}_{\text{aug}}}\right)$$
$$+ \left(\underbrace{\begin{bmatrix} \mathbb{O} & -\mathbb{I} \\ \mathbb{O} & \bar{C} \end{bmatrix}}_{\bar{C}_{\text{aug}}} + \underbrace{\begin{bmatrix} \mathbb{O} & \mathbb{O} \\ \mathbb{O} & \tilde{C}_o \end{bmatrix}}_{\tilde{C}_{\text{aug}}}\right)\underbrace{\begin{bmatrix} x_o \\ \dot{x}_o \end{bmatrix}}_{x} \tag{15.7}$$
$$= \underbrace{\begin{bmatrix} \mathbb{O} \\ \mathbb{I} \end{bmatrix}}_{B_F} F,$$

where $\mathbb{I}, \mathbb{O} \in \mathbb{R}^{n_x \times n_x}$. Eq. (15.7) can be rewritten as

$$\dot{x} = -\bar{M}_{\text{aug}}^{-1}\bar{C}_{\text{aug}}x - \bar{M}_{\text{aug}}^{-1}\bar{N}_{\text{aug}} + \bar{M}_{\text{aug}}^{-1}B_F F - \bar{M}_{\text{aug}}^{-1}\tilde{M}_{\text{aug}}\dot{x} - \bar{M}_{\text{aug}}^{-1}\tilde{C}_{\text{aug}}x - \bar{M}_{\text{aug}}^{-1}\tilde{N}_{\text{aug}}. \tag{15.8}$$

In 3D manipulation, the parameters of Eq. (15.8) can be decomposed as

$$-\bar{M}_{\text{aug}}^{-1}\tilde{M}_{\text{aug}} = L_1 \boldsymbol{\Delta} R_1, \quad -\bar{M}_{\text{aug}}^{-1}\tilde{C}_{\text{aug}} = \sum_{j=1}^{2} L_{2j} \boldsymbol{\Delta} R_{2j} \tag{15.9}$$

when parameterizing the rotation matrix R of the object by Z-Y-X Euler angles E, with

$$\begin{aligned}
L_1 &= L_{21} = [0_{6\times6}; \bar{M}^{-1}] \times \text{diag}(I_{3\times3}, Q_E^T),\\
\boldsymbol{\Delta} &= \text{diag}(\delta_m I_{3\times3}, \delta_{\mathcal{I}_1}, \ldots \delta_{\mathcal{I}_3}) \quad \text{with} \|\boldsymbol{\Delta}\|_\infty \le 1,\\
R_1 &= -\text{diag}(\Delta m I_{3\times3}, \Delta\mathcal{I}) \times [0_{6\times6}, \text{diag}(I_{3\times3}, Q_E)],\\
R_{21} &= -\text{diag}(\Delta m I_{3\times3}, \Delta\mathcal{I}) \times [0_{6\times6}, \text{diag}(0_{3\times3}, \dot{Q}_E)],\\
L_{22} &= [0_{6\times6}; \bar{M}^{-1}] \times \text{diag}(I_{3\times3}, R(Q_E\dot{E})\hat{}\,),\\
R_{22} &= -\text{diag}(\Delta m I_{3\times3}, \Delta\mathcal{I}) \times [0_{6\times6}, \text{diag}(0_{3\times3}, Q_E)],
\end{aligned}$$

where $\Delta m \in \mathbb{R}$ and $\Delta\mathcal{I} = \text{diag}(\Delta\mathcal{I}_1, \ldots \Delta\mathcal{I}_3)$ are the maximal mass and inertia uncertainties, $Q_E \in \mathbb{R}^{3\times3}$ is a Jacobian matrix from Euler angle rate $\dot{E}$ to angular velocity of the object in body frame, and $(\bullet)^\wedge$ denotes the matrix representation of the cross product.

With Eq. (15.9), the uncertainty term $-M_{\text{aug}}^{-1}(\tilde{M}_{\text{aug}}\dot{x} + \tilde{C}_{\text{aug}}x)$ in Eq. (15.8) can be represented by

$$L_1 \mathbf{\Delta} \underbrace{(R_1\dot{x} + R_{21}x)}_{z_1} + L_{22}\mathbf{\Delta} \underbrace{R_{22}x}_{z_2} = L_1 \underbrace{\mathbf{\Delta} z_1}_{w_1} + L_{22} \underbrace{\mathbf{\Delta} z_2}_{w_2} = L_1 w_1 + L_{22} w_2. \tag{15.10}$$

2D manipulation is introduced below for illustration and comparison purposes. The Coriolis uncertainty can be eliminated by choosing the local parametrization as body frame translation and rotation angle. Thus L_{21}, R_{21} and L_{22}, R_{22} are removed, and

$$\begin{aligned} L_1 &= [0_{3\times 3}; \bar{M}^{-1}], \\ \mathbf{\Delta} &= \text{diag}(\delta_m I_{2\times 2}, \delta_{\mathcal{I}_3}) \quad \text{with} \|\mathbf{\Delta}\|_\infty \le 1, \\ R_1 &= [0_{3\times 3}, -\text{diag}(\Delta m I_{2\times 2}, \Delta \mathcal{I}_3)]. \end{aligned} \tag{15.11}$$

In general 3D manipulation, the Coriolis term is typically ignored due to the low-speed operation condition, as shown in [10].

The control input u is F, and the augmented gravity $\bar{N}_{\text{aug}}$ can be compensated by an additional control input $u_0 = \bar{N}_{\text{aug}}$. The gravity uncertainty $\tilde{N}_{\text{aug}}$ is considered as part of the disturbance u_{dis}. Then the uncertain state space model is represented as

$$\begin{aligned} \dot{x} &= \underbrace{-\bar{M}_{\text{aug}}^{-1}\bar{C}_{\text{aug}}}_{A} x + \underbrace{L_1}_{B_1} w_1 + \underbrace{\bar{M}_{\text{aug}}^{-1} B_F}_{B_2}(u - u_0 + u_{\text{dis}}), \\ z_1 &= C_1 x + \underbrace{R_1 L_1}_{D_{11}} w_1 + \underbrace{R_1 \bar{M}_{\text{aug}}^{-1} B_F}_{D_{12}}(u - u_0 + u_{\text{dis}}), \\ y &= \underbrace{[I_{3\times 3}, 0_{3\times 3}]}_{C_2} x, \qquad w_1 = \mathbf{\Delta} z_1, \end{aligned} \tag{15.12}$$

where $C_1 = -R_1 \bar{M}_{\text{aug}}^{-1} \bar{C}_{\text{aug}}$. Eq. (15.12) describes uncertainties by linear fractional transformation (LFT). Note though that the system is nonlinear, due to the state dependencies of the dynamics parameters.

15.3.2 Combining feedback linearization with modeling

A challenge in robust control is the implementation on nonlinear systems. Although some extensions have been made for LPV systems, the application of robust control to a general nonlinear system is still challenging.

To reduce the influence of nonlinearities, feedback linearization is applied to linearize the model. More specifically, for Eq. (15.8), the command force may be

$$F = \left(\bar{M}_{\text{aug}}^{-1} B_F\right)^{\dagger} \left[\bar{M}_{\text{aug}}^{-1} \bar{C}_{\text{aug}} x + \bar{M}_{\text{aug}}^{-1} \bar{N}_{\text{aug}} + B_F u\right]. \tag{15.13}$$

Note $\left(\bar{M}_{\text{aug}}^{-1} B_F\right)\left(\bar{M}_{\text{aug}}^{-1} B_F\right)^{\dagger} = [0_{3\times3}, 0_{3\times3}; 0_{3\times3}, I_{3\times3}]$, rather than identity. Therefore, Eq. (15.8) becomes

$$\begin{aligned} \dot{x} &= Ax + B_F u - \bar{M}_{\text{aug}}^{-1} \tilde{M}_{\text{aug}} \dot{x} - \bar{M}_{\text{aug}}^{-1} \tilde{C}_{\text{aug}} x - \bar{M}_{\text{aug}}^{-1} \tilde{N}_{\text{aug}}, \\ A &= \begin{bmatrix} 0_{3\times3} & I_{3\times3} \\ 0_{3\times3} & 0_{3\times3} \end{bmatrix}. \end{aligned} \tag{15.14}$$

Following the similar procedure as Eq. (15.9) and Eq. (15.11),

$$\begin{aligned} \dot{x} &= Ax + L_1 w + B_F(u + d), \\ z_1 &= R_1 A x + R_1 L_1 w_1 + R_1 B_F (u + d), \\ y &= [I_{3\times3}, 0_{3\times3}]x, \qquad w_1 = \boldsymbol{\Delta} z_1. \end{aligned} \tag{15.15}$$

The model would be an LTI system if there were no uncertainties. However, due to the parametric uncertainties, the feedback linearization based on nominal parameters will not be able to eliminate all the nonlinearities. Therefore, the remaining nonlinear uncertainties after feedback linearization are approximated by an LTI system evaluated around an equilibrium point. The resultant system has the same form as Eq. (15.15), except that L_1 and R_1 are evaluated at the equilibrium point. The feasibility of this approximation is validated in Section 15.6.

The linearized plant described by Eq. (15.15) is controllable and observable. The robust controller will be designed based on this linearized plant.

15.4 Robust manipulation controller design

An RMC consists of a robust controller and a manipulation controller. We first present the design process of the robust controller.

15.4.1 Design scheme

A robust controller is used to obtain the desired Cartesian force of the object for motion tracking with guaranteed robust stability and performance. The generalized plant P_{general} that the robust controller will work on is shown in Fig. 15.4. G_{NL} and $\boldsymbol{\Delta}$ define an upper LFT with respect to $\boldsymbol{\Delta}$ (denoted as $F_u(G_{NL}, \boldsymbol{\Delta})$) to represent the nonlinear uncertain dynamics, as shown in the red dash box. The feedback linearization described by $\alpha(x) + \beta(x)u$ is connected with the nonlinear uncertain plant to linearize the nominal model, as shown in the blue dash-dot box. Eq. (15.15) is the combination of two boxes.

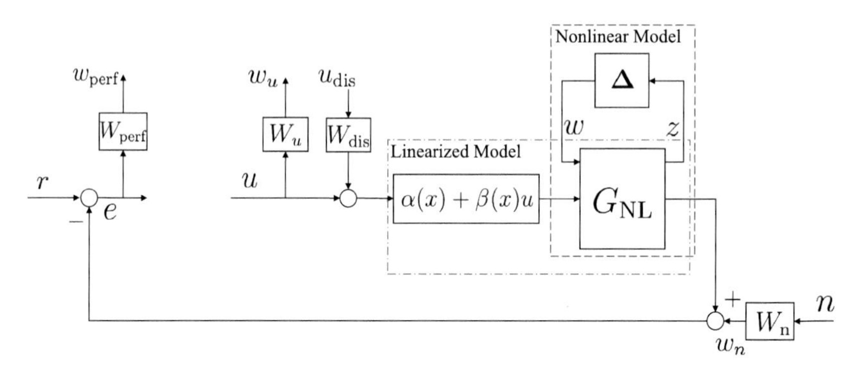

FIGURE 15.4

The inputs to the generalized plant P_{general} are $\{r, u_{\text{dis}}, n, u\}$, which denote the pose reference, the input disturbance, the noise, and the control input to the plant. The outputs of the plant are $\{w_{\text{perf}}, w_u, e\}$, which denote the tracking performance, the action magnitude, and the pose error; W_{perf} suppresses tracking errors at different frequencies, W_u regulates the control input, W_{dis} shapes the input disturbance., and W_n shapes the measurement noise. The structures of the weighting functions will be described in Section 15.4.2.

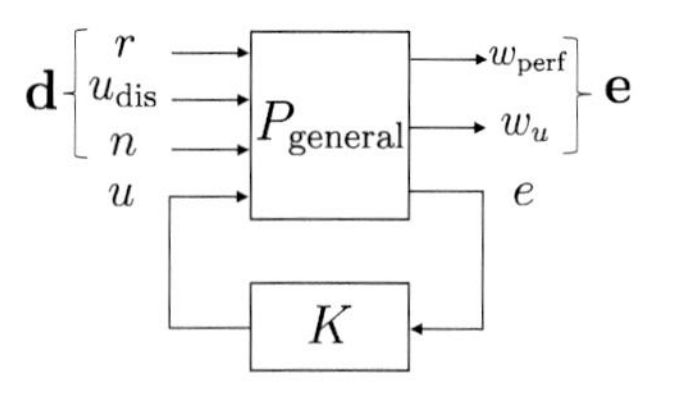

FIGURE 15.5

The connection between the generalized plant P_{general} and the controller K is described by Fig. 15.5; P_{general} and K define a lower LFT with respect to K as $F_l(P_{\text{general}}, K)$, to denote the closed-loop system. The closed-loop system concatenates the inputs $\{r, u_{\text{dis}}, n\}$ as $\mathbf{d}$ and the outputs $\{w_{\text{perf}}, w_u\}$ as $\mathbf{e}$.

The objective of the robust controller design is to synthesize K to keep $\mathbf{e}$ small for all reasonable $\mathbf{d}$, where the word "small" is used in the sense of infinity norm, i.e.,

$$\begin{aligned} &K = \underset{K}{\operatorname{argmin}} \, \| F_L(P_{\text{general}}, K) \|_\infty \\ &\text{with} \\ &\mathbf{e} = F_L(P_{\text{general}}, K)\mathbf{d}, \\ &\| F_L \|_\infty := \max_{\omega \in \mathbb{R}} \bar{\sigma}(F_L(j\omega)). \end{aligned} \tag{15.16}$$

D-K iteration is applied to solve Eq. (15.16):

$$\min_{K} \inf_{D} \| D F_L(P_{\text{general}}, K) D^{-1} \|_\infty < 1. \tag{15.17}$$

Readers can refer to [17] for implementation details.

The designed controller K will be used to calculate u based on the pose error e. Then the output of the controller is combined with feedback linearization equation (15.13) to obtain the desired Cartesian space force F for the object.

15.4.2 Design of weighting functions

The general form of a weighting function $W(s)$ in P_{general} can be written as

$$W(s) = \text{diag}([a_1 W_{1,1}(s), a_2 W_{2,2}(s), a_3 W_{3,3}(s)]),$$

where a_i is the weight of the ith channel and $W_{i,i}$ is a SISO filter determined by high-frequency gain G_h, low-frequency gain G_l, cross-over frequency ω_c, and order n. In this section, the principle of parameter selection is introduced. The concrete values for these parameters will be shown in Section 15.6.

15.4.2.1 Design of performance weighting function W_{perf}

The term W_{perf} penalizes the tracking error caused by the general disturbance $\mathbf{d}$. A high cross-over frequency w_c penalizes the disturbance with large bandwidth. With larger w_c, the system takes shorter time to settle down, and the desired force tends to change at higher frequencies. Consequently, the error oscillates at higher frequencies. The low-frequency gain G_l penalizes the magnitude of low-frequency disturbance. When G_l is very small, the low-frequency error is large, but the high-frequency error is small, which means that the system takes a shorter time to converge. On the other hand, the high-frequency gain G_h penalizes the magnitude of high-frequency disturbances. Increasing G_h will speed up the convergence. However, the oscillation will be enlarged, and the low-frequency performance will be compromised. As for the order n, high values give the system more freedom to choose the best controller, while an excessively large n increases the order of the controller. The motivation for tuning a_i is the fact that the behavior in translation directions and the rotation direction is usually different because of different parameter scales.

15.4.2.2 Design of action weighting function W_u

The actions at different frequencies are penalized equally. This is a special case when $G_l = G_h$, which means the weighting function is a constant. Similar to before, large $G_{l/h}$ penalizes the magnitude of action. A larger $G_{l/h}$ results in more penalty to control effort, thus the force generated by the controller is smaller. The smaller force can result in a slower convergence speed and poor disturbance rejection. On the contrary, a small $G_{l/h}$ can make the controller generate excessively large forces and damage the object. The influences of a_i and n can be reflected into changing $G_{l/h}$.

15.4.2.3 Design of disturbance weighting function W_{dis}

The disturbance weighting function is used to shape the exogenous disturbance in the generalized plant P_{general}. The cross-over frequency ω_c indicates the shaping bandwidth. Generally, it enlarges the magnitude of low-frequency disturbances and shrinks the magnitude of high-frequency disturbances. A large G_l will create a virtual disturbance with large low-frequency gain. Therefore, the controller would concentrate on reducing the low-frequency disturbance. In our implementation, the gravity is treated as static disturbance. Therefore, increasing G_l makes the actual system response faster by using the larger control effort. The high-frequency gain G_h specifies the shaping factor to high-frequency disturbances. A large value makes the system consider the disturbance rejection in full scale, and the low-frequency disturbance response will be compromised. Similar to W_{perf}, a_i specifies the scales of shaping for different channels, and n specifies the freedom of designing W_{dis}.

15.4.2.4 Design of noise weighting function W_n

The W_n is designed to be a high pass filter to shape the noise to the generalized plant P_{general}. The reason is twofold. First, the vision sensor used for object pose detection has high-frequency noises. Second, the manipulation controller used for desired force approximation is a low pass filter, which may result in additional high-frequency approximation error. The tuning of noise weighting is similar to disturbance weighting tuning.

15.4.3 Manipulation controller design

The manipulation controller is utilized to generate torque commands for the hand to track the desired force generated by the robust controller. The manipulation controller consists of a force optimizer, which searches the desired contact force f on fingertips from the desired force F on the object, and a joint-level torque controller, which generates an appropriate joint torque vector τ to reproduce f. The force optimizer is formulated into a quadratic programming (QP):

$$\min_{\beta} \quad \alpha_1 \|f\|_2^2 + \alpha_2 \|f - f_{\text{prev}}\|_2^2 + \alpha_3 \|\Psi\|_2^2 \tag{15.18a}$$

$$s.t. \quad \Psi = F - G(q, x_o) f, \tag{15.18b}$$

$$f = B\beta, \tag{15.18c}$$

$$\beta \geq 0, \tag{15.18d}$$

$$\tau_{\min} \leq J_h^T(q, x_o) f \leq \tau_{\max}, \tag{15.18e}$$

where $f = [f_1^T, ..., f_{n_c}^T]^T$ is a concatenated contact force vector in contact frame, f_{prev} is the contact force of the previous time step, $B = \text{diag}\{B_1, ..., B_{n_c}\}$ and B_i is a conservative pyramid approximation of friction cone [18], and $\beta \geq 0$ is the nonnegative coefficient of columns of B. A slack variable Ψ is introduced to relax the hard constraint $F = Gf$, since $F = Gf$ might result in an infeasible solution, and the location measurements of contact points might be noisy. The constraints (15.18c) and (15.18d) together ensure that the contact force remains within positive $colspan(B)$ (i.e., friction cone). Constraint (15.18e) guarantees that the contact force f is realizable.

The weights α_1, α_2, and α_3 are used to balance different cost terms. They are tuned to penalize the magnitude of the contact force, the change of the contact force, and the force tracking error, respectively. The tuning process considers the response speed of the real-world hand actuators and the force tracking performance.

The joint-level (torque) tracking controller takes the optimal contact force f^* from the force optimization as input and yields the control torque by $\tau = J_h^T(q, x_o) f^*$.

15.5 Real-time finger gaits planning

Sections 15.3 and 15.4 introduce the RMC for small-range object manipulation without considering the manipulability of the hand. In this section, we further consider the case where fingers need to adjust their grasping gestures to address the workspace limits and guarantee the object stability.

15.5.1 Grasp quality analysis

Grasp quality has been well explored in [2,19]. For grasping with multifingered hands, we aim to simplify the complex algebraic or force-based qualities into geometric types to accelerate grasp optimization [20,21]. Meanwhile, it is desired that both hand manipulability and object grasp quality are considered during the finger gaits planning. The hand manipulability describes the ability for a hand to manipulate the object to realize arbitrary object motions. The object grasp quality describes the capacity to resist external disturbances given a group of contact points on the object. This chapter adopts a quality metric in [22] to represent the hand manipulability Q_h: $Q_h = -0.5 \sum_{j=1}^{N_{\text{finger}}} \sum_{i=1}^{N_{\text{joint}}} \left((q_j^i - \bar{q}_j^i)/(q_{\text{max,j}}^i - q_{\text{min,j}}^i) \right)^2$, where q_j^i is the ith joint angle of the jth finger, $q_{\text{min,j}}^i$ and $q_{\text{max,j}}^i$ are the limits of q_j^i, $\bar{q}_j^i = (q_{\text{max,j}}^i + q_{\text{min,j}}^i)/2$ is the middle position of the corresponding joint, and N_{joint} is the number of joints per finger. The object grasp quality Q_o is represented as $Q_o = 2\text{Area}\left(\{p_j\}_{j \in I_c}, \text{proj}(p_k)\right)$ [23], where I_c is the set of indices of all fingertips that are in contact with the object, p_j is the contact position in Cartesian space for the jth fingertip, and $\text{proj}(p_k)$ denotes the projection operation of p_k onto the plane specified by $\{p_j\}_{j \in I_c}$.

Similar to [24], the overall quality Q can be obtained by combining Q_o and Q_h:

$$Q = w_1 Q_o + w_2 Q_h, \tag{15.19}$$

where $w_i > 0$ is the weight for the corresponding term.

Once the overall grasp quality Q drops below a threshold, the finger gaits should be replanned to adjust contact points on the object. It is observed that humans tend to relocate their fingers one by one during finger gaiting. This idea is adopted and the finger gaits are sequentially planned. Thus, the proposed algorithm will compare all the fingers and choose one of them to initialize finger gaits planning if all fingertips are in static contacts and $Q < \delta_Q$, where δ_Q is a threshold. The free finger is selected based on the finger manipulability of itself and the grasp quality of the remaining fingers to the object. To be more specific, the finger manipulability for the kth finger is $-0.5\sum_{i=1}^{N_{\text{joint}}}\left((q_k^i - \bar{q}_k^i)/(q_{\text{max,k}}^i - q_{\text{min,k}}^i)\right)^2$. The grasp quality of remaining fingers to the object is the area of convex hull spanned by the remaining fingertips. The candidate free finger for gaiting is the one with small finger manipulability and large remaining grasp quality. If there is already one free finger that is not in contact with the object, that finger will continue its gaiting.

In this section, the related position-level finger gaits planner is first presented and then the velocity-level finger gaits planner is proposed to resolve the problems in the position-based planner.

15.5.2 Position-level finger gaits planning

The position-level finger gaits planner consists of a contact optimization and a trajectory planning. The contact optimization searches an optimal contact point to maximize the overall grasp quality in Eq. (15.19), and the trajectory planning generates trajectories to relocate the finger to the optimal contact point. The contact optimization can be formulated in the following form:

$$\max_{p_k, q_k} \quad Q \tag{15.20a}$$

$$\text{s.t.} \quad p_k \in \partial O, \tag{15.20b}$$

$$\|p_k - p_o\| \le \epsilon, \tag{15.20c}$$

$$p_k = \text{FK}(q_k), \tag{15.20d}$$

$$q_{l,k} \le q_k \le q_{u,k}. \tag{15.20e}$$

Constraint (15.20b) indicates that the fingertip position p_k of the free finger k should be on the surface of object ∂O. Constraint (15.20c) means that the searching region should be constrained in a certain region ϵ from the original position p_o to keep the stability of the object. Constraint (15.20d) is the forward kinematics of the finger. Constraint (15.20e) means that the joint space searching should be in the feasible region. After finding the optimal contact point, the trajectory planning algorithm is required to generate a feasible trajectory.

The position-level finger gaits planning has the following drawbacks. First, the problem has nonlinear equality constraints. Therefore, it is difficult for real-time computation. In addition, this method requires a trajectory planning to avoid collision with the object and reach the planned optimal point. Moreover, the trajectory planning should consider the collision avoidance. Furthermore, the equality constraint (15.20b) corresponding to the object surface is usually unknown in advance. Lastly, the contact optimization in Eq. (15.20) uses current contacts $\{p_j\}_{j \in I_c}$ to find the optima, while $\{p_j\}_{j \in I_c}$ actually keep moving during the contact optimization, trajectory planning, and execution. With all aforementioned issues, an efficient velocity-level gaits planner is proposed below.

15.5.3 Velocity-level finger gaits planning

The task of the finger gaits planner is to generate commands to change the contact location of the free finger in order to achieve better object grasp quality and finger manipulability in real-time. However, searching contact position by maximizing the quality in Eq. (15.19) is challenging. First, the search of p_k should be conducted on the surface of the object, and the formulation of the object surface requires 3D reconstruction and surface modeling. Second, the searching of q_k should be constrained within the joint limits, and q_k is coupled with p_k by forward kinematics. Third, after finding the optimal contact point, a trajectory planning algorithm is required to generate a feasible trajectory. In our previous work [25], a velocity level finger gaits planner is proposed to overcome the aforementioned challenges.

In this planner, the contact optimization is modified into a short-term optimization. To be more specific, rather than finding an optimal contact point, an optimal moving velocity of the fingertip of the free finger is calculated at each time step, and the finger is actuated by a velocity-force controller to achieve that velocity. Formally, instead of optimizing Q in Eq. (15.19), we optimize $\dot{Q}$ with joint velocity of the kth finger $\dot{q}_k$ as the decision variable. The solution $\dot{q}_{\mathrm{des,k}}$ is used to control the robotic hand in each time step.

The intuition behind it is the Taylor series expansion; Q is a function of states $\{q_k, p_k\}$, and the states are the functions of time t. Therefore, Q is a function of t. By this interpretation, Q in time instant $t + T_s$ can be written as

$$Q(t + T_s) \approx Q(t) + \dot{Q}(t)T_s, \tag{15.21}$$

where T_s is the time step. In this equation, higher-order terms have been omitted, because T_s is usually a small period. Therefore, designing control policy to maximize $Q(t + T_s)$ is equivalent to maximizing $\dot{Q}(t)$.

With the short-term approximation, $\dot{Q}$ becomes

$$\begin{aligned}
\dot{Q} &= w_1\dot{Q}_o + w_2\dot{Q}_h, \\
\dot{Q}_o &= \|p_{j_2} - p_{j_3}\|_2 n_{j_1}^T v_{p_k}, \\
\dot{Q}_h &= \sum_{i=1}^{N_{\text{joint}}} \left(\frac{\bar{q}_k^i - q_k^i}{(q_{\text{max,k}}^i - q_{\text{min,k}}^i)^2} \dot{q}_k^i \right).
\end{aligned} \tag{15.22}$$

We assume that $\{p_j\}_{j\in I_c} = \{p_{j_1}, p_{j_2}, p_{j_3}\}$, n_{j_1} is a normal vector of line segment $\overline{p_{j_2}p_{j_3}}$ in the plane specified by $\{p_j\}_{j\in I_c}$, and v_{p_k} is the velocity of contact point p_k; $\dot{q}_k^i$ is the joint velocity of the ith joint for the kth finger. In this optimization, the states v_{p_k} and $\dot{q}_k$ in $\dot{Q}_o$ and $\dot{Q}_h$ are coupled linearly by $v_{p_k} = J(q_k)\dot{q}_k$, where $J(q_k)$ is the Jacobian matrix of the kth finger. By plugging in the coupled term, $\dot{Q}$ becomes

$$\dot{Q} = w_1\|p_{j_2} - p_{j_3}\|_2 n_{j_1}^T J\dot{q}_k + w_2 \sum_{i=1}^{N_{\text{joint}}} c_k^i \frac{\bar{q}_k^i - q_k^i}{(q_{\text{max,k}}^i - q_{\text{min,k}}^i)^2} \dot{q}_k^i, \tag{15.23}$$

where c_k^i is a weighting function added to Eq. (15.23) to address the influence of joint limits:

$$c_k^i = \begin{cases} \ln\left(\frac{\bar{q}_k^i - q_{\text{min},k}^i - q_{\text{thres}}^i}{q_k^i - q_{\text{min},k}^i}\right) + 1, & q_k^i - \bar{q}_k^i < -q_{\text{thres}}^i, \\ 1, & |q_k^i - \bar{q}_k^i| \le q_{\text{thres}}^i, \\ \ln\left(\frac{q_{\text{max},k}^i - \bar{q}_k^i - q_{\text{thres}}^i}{q_{\text{max},k}^i - q_k^i}\right) + 1, & q_k^i - \bar{q}_k^i > q_{\text{thres}}^i, \end{cases}$$

where q_{thres}^i is a threshold where the weighting should start to increase. In the meantime, constraints (15.20b) and (15.20d) become $n_{p_k}^T J(q_k)\dot{q}_k = 0$, where n_{p_k} is the surface normal of the object at p_k and $\dot{q}_k = [\dot{q}_k^1, ..., \dot{q}_k^{N_{\text{joint}}}]^T$. The surface normal can be inferred by the tactile sensor on the fingertip. The constraint (15.20c) is eliminated because we are working on short-term optimization, and the optimization would be solved in each time step.

With the above analysis, a new optimization can be formulated to approximate the original contact optimization:

$$\dot{q}_{\text{des,k}} = \arg\max_{\dot{q}_k} \ \dot{Q} \tag{15.24a}$$

$$s.t. \quad \dot{q}_{\text{min,k}} \le \dot{q}_k \le \dot{q}_{\text{max,k}}, \tag{15.24b}$$

$$n_{p_k}^T J(q_k)\dot{q}_k = 0, \tag{15.24c}$$

$$\|\dot{q}_k - \dot{q}_{\text{des,prev}}\|_\infty \le \sigma, \tag{15.24d}$$

where constraint (15.24b) means that the desired joint velocity $\dot{q}_k$ should be bounded in $[\dot{q}_{\text{min,k}}, \dot{q}_{\text{max,k}}]$. Constraint (15.24c) indicates that p_k must move perpendicular to the current surface normal n_{p_k}. Constraint (15.24d) limits the joint acceleration by σ/T_s, where T_s denotes the sampling time of the system; $\dot{q}_{\text{des,prev}}$ is the desired joint velocity in the previous time step. Eq. (15.24) is an LP, which can be solved in real-time.

After obtaining the desired joint velocity $\dot{q}_{\text{des,k}}$ by solving Eq. (15.24), a velocity-force controller is implemented to calculate the desired torque for the kth finger:

$$\tau_{\text{des,k}} = K_v \dot{q}_{\text{des,k}} + K_f J(q_k)^T (f^n_{\text{des}} - f^n_{\text{act,k}}), \tag{15.25}$$

where $\tau_{\text{des,k}}$ and f^n_{des} are the desired torque and desired contact force in the normal direction, f^n_{des} is set to be a constant small force during finger gaiting, and $f^n_{\text{act,k}}$ is the actual contact force in the normal direction and can be measured by a 1D tactile sensor. The force component $K_f J(q_k)^T (f^n_{\text{des}} - f^n_{\text{act,k}})$ in Eq. (15.25) attempts to maintain the contact between the fingertip and the surface, which makes the normal vector n_{p_k} measured from the tactile sensor updated.

The velocity-level gaits planner that is composed of Eq. (15.24) and Eq. (15.25) has several advantages. First, the proposed planner is computationally efficient. The optimization equation (15.24) is an LP that can be solved in each time step. Second, the 3D object model is not required. Instead, 1D tactile sensors are employed to detect contact points on the hand and infer the surface normals by the known fingertip structures, and the sensor update can be accomplished by the force component in the velocity-force controller.

The grasp quality is expected to be improved at the beginning of gaits planning. The velocity-level gaits planner can be terminated when there is little grasp quality improvement (i.e., $\dot{Q} < \delta$, where δ is a small positive number), or when the grasp quality is above the threshold (i.e., $Q > \delta_Q$).

15.5.4 Similarities between position-level and velocity-level planners

This section shows the similarity of the performance between the velocity-level planner Eq. (15.24) and Eq. (15.25) and one step of Eq. (15.20) if solved by a gradient projection method [26].

An abbreviated form of Eq. (15.20) is formulated for notational convenience:

$$\max_x Q \quad \text{s.t. } h(x) = 0, g(x) \leq 0, \tag{15.26}$$

where $x = \left[c_k^T, q_k^T\right]^T$, $h(x) = 0$ represents the equality constraints (15.20b) and (15.20d), and $g(x) \leq 0$ represents inequality constraints (15.20c) and (15.20e).

In each step of the gradient projection method, the search direction d is found by projecting ∇Q onto the tangent space of equality constraints $T =$

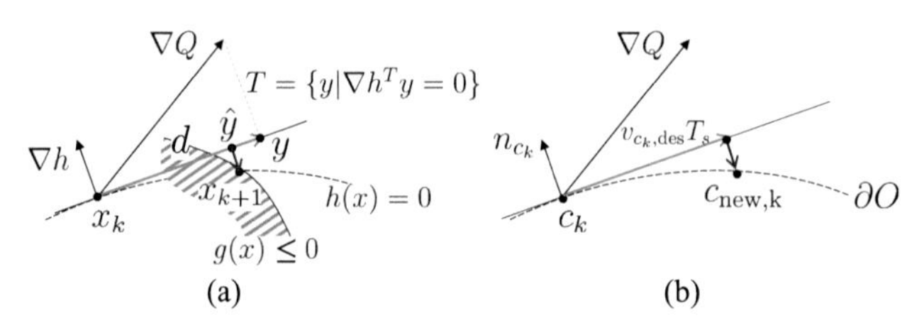

FIGURE 15.6

$\{y|\nabla h^T y = 0\}$, as shown in Fig. 15.6 (A). Then, an iterative technique is employed to project the points along d onto $h(x) = 0$, until the projected point x_{k+1} lies in the set $\{x|h(x) = 0, g(x) \leq 0\}$.

Instead of projecting the gradient into the tangent space, the proposed planner searches the optimal direction $v_{c_k,\text{des}}$ in tangent space by LP (Eq. (15.24)) with consideration of the feasibility of the motion (15.24d) and (15.24b), as shown in Fig. 15.6 (B). Moreover, the velocity-force controller (Eq. (15.25)) is a physical actualization of projecting $\hat{y}$ onto $h(x) = 0$ by maintaining the contact force between the fingertip and the surface. As the LP (Eq. (15.24)) is solved in each time step T_s, the search step $\Delta x_k = x_{k+1} - x_k$ is quite small. Thus, the planned $\dot{q}_{\text{des,k}}$ in Eq. (15.24) is usually smooth.

15.5.5 Finger gaiting with jump control

The above finger gaits planner achieves short-range sliding along convex or concave object surfaces. However, the finger gaits planner would not able to slide through the sharp edges or hollows where the curvatures are excessively large, since the normal direction estimation from tactile sensors might be noisy due to the possible multiple-point contact conditions.

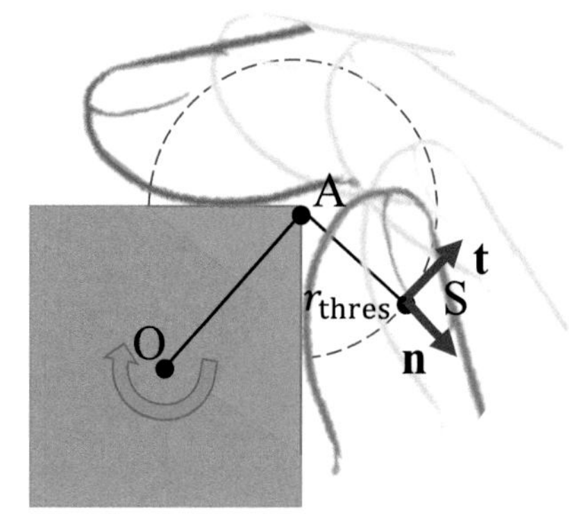

FIGURE 15.7

To avoid the failure of the finger gaits planner around the sharp edge or hollows, we observe the human behavior and propose a strategy called short-range jump control. First, the rough pose of the edge is estimated either by vision or a tactile sensor. The jump is triggered by checking whether (1) the distance between the fingertip and edge is within a certain threshold and (2) the desired velocity of the fingertip generated by the finger gaiting is towards the edge. Once the jump is triggered, the jump controller would choose a reference point and jump around the point, as shown in Fig. 15.7. In this figure, A and S denote the reference jump point and the fingertip and $\mathbf{n}$ and $\mathbf{t}$ are normal and tangent vectors computed from the current configuration. The red dash circle with radius r_{thres} is the desired path for jumping. An LP is applied to generate the velocity command for the fingertip:

$$\max_{v_t, \dot{q}_k} \quad v_t \tag{15.27a}$$

$$\text{s.t.} \quad J_{v,k}\dot{q}_k = \mathbf{t}v_t + \mathbf{n}k_n(r_{\text{thres}} - \|AS\|), \tag{15.27b}$$

$$\dot{q}_{\text{min,k}} \leq \dot{q}_k \leq \dot{q}_{\text{max,k}}, \tag{15.27c}$$

$$v_t \geq 0, \tag{15.27d}$$

where v_t is the fingertip velocity in tangent direction $\mathbf{t}$ and $\|AS\|$ is the current distance between the fingertip and the reference point. Eq. (15.27) aims to maximize the moving speed along the tangent direction. Constraint (15.27b) regulates the fingertip along the desired circular path to avoid unexpected collision with the edge. The object velocity is not included due to the measurement noise and the assumption that the object moves slow. Constraints (15.27c) and (15.27d) ensure that the jump is feasible and progressive.

The desired joint velocity $\dot{q}_{\text{des,k}}$ generated from Eq. (15.27) is converted into the joint torque by $\tau_{\text{des,k}} = K_v\dot{q}_{\text{des,k}}$. During the jump, the finger gaiting is bypassed. The jump control is reset once the fingertip makes contact with the object again; r_{thres} is a small value, thus the grasp quality (Eq. (15.19)) is assumed not to be affected by the contact change due to the short-range jump control.

15.6 Simulation and experiment studies

15.6.1 Simulation setup

The controller was implemented in the Mujoco physics engine [27]. The simulation time step was set to be 2 ms. Our platform was a desktop with 4.0 GHz Intel Core Quad CPU, 32 GB RAM, and running Windows 10 operating system.

The hand models used in the simulation are shown in Fig. 15.8. The general hand model used in 3D manipulation is shown in Fig. 15.8 (A). It has four identical fingers and 12 DoFs. Each finger has three revolute joints, J1, J2, and J3. The joint angles of J1, J2, and J3 are constrained in $[-10°, 135°]$, $[-45°, 45°]$, and $[-10°, 170°]$, respectively. A three-fingered hand for the box flipping task is shown in Fig. 15.8 (B). The three-fingered hand is set up with nine DoFs and each finger has three revolute joints, J1, J2, and J3, and each joint is constrained

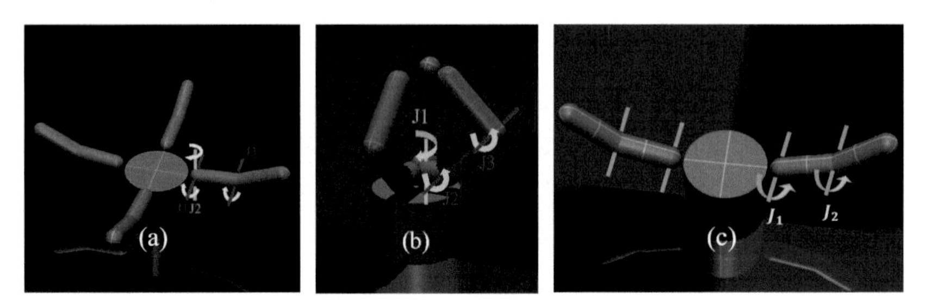

FIGURE 15.8

in [−90°, 90°]. For illustration purposes, a planar hand with two identical fingers and four DoFs was set up, as shown in Fig. 15.8 (C). All hands are equipped with joint encoders, motor torque sensors, and 1D distributive tactile sensors. The manipulated object is approximately 0.5 kg. To mimic an actual real-world finger, the density of each finger link is set to 10,000 kg/m^3. The manipulated objects for a four-fingered hand and a three-fingered hand are approximately 0.5 kg and 0.3 kg. The 3D mesh models of objects are unknown to the planner. Rather, a vision system can be employed to obtain the pose of the object by tracking the features on it. In the simulation test, the object pose is obtained from Mujoco.

15.6.2 Experimental setup

We also present the experimental setup using a BarrettHand BH8-282 to validate the proposed RMC algorithm. The BarrettHand has three fingers, four DoFs, and eight joints. Each finger has two joints driven dependently by one motor through gears and wires. An additional motor controls the spread joints of both finger 1 and finger 2, causing dependent motion of two spread joints. It is equipped with a joint encoder in each joint, pressure profile system (PPS) tactile sensors on fingertips and palms, and a strain gauge joint torque sensor at the distal link of each finger. It is worth noting that the BarrettHand is not used to validate the finger gaiting algorithm due to the limited independent DoFs.

Fig. 15.9 shows the experimental setup for RMC validation. A Logitech C270 webcam was used to perceive the scene and capture the motion of ArUco markers. Marker detection and pose estimation algorithms from OpenCV were applied to estimate the object 3D position and orientation. The pose error was then sent to the proposed RMC algorithm to produce the desired contact force for hand–object interaction. Since the wire-driven property introduces excessive frictions for direct open-loop force implementation, tactile and torque signals were fused to estimate the actual contact force and provide feedback for a low-level contact force tracking controller. The introduction of the tracking controller is neglected due to limited space.

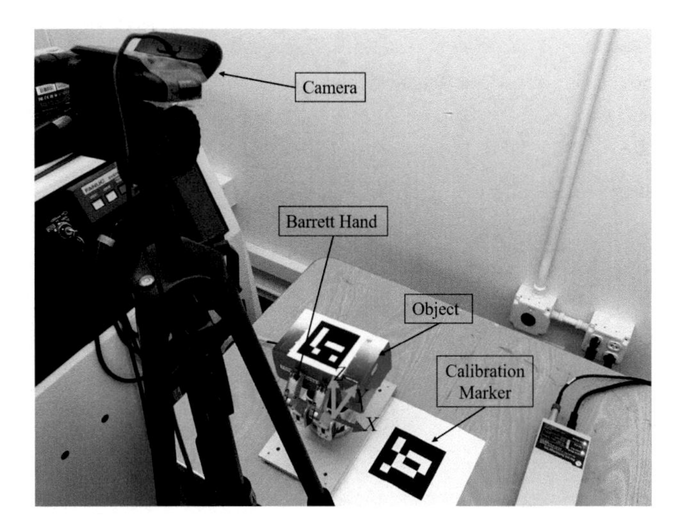

FIGURE 15.9

15.6.3 Parameter lists

15.6.3.1 RMC parameters for simulation test

The parameters of the weighting functions in Section 15.4.2 are shown in Table 15.1.

TABLE 15.1 Parameters of weighting functions for the simulation test.

Weighting	ω_c	G_l	G_h	(a_1, a_2, a_3)	n
W_{perf}	2π	1100	0.9	$(1, 1, 2)$	2
W_u	N/A	0.0001	0.0001	$(1, 1, 0.5)$	1
W_{dis}	200π	80	0.1	$(1, 1, 10)$	2
W_n	20π	0.1	10	$(1, 1, 1)$	1

The parameters of the manipulation controller were set as follows: the joint torques were constrained by $\tau_{\text{min}} = -0.5$ Nm and $\tau_{\text{max}} = 0.5$ Nm. The weights for different cost terms in Eq. (15.18) were $\alpha_1 = 0.1,\ \alpha_2 = 0.1,\ \alpha_3 = 1000$. The dimensionality of the contact space in the simulation was set as 6, with sliding, torsional, and rolling friction coefficients of 1, 0.005, and 0.0001, respectively. In manipulation controller design, we used the point contact with friction model [2] and assumed a conservative sliding friction coefficient of 0.5774. Moreover, $f_{\text{min}} = 2$ N and $f_{\text{max}} = 20$ N.

15.6.3.2 RMC parameters for BarrettHand experiment

The object to be manipulated is 0.29 kg, while the estimated mass is 0.2 kg (31% mass uncertainty). RMC was designed to resist 40% mass and 50% moment of inertia uncertainties. Due to the large sampling time of the force tracking control, the continuous plant was first discretized with $T_s = 0.033$ s before RMC design. With the plant discretization, the parameters of weighting functions in Section 15.4.2 are redesigned and shown in Table 15.2

TABLE 15.2 Weighting functions of the BarrettHand experiment.

Weighting	ω_c	G_l	G_h	(a_1, a_2, a_3)	n
W_{perf}	8π	20	0.9	$(1, 1, 1, 1, 1, 1)$	2
W_u	N/A	0.0003	0.0003	$(1, 1, 1, 0.1, 0.1, 0.1)$	1
W_{dis}	8π	32	0.1	$(1, 1, 1, 1, 1, 1)$	2
W_n	8π	0.1	10	$(1, 1, 1, 1, 1, 1)$	1

The closed-loop system with the designed RMC has a robust stability margin of around 1.13, which means that the system can withstand about 13% more uncertainty than is specified in the uncertain elements without becoming unstable.

15.6.3.3 Parameters for finger gaits planner simulation

The parameter values for the LP (Eq. (15.24)) are $w_1 = 0.99$, $w_2 = 0.01$, $q^i_{\text{thres}} = 0.25(q^i_{\text{max},k} - q^i_{\text{min},k})$, $\dot{q}_{\text{min,k}} = -1$ rad/s, $\dot{q}_{\text{max,k}} = 1$ rad/s, $\sigma = 0.002$ rad/s, $\delta = 10^{-5}$. The parameter values for the velocity-force controller are $K_v = 0.1 \sim 0.25 \times \text{diag}([1, 1, 1])$, $K_f = 1.5 \sim 3.4 \times \text{diag}([1, 1, 1])$. The planner time step $T_s = 2$ ms, and the simulation time step $t_s = 0.5$ ms. Again, the BarrettHand cannot be used for validating finger gaits planner due to the limited DoFs. Therefore, the proposed gaits planner is tested by simulations only.

15.6.4 RMC simulation results

15.6.4.1 Comparison with different methods

To simplify comparison, a 2D manipulation task is first employed to compare the proposed RMC with other methods. The desired object motion is to move to (150 mm, −10 mm, 5 degrees) from (139 mm, 0 mm, 0 degrees). The equilibrium point is chosen as the configuration upon contact, which can be planned by grasp planning. In this chapter, the equilibrium point is prerecorded for simplicity, and the nominal parameters required for modeling can be calculated accordingly.

The controller was designed to be robust to 40% mass and 50% moment of inertia uncertainties. The resultant robust stability margin was 1.73, which means that the system could withstand about 73% more uncertainties than were specified in the uncertain elements without going unstable. The proposed method is compared with the modified impedance control (MIC) from [25] and the DOB-based tracking in [6], as shown in Fig. 15.10. All these three methods

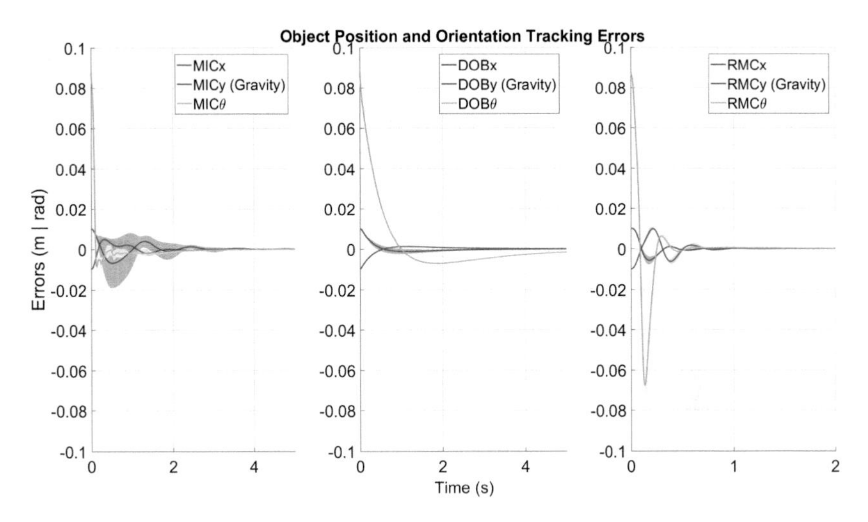

FIGURE 15.10

assume the variations of mass and moment of inertia are within 20% and 50%, respectively.

The object motion tracking result using MIC is shown in Fig. 15.10 (A). The solid lines are average convergence profiles while the shaded batches are associated variations.[1] The average settling time[2] for different uncertainties is 3.14 s. The convergence in the gravitational direction exhibits larger variation for different uncertainties, since it takes longer time to accumulate force when the believed gravity is lighter than the actual gravity.

The convergence behavior using DOB is shown in Fig. 15.10 (B). The average settling time is 3.28 s and the rotational direction is the critical direction that affects the convergence. The tuning of the parameters is described in [28]. The object pose tracking using the proposed RMC is shown in Fig. 15.10 (C). The average settling time is 0.72 s. Consequently, the oscillation is large compared with other methods. The error profiles have small variations under different mass and moment of inertia uncertainties.

The feedback linearization is applied in Section 15.3 to eliminate the nonlinearities of the nominal system. However, the nonlinearities still exist in the uncertain terms, as shown in Eq. (15.14). The remaining uncertainties are approximated as linear by evaluating parameters at an equilibrium point. The error introduced by this approximation could be treated as a disturbance and is called the disturbance from LTI approximation d_{LTI}:

$$d_{\text{LTI}} = (I - \bar{M}_{\text{eq}}\bar{M}^{-1})(\tilde{M}_o\ddot{x}_o + \tilde{N}_o) - \bar{M}_{\text{eq}}\bar{M}^{-1}\tilde{C}_o\dot{x}_o, \tag{15.28}$$

[1] ± standard derivation is used as the boundary of variation.

[2] 5% threshold is used for all settling time calculations.

where $\bar{M}_{\text{eq}}$ is the nominal inertia matrix at the equilibrium point. The magnitudes of d_{LTI} in both time and frequency domain are shown in Fig. 15.11.

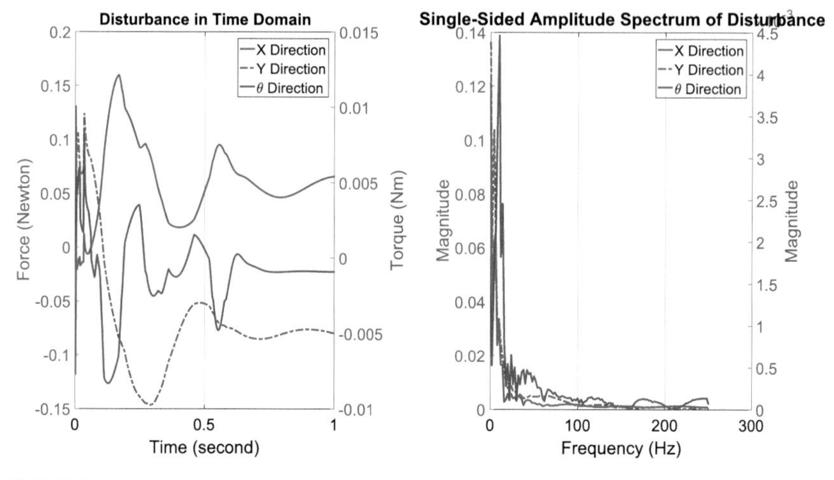

FIGURE 15.11

The magnitude and spectrum of d_{LTI} are shown in Fig. 15.11. The disturbance caused by LTI approximation has small magnitude compared with disturbance rejection introduced later, and mainly lies in the low-frequency region ($<$12 Hz), and thus can be suppressed by the proposed robust controller.

15.6.4.2 Robustness to uncertainties

A general 3D manipulation task using the four-fingered hand introduced in Section 15.6.1 is presented to demonstrate the robustness of the proposed RMC to various uncertainties. The desired object displacements are (4, 4, 11) mm and (0, 0, 0.5) rad. The velocity measurement is not required since the Coriolis force in Eq. (15.3) is neglected.

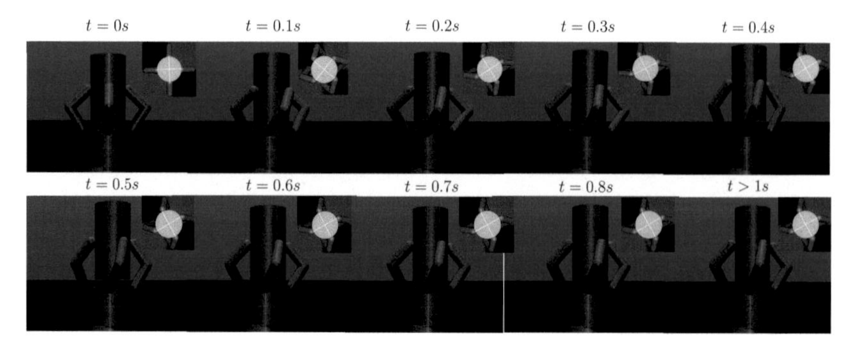

FIGURE 15.12

The snapshot of this type of manipulation is presented in Fig. 15.12. The object is a cylinder with mass 0.535 kg and radius 35 mm.

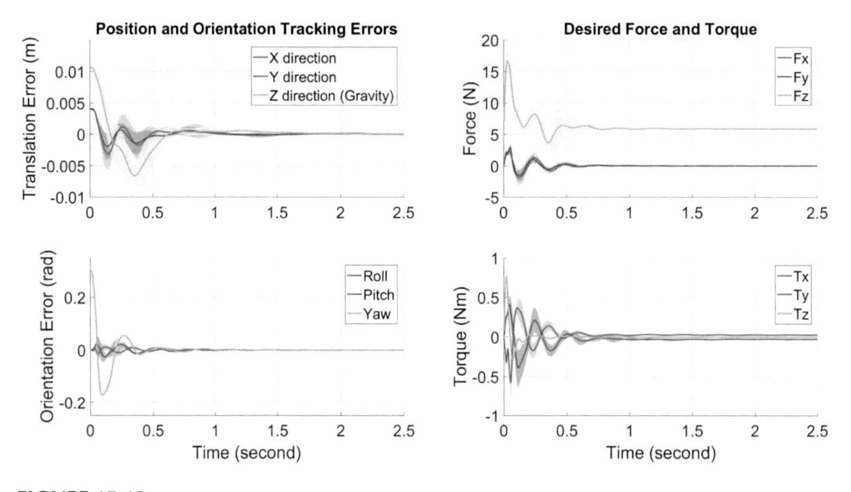

FIGURE 15.13

The proposed RMC is robust to object dynamics uncertainties. The robust controller was designed to be robust to 40% mass and 50% moment of inertia uncertainties, while the mass and moment of inertia uncertainties in the controller vary from $\{-40\%, -80\%\}$ to $\{+50\%, +80\%\}$ of their true values. The tracking errors and the corresponding desired force on the object of RMC with the sampled uncertainties are shown in Fig. 15.13 (left) and Fig. 15.13 (right). As before, the solid lines represent the average values and the shaded patches represent the variations for different uncertainties. The average settling time for all the samples is 1.31 s, while the largest settling time (2.25 s) appears in the x-direction when the uncertainties are $\{-40\%, -80\%\}$.

In addition, the proposed RMC is also robust to COM uncertainty. The maximum position deviation from the actual COM of the object is $\pm\{10, 10, 15\}$ mm, while the maximum orientation derivation from the actual principal axes of the object is $\pm\{0.3, 0.3, 0.3\}$ rad in Z-Y-X Euler angle. The object has 20% mass and 50% moment of inertia uncertainties at the same time. The tracking errors and the associated desired forces on the object are shown in Fig. 15.14 (left) and Fig. 15.14 (right). The average settling time for all the samples is 2.13 s, while the largest settling time (2.46 s) appears in the x-direction when the uncertainties are $\{5, 5, -15\}$ mm and $\{0.3, 0.3, 0.3\}$ rad. The Cartesian force converges to effective gravitational force to compensate the gravity for the composite system.

Thirdly, we demonstrate the robustness of the RMC to contact dynamics uncertainties. In Mujoco, the reference acceleration a_{ref} after contact is modeled by a virtual spring with stiffness and damping $\{k_c, b_c\}$, with $a_{\text{ref}} = -k_c r_c - b_c v_c$, where r_c, v_c are residual and velocity, and the implemented acceleration a_{imp} is interpolated by $a_{\text{imp}} = d a_{\text{ref}} + (1 - d) a_0$, where d is an interpolation factor and

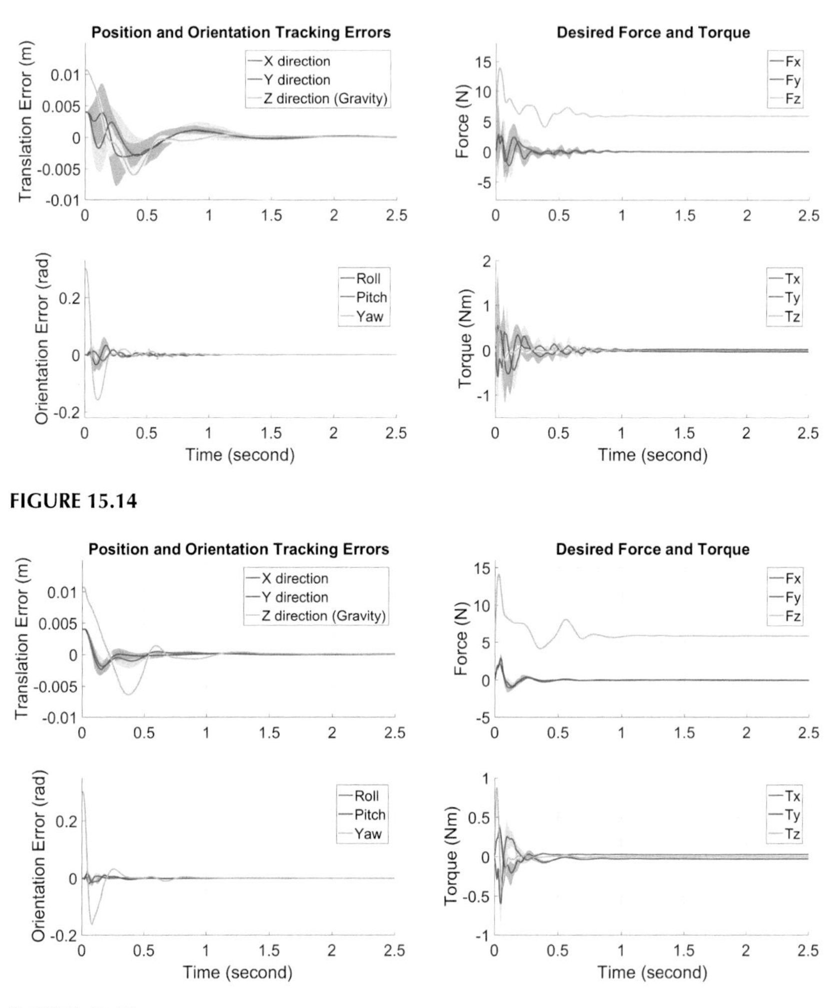

FIGURE 15.14

FIGURE 15.15

a_0 is the acceleration in the absence of constraints. In the simulation, $\{k_c, b_c\}$ vary from $\{4440.9, 133.3\}$ to $\{63131.5, 505.2\}$, and the object has 20% mass and 50% moment of inertia uncertainties in the meantime. The tracking errors and the corresponding desired forces under the sampling of these parameters are shown in Fig. 15.15. The average settling time for all samples is 1.29 s, and the largest settling time (2.28 s) appears in the x-direction when $k_c = 4440.9$ and $b_c = 133.3$. The Cartesian force converges to effective gravitational force to compensate the gravity for the composite system. The robustness to friction uncertainties will be described in Section 15.6.6.

Finally, the proposed RMC is robust to tactile uncertainty. The nominal contact position $\bar{c}_i$ and rotation $\bar{R}_{c_i}$ for the ith contact are computed by $\bar{c}_i =$

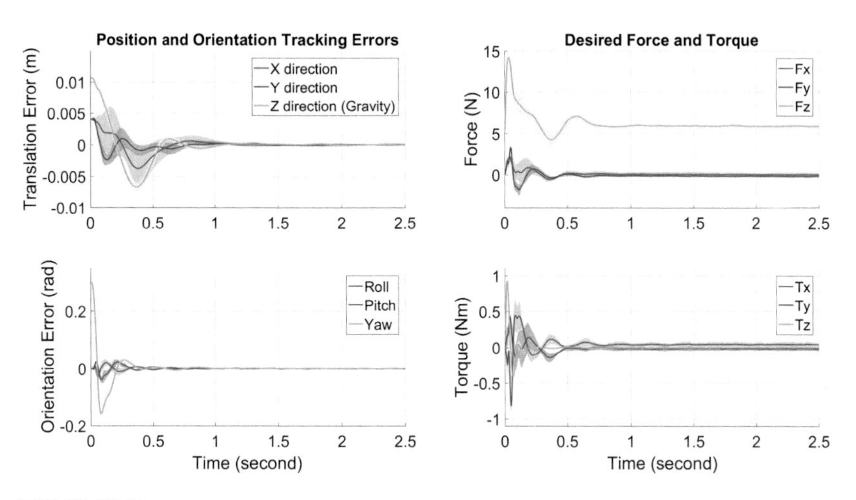

FIGURE 15.16

$c_i + \delta c$, $\bar{R}_{c_i} = R_{c_i} \cdot \delta R_c$, where c_i and R_{c_i} are actual contact position and rotation and δc and δR_c are uncertainties added to the contact, with $\delta c \sim \mathcal{N}(m_{\delta c}, \sigma_{\delta c}^2)$ and δR_c generated from Euler angle uncertainty $\delta E_c \sim \mathcal{N}(m_{\delta E}, \sigma_{\delta E}^2)$. The nominal contact positions for finger 1 and finger 3 $\bar{c}_{1/3}$ were set to $c_{1/3} - \delta c$ to avoid the influence of symmetry. In the simulation, $m_{\delta c}$ varies from $[-5, -5, -5]$ to $[5, 5, 5]$ mm, $\sigma_{\delta c} = 3\text{diag}([1, 1, 1])$ mm, and $m_{\delta E}$ varies from $[-0.2, -0.2, -0.2]$ to $[0.3, 0.3, 0.3]$ rad and $\sigma_{\delta E} = 0.1\text{diag}([1, 1, 1])$ rad. In the meantime, the object has 20% mass and 50% moment of inertia uncertainties. The tracking errors and the associated force on object under the above uncertainties are shown in Fig. 15.16. The average settling time for all samples within the range is 1.39 s, and the largest settling time (2.58 s) appears in the x-direction when $m_{\delta c} = [5, 5, 5]$ mm and $m_{\delta E} = [0.2, 0.2, 0.2]$ rad.

15.6.5 RMC experiment results

This section shows the experimental results using the proposed RMC with a low-level contact force tracking controller. Readers may refer to [29] for the details of the low-level tracking controller. Since the BarrettHand has four DoFs, it cannot span the whole 6-DoF Cartesian space. Therefore, we demonstrate the performance of RMC on tracking a desired trajectory that is within the feasibility space of object motion.

Without considering the spread motion, the desired trajectory was prerecorded during tracking to fixed position $(0, 0.03, 0.16)$ m and rotation $(0, 0, 0)$ rad, though it could be computed analytically given the initial contacts. Readers may refer to [29] for the results of the fixed position tracking. Fig. 15.17 and Fig. 15.18 illustrate the execution snapshots and the error profile for the trajectory tracking. RMC tracked the target motion in the Y- and Z-

directions and maintained a firm grasp to keep the stability of other directions. It can be seen that the desired motion on the Y-axis can be accurately tracked, though motion in the Z-direction exhibits a 0.005-m error.

The error may come from the difference of the contact locations in recording the desired trajectory and in performing the actual experiment. This implies that the 6-DoF Cartesian space cannot be spanned by the object held by a 4-DoF gripper.

The robustness of the proposed algorithm to external disturbances is demonstrated at last from 31 $\sim$ 35 s. RMC is able to comply with the external disturbance to avoid excessive large contact forces while maintaining the firm grasp without losing stability, as shown in 31 $\sim$ 35 s of Fig. 15.18.

FIGURE 15.17

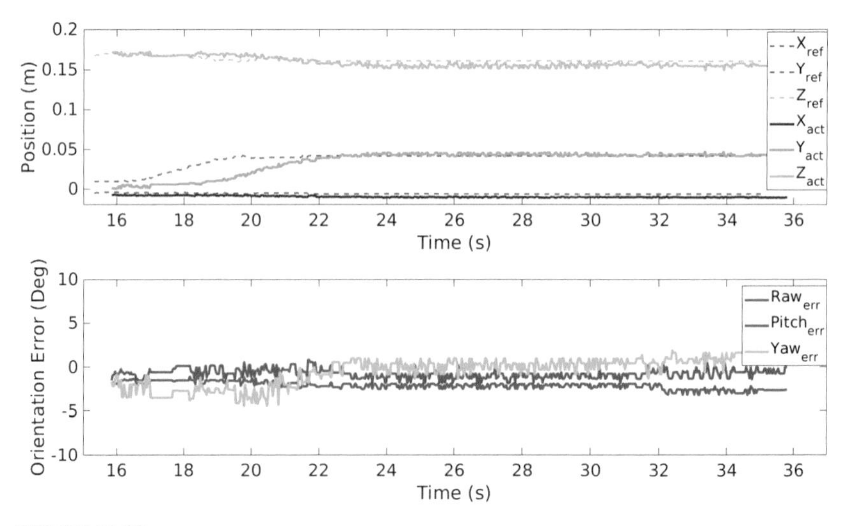

FIGURE 15.18

15.6.6 Finger gaiting simulation results

Finger gaiting cannot be validated by the current physical setup due to the limited DoFs of the BarrettHand. Therefore, we conduct several simulation tests to verify the robust finger gaiting that is composed of RMC and velocity-based finger gaits planning.

15.6.6.1 Finger gaiting on smooth surfaces under uncertainties

The proposed finger gaits planning and the RMC algorithm are first validated by a lifting and rotating task. The desired object motion is to move along the Z-axis by 11 mm, rotate continuously around the Z-axis with 0.2 rad/s, and rotate sinusoidally around the Y-axis with 0.4 rad/s. The manipulation surface of the object is smooth. The following uncertainties are included in **every** simulation of this section: (1) object dynamics uncertainty: the object has 20% mass and 50% moment of inertia uncertainties; (2) COM uncertainty: the COM position has $(3, 3, -10)$ mm offset and the principal axes have $(0.1, 0.1, 0.1)$ rad offset represented by Euler angle; (3) tactile uncertainty: the contacts measured by tactile sensors have $(2, 2, 2)$ mm position offset and $(0.05, 0.05, 0.05)$ rad orientation offset, plus additional noises; (4) contact dynamics uncertainty: the planner uses Coulomb friction and point contact with the friction model, while the contacts in the simulator also contain torsional friction and rolling friction. The stiffness $k_c = 15{,}783$ and damping $b_c = 253$.

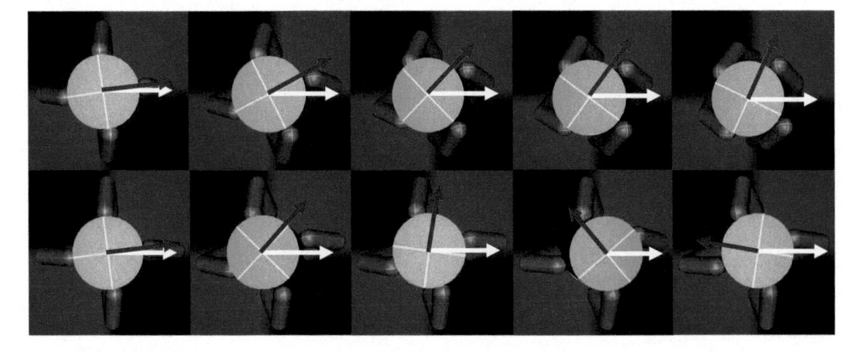

FIGURE 15.19

The performances of manipulating a cylindrical object without/with the proposed finger gaits planner are presented in Fig. 15.19. The initial and the current pose of the object are shown by white and red arrows. Without the proposed finger gaits planner, the object cannot be rotated over 90 degrees due to the decreasing finger manipulability, as shown in Fig. 15.19 (top). With the proposed finger gaits planner, the object can be rotated continuously with the desired velocity, as shown in Fig. 15.19 (bottom). The finger gaits planner ensures the grasp quality (Eq. (15.24)) is kept above a threshold, and the RMC guarantees robust stability and robust performance. The computation time for solving Eq. (15.24) and Eq. (15.18) is less than 1 ms for each planning step.

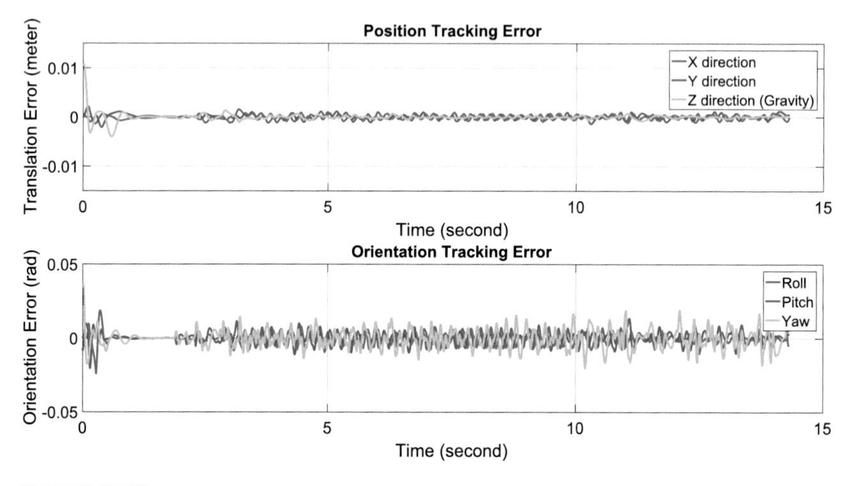

FIGURE 15.20

Fig. 15.20 shows the pose tracking errors during the lifting and rotation. RMC is able to drive the object to track the desired motion in $0 \sim 2$ s. The finger gaiting is triggered at 2 s since the grasp quality drops below a threshold. The pose errors do not converge to zero due to the disturbance introduced by contact allocating. The maximum position error is 1.6 mm in the x-direction, and the maximum orientation error is 0.019 rad (1.09 degrees) around the z-axis.

The disturbance rejection of the proposed two-level planner is shown in Fig. 15.21. Besides the disturbances from various uncertainties, external perturbation force/torque are exerted on the object. The force perturbation and the associated position errors are shown in Fig. 15.21 (A and B), and the torque perturbation and the associated orientation errors are shown in Fig. 15.21 (C and D). The system can resist at least 5 N force and 0.2 Nm torque in different directions without becoming unstable.

The robustness of the proposed two-level planner to different shapes is demonstrated by lifting and rotating an ellipsoid, as shown in Fig. 15.22. Fig. 15.22 (top) is the top view, and Fig. 15.22 (bottom) is the lateral view. The mass of the ellipsoid is 0.34 kg. An identical two-level planner (finger gaits planner + RMC) is used for the ellipsoid manipulation, though the ellipsoid has different geometries and dynamics with cylinder.

The quality rate $\dot{Q}$ in (Eq. (15.24)) during a typical contact relocation period is shown in Fig. 15.23. The rate is above δ, which means that the proposed finger gaits planner is able to continuously improve the grasp quality.

The robustness of RMC to different contact conditions has been presented in Section 15.6.4.2. However, RMC requires a conservative pyramid approximation of the friction cone specified by the Coulomb friction coefficient μ_c.

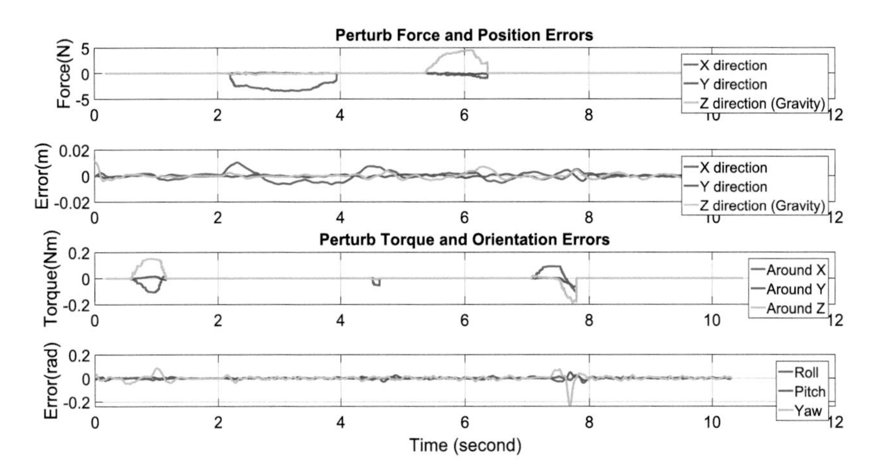

FIGURE 15.21

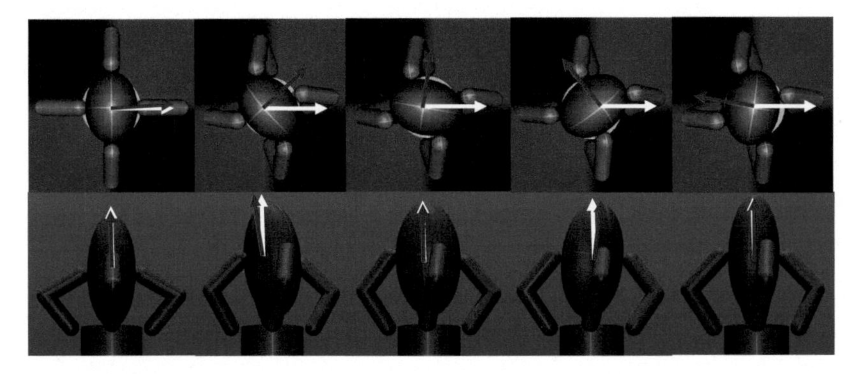

FIGURE 15.22

However, this approximation might be too conservative if the contact dynamics are uncertain. One potential solution is to detect slippage and adaptively adjust μ_c used in RMC, and slippage detection can be difficult by 1D tactile sensors [30]. In this chapter, the finger gaits planner is employed to relocate the slipping finger if the quality drops below a threshold, instead of developing a complex algorithm to prevent slippage. Besides other uncertainties, the nominal Coulomb friction coefficient $\bar{\mu}_c$ in the planner is set as 0.5236, while the actual $\mu_c = 0.4$. The unexpected slippage of the fingers is compensated by the finger gaits planner. The tracking errors of the two-level planner under friction overestimation are shown in Fig. 15.24.

15.6.6.2 *Finger gaiting of a three-fingered hand*

Finally, we demonstrate the proposed two-level planner on different hands and tasks. A box flipping task using a three-fingered hand shown in Section 15.6.1

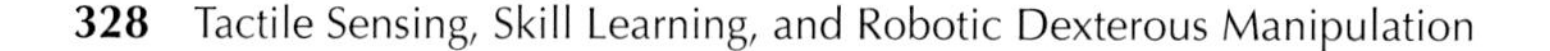

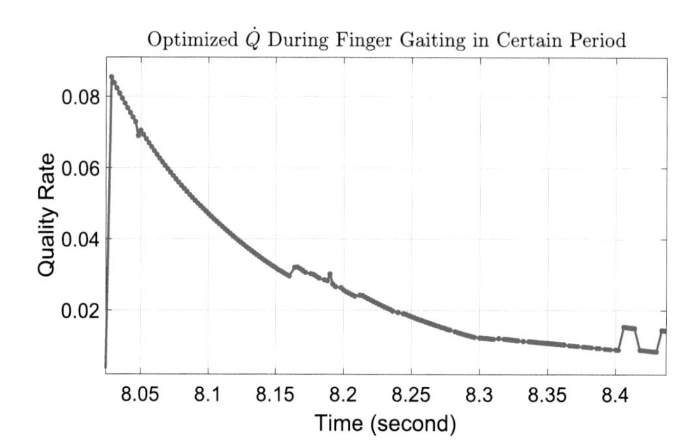

FIGURE 15.23

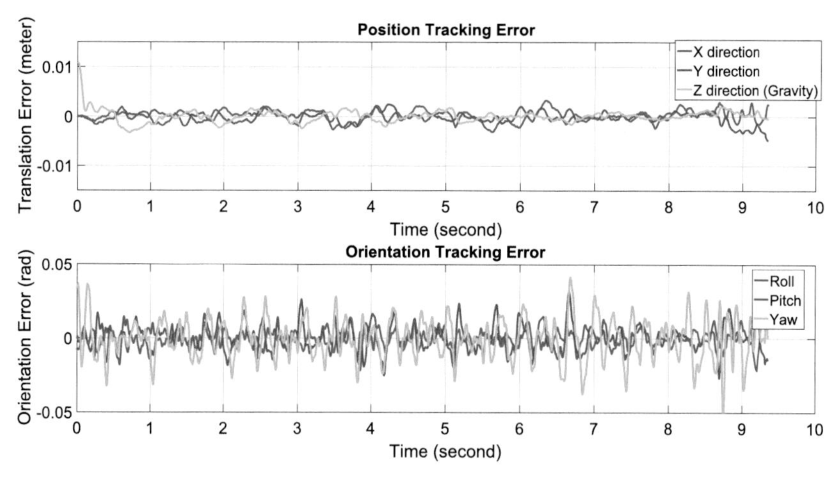

FIGURE 15.24

is illustrated. The desired flipping speed is 0.5 rad/s. The box to be manipulated has 20% mass and 50% moment of inertia uncertainties. Compared with the above simulations, the box has a sharp edge, and the task cannot be finished without changing manipulation surfaces. Therefore, a jump controller is desired to drive the middle finger through the edge, as shown in Fig. 15.25 and Fig. 15.26. Fig. 15.25 shows the snapshots of the box flipping process. The finger gaiting starts whenever losing contact or the quality drops below the threshold, and the jumping starts from 1.78 s and rotates along the edge until contacting with the other surface. Compared with sliding, the recontacting after jumping results in larger disturbance, thus the pose tracking errors oscillate to some extent, as shown in Fig. 15.26. The fingers on two sides manipulate the box by RMC. Note though that RMC is not able to maintain the robust performance

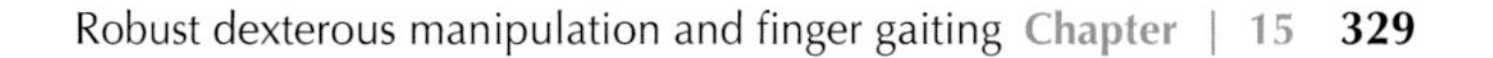

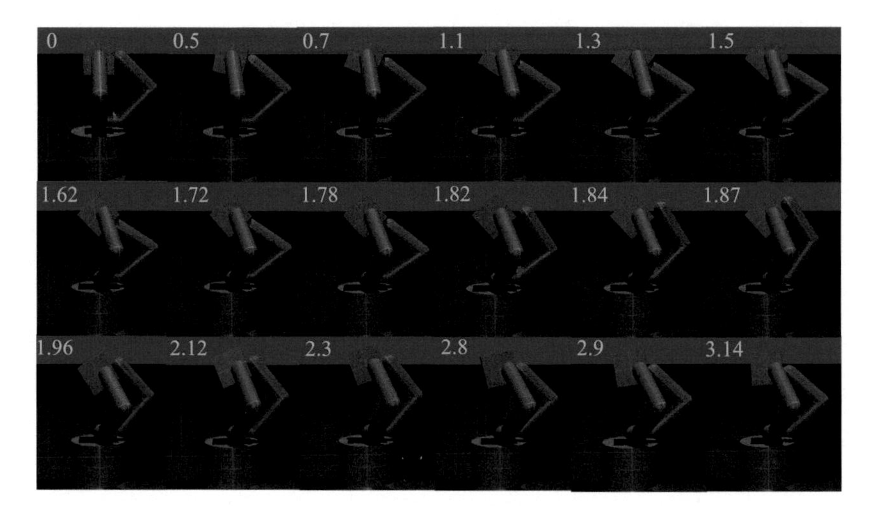

FIGURE 15.25

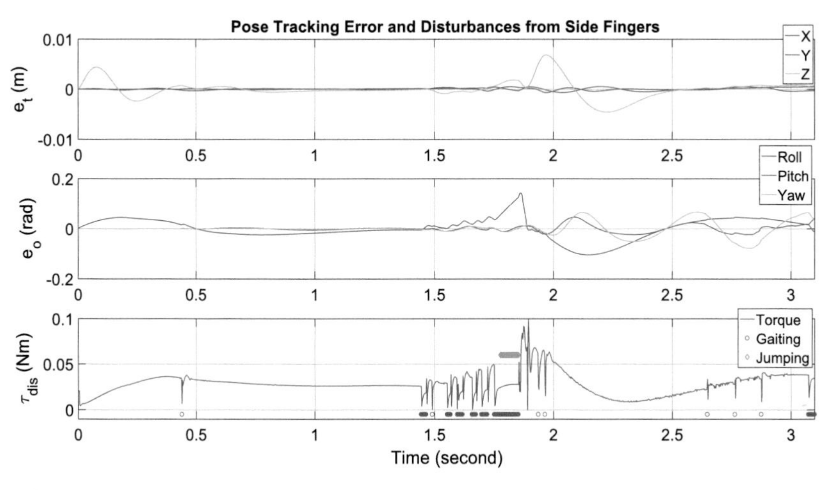

FIGURE 15.26

during jumping, since the remaining six DoFs cannot form force-closure grasps and the box cannot track the desired pitch motion, as shown in Fig. 15.26 (B). The manipulation employs point contact with the friction model, while the contact in Mujoco is set up with both torsional friction and rolling friction. The additional frictions act as a disturbance and the value is shown in Fig. 15.26 (C).

15.7 Chapter summary

This chapter proposed a two-level optimization-based planner, which includes a velocity-based finger gaits planner and an RMC, to achieve real-time finger gaiting under different types of uncertainties. The finger gaits planner searches optimal velocities to improve the object grasp quality and the hand manipulability, rather than directly finding optimal contact points by nonlinear programming methods. The proposed planner is computationally efficient and can be solved in real-time. Besides, the planner does not rely on precise 3D reconstruction for surface modeling and high-resolution encoders for velocity measurements. The presented two-level optimization-based planner can handle 50% mass and 80% moment of inertia uncertainties of the object, robust to the friction overestimation and unexpected slippage, and address the complex dynamically critical tasks such as box flipping with low-DoF hands. Simulations showed that the proposed method can achieve real-time finger gaiting and realize long-range object motions that are infeasible without the proposed finger gaits planner.

References

[1] A. Bicchi, Hands for dexterous manipulation and robust grasping: a difficult road toward simplicity, IEEE Transactions on Robotics and Automation 16 (6) (2000) 652–662.

[2] R.M. Murray, Z. Li, S.S. Sastry, A Mathematical Introduction to Robotic Manipulation, CRC Press, 1994.

[3] A. Caldas, A. Micaelli, M. Grossard, M. Makarov, P. Rodriguez-Ayerbe, D. Dumur, Object-level impedance control for dexterous manipulation with contact uncertainties using an lmi-based approach, in: 2015 IEEE International Conference on Robotics and Automation (ICRA), IEEE, 2015, pp. 3668–3674.

[4] T. Takahashi, T. Tsuboi, T. Kishida, Y. Kawanami, S. Shimizu, M. Iribe, T. Fukushima, M. Fujita, Adaptive grasping by multi fingered hand with tactile sensor based on robust force and position control, in: Robotics and Automation, 2008. ICRA 2008. IEEE International Conference on, IEEE, 2008, pp. 264–271.

[5] H. Koc, D. Knittel, M. de Mathelin, G. Abba, Modeling and robust control of winding systems for elastic webs, IEEE Transactions on Control Systems Technology 10 (2) (2002) 197–208.

[6] C.-S. Liu, H. Peng, Disturbance observer based tracking control, Journal of Dynamic Systems, Measurement, and Control 122 (2) (2000) 332–335.

[7] A. Mokhtari, N.K. M'Sirdi, K. Meghriche, A. Belaidi, Feedback linearization and linear observer for a quadrotor unmanned aerial vehicle, Advanced Robotics 20 (1) (2006) 71–91.

[8] S. Andrews, P.G. Kry, Goal directed multi-finger manipulation: control policies and analysis, Computers & Graphics 37 (7) (2013) 830–839.

[9] N. Furukawa, A. Namiki, S. Taku, M. Ishikawa, Dynamic regrasping using a high-speed multifingered hand and a high-speed vision system, in: Proceedings 2006 IEEE International Conference on Robotics and Automation, 2006. ICRA 2006, IEEE, 2006, pp. 181–187.

[10] M. Li, H. Yin, K. Tahara, A. Billard, Learning object-level impedance control for robust grasping and dexterous manipulation, in: 2014 IEEE International Conference on Robotics and Automation (ICRA), IEEE, 2014, pp. 6784–6791.

[11] P. Vinayavekhin, S. Kudoh, K. Ikeuchi, Towards an automatic robot regrasping movement based on human demonstration using tangle topology, in: Robotics and Automation (ICRA), 2011 IEEE International Conference on, IEEE, 2011, pp. 3332–3339.

[12] R.J. Platt Jr, Learning and generalizing control-based grasping and manipulation skills, PhD thesis, Citeseer, 2006.

[13] R. Platt, A.H. Fagg, R.A. Grupen, Manipulation gaits: sequences of grasp control tasks, in: Robotics and Automation, 2004. Proceedings. ICRA'04. 2004 IEEE International Conference on, vol. 1, IEEE, 2004, pp. 801–806.

[14] I. Mordatch, Z. Popović, E. Todorov, Contact-invariant optimization for hand manipulation, in: Proceedings of the ACM SIGGRAPH/Eurographics Symposium on Computer Animation, Eurographics Association, 2012, pp. 137–144.
[15] J. Xu, T.-K.J. Koo, Z. Li, Sampling-based finger gaits planning for multifingered robotic hand, Autonomous Robots 28 (4) (2010) 385–402.
[16] Q. Li, R. Haschke, B. Bolder, H. Ritter, Grasp point optimization by online exploration of unknown object surface, in: 2012 12th IEEE-RAS International Conference on Humanoid Robots (Humanoids 2012), IEEE, 2012, pp. 417–422.
[17] G.J. Balas, J.C. Doyle, K. Glover, A. Packard, R. Smith, μ-Analysis and Synthesis Toolbox, MUSYN Inc. and The MathWorks, Natick, MA, 1993.
[18] C.K. Liu, Dextrous manipulation from a grasping pose, in: ACM Transactions on Graphics (TOG), vol. 28, ACM, 2009, p. 59.
[19] M.A. Roa, R. Suárez, Grasp quality measures: review and performance, Autonomous Robots 38 (1) (2015) 65–88.
[20] Y. Fan, X. Zhu, M. Tomizuka, Optimization model for planning precision grasps with multi-fingered hands, arXiv preprint, arXiv:1904.07332, 2019.
[21] Y. Fan, M. Tomizuka, Efficient grasp planning and execution with multifingered hands by surface fitting, IEEE Robotics and Automation Letters 4 (4) (2019) 3995–4002.
[22] A. Liegeois, Automatic supervisory control of the configuration and behavior of multibody mechanisms, IEEE Transactions on Systems, Man, and Cybernetics 7 (12) (1977) 868–871.
[23] T. Supuk, T. Kodek, T. Bajd, Estimation of hand preshaping during human grasping, Medical Engineering & Physics 27 (9) (2005) 790–797.
[24] Y. Fan, T. Tang, H.-C. Lin, M. Tomizuka, Real-time grasp planning for multi-fingered hands by finger splitting, in: 2018 IEEE/RSJ International Conference on Intelligent Robots and Systems (IROS), IEEE, 2018, pp. 4045–4052.
[25] Y. Fan, W. Gao, W. Chen, M. Tomizuka, Real-time finger gaits planning for dexterous manipulation, IFAC-PapersOnLine 50 (1) (2017) 12765–12772.
[26] J. Rosen, The gradient projection method for nonlinear programming. Part II. Nonlinear constraints, Journal of the Society for Industrial and Applied Mathematics 9 (4) (1961) 514–532, http://epubs.siam.org/doi/pdf/10.1137/0109044.
[27] E. Todorov, T. Erez, Y. Tassa, Mujoco: a physics engine for model-based control, in: 2012 IEEE/RSJ International Conference on Intelligent Robots and Systems, IEEE, 2012, pp. 5026–5033.
[28] Y. Fan, L. Sun, M. Zheng, W. Gao, M. Tomizuka, Robust dexterous manipulation under object dynamics uncertainties, in: 2017 IEEE International Conference on Advanced Intelligent Mechatronics (AIM), IEEE, 2017, pp. 613–619.
[29] Y. Fan, Dexterity in robotic grasping, manipulation and assembly, PhD thesis, UC Berkeley, 2019.
[30] R. Fernandez, I. Payo, A.S. Vazquez, J. Becedas, Micro-vibration-based slip detection in tactile force sensors, Sensors 14 (1) (2014) 709–730.

Appendix A

Key components of dexterous manipulation: tactile sensing, skill learning, and adaptive control

Qiang Li[a], Shan Luo[b], Zhaopeng Chen[c], Chenguang Yang[d], and Jianwei Zhang[c]
[a]*Center for Cognitive Interaction Technology (CITEC), Bielefeld University, Bielefeld, Germany,* [b]*Department of Computer Science, University of Liverpool, Liverpool, United Kingdom,* [c]*University of Hamburg, Faculty of Mathematics, Informatics and Natural Science, Department Informatics, Group TAMS, Hamburg, Germany,* [d]*Bristol Robotics Laboratory, University of the West of England, Bristol, United Kingdom*

A.1 Introduction

Dexterous manipulation is a fundamental capability for robots that go outside the laboratory and into our real life. Although this looks like a trivial capability for humans, it is still a big challenge even for the most advanced robots. Over several billion years of evolution, humans have developed a rich multimodal perception system. After a baby's birth, he or she needs many years to train the sensor–motor loop, learn manipulation skills, and learn to make decisions while facing complex situations. From this perspective, robots still have a long way to go before reaching human-like manipulation skills. It is impossible to cover everything within one book, reveal all "secrets" behind manipulation capabilities, and report all robotic manipulation progress. Therefore, we decided to present such progress from several vital perspectives. We believe that research progress in these fields will reshape the future of robotic manipulation, especially when the robots start to make "contact" with external unstructured environments physically. These components make up the title of this chapter: tactile sensing, skill learning, and adaptive control.

A.2 Why sensing, why tactile sensing

Traditionally, robots were exploited and deployed in industrial factories. In such well-structured environments, the programmers can model everything; all robots need to do is plan and then follow that plan. This, however, cannot fulfill the requirements of work in an unstructured environment. Service robots that work in the hospital and take care of patients need to grasp an object and deliver it to a patient. They are even required to help and lift patients and put them in bed. Collaborative robots that work with the workers side by side need to help workers perform fussy tasks in order for the workers to focus on thinking and decision making. Agricultural robots that work in the farm need to help farmers plant and reap. The environment cannot be entirely defined in advance by the robotic experts; robots have to use their sensing system to understand the environment and implement proper decisions on their own.

Vision sensing is a well-developed research domain that has achieved lots of promising results based on the scientific publications in the fields of robotics, artificial intelligence, control, and neuroscience. Via vision, robots can understand the environment, make decisions, take actions, and implement complex manipulation tasks even in cluttered environments, because vision provides a global dynamic image sequence, yielding huge amounts of available data, which are processed by advanced deep neural network algorithms running on powerful computational hardware. However, more and more researchers are beginning to realize that physical contact cannot be modeled well purely based on vision, and modeling and predicting information in the force domain is especially difficult. Another downside of vision is the unavoidable occlusion when robots physically interact with objects. One can easily imagine human beings who exploit touch to recognize the context of contact and make decisions via tactile sensing and force feedback.

The first part of this book includes four chapters. The authors present and discuss several important aspects of tactile sensing: tactile sensors, tactile perception, and cross-modality learning. Mainly two types of tactile sensors are presented: optical-based and capacitive-based sensors. In Chapter 1, apart from introducing the evolution of optical tactile sensors in detail, the authors also present their GelTip tactile sensor. One advantage of this sensor is that it can detect contacts inside and outside of the grasp closure and render the robotic gripper's grasp more robustly. The authors also compare the proposed sensor with other state-of-the-art sensors (GelSight, GelSlim, DIGIT, RoundFingertip, OmniTact). In Chapter 4, the authors explain the principle of capacitive proximity sensing and propose using the sensor's measurement to classify the objects. The advantage of the proposed sensor is that it can catalog the materials in a granular way, e.g., not only recognize the object as made of stone, wood, or plastic, but also identify which kind of wood. All measuring and classification is done without any contact. The authors of Chapter 2 focus on the processing of tactile sensing, and they do a survey on how to extract material properties, grasping stability, and object pose and shape. Besides that, the authors also introduce a visually guided tactile perception approach and its application in a crack detection and reconstruction task. Except for tactile sensing, other modalities are also considered. This stimulates the work in Chapter 3. The authors' study in this chapter is inspired by the concept of synesthesia. Technically, this can be represented as the same features from various sensory modalities sharing a common subspace. Through the common subspace, it is possible to have a unitary representation from multiple modalities and transfer the features from one modality to another. The authors study synesthesia by cross-model learning. After the learning procedure, sensing data from one modality, e.g., vision, can be mapped to the tactile domain, which makes up realistic pseudodata to replace the inaccessible real data. The authors also examine the attention mechanism in multimodality feedback, which is helpful to improve the learning efficiency, e.g., a higher learning weight is given to the vital modality features.

Tactile sensing is only half the story for robots interacting with unknown objects; the other half is the actions. The goal of action is twofold. An action is taken to change an object to its goal state, and it is also taken so that the robot can measure the object better and improve its confidence in itself and its external environment. Research on the second half of the story is distributed over the following two components.

A.3 Why skill learning

A robot's skill is a widely used term in the robot learning domain. Generally speaking, the skill is the capability of a robot to perform a given task. Some examples of robotic skills are locomotion, peg-in-hole, grasping, active object categorization, and joint manipulation of an object with a human. A critical feature of owning a skill is that skillful manipulation can be extended to an unseen scenario and used for implementing complex manipulation if the robot has acquired primitive skills. Skill acquisition can be achieved through two approaches: autonomous self-exploration or guidance from a human teacher.

From the aspects of robotic imitation, robotics experts expect to gain research insights from a subject's skill demonstration. In Chapter 6, the author tries to find the neurophysiological mechanisms underpinning the sensorimotor control of human grasping – how the different sensory modalities like visual or tactile sensations influence the kinematic performance of a daily manual task that usually involves multiple consecutive subgoals. To answer this question, the author firstly studies human aspects to find the functionality of multimodality (vision, force) for a grasp-to-pour task of a glass of water and investigates the effects of sensorimotor control on the kinematic performance. In order to understand the human's grasping dynamics while the object's center of mass is changed, the author uses a sensorized object to measure and record each digit force of the grasping hand. Statistical analysis is conducted to study how the human changes his digit force (skill) to keep the grasp stable without pronounced tilt or vibration. In Chapter 7, the authors study the analysis of human data from kinematic and dynamic aspects and transfer the findings, e.g., human grasping skills, to different robotic hands. For dynamic grasping, authors deploy the human grasping skill – fixing the stiffness of the hand as a whole instead of controlling every digit force – to a Shadow hand. From kinematic grasping, the authors study the human hand and the underactuated prosthetic hand Hannes. A comparison of the principal components shows that the synergistic concept is quite an excellent approach to building a prosthetic hand, which was designed based on anthropomorphism and biomimicry and replicates human hand-like grasping. A more detailed description and dexterity evaluation of the Hannes hand are presented in Chapter 11, where its high effectiveness and usability are compared to advanced prostheses. Imitation skills can be obtained from the insight of the subject's experiment and robotic demonstration. In Chapter 5, the authors study force-related dexterous tasks, e.g., human–robot collaboration for sawing, and their proposed solution is to learn the variable impedance profile. The data are extracted by kinesthetic teaching.

Except for imitation, peg-in-hole, grasping, and in-hand manipulation skills can be extracted from exploration learning. The authors present this methodology in several chapters. In Chapter 10, the authors study an insertion task using a robot arm. Even in an arm with six degrees of freedom (DoFs), a reinforcement learning (RL) algorithm generally requires excessive data to explore the entire space and locate the optimal policy. The authors evaluate their robotic insertion idea with specified insertion tasks. It is unnecessary to explore the whole action space because the insertion action is limited in narrower action space which can be initialized well with the trajectories generated from a fitted dynamics model. When the robot is fitted with a hand with even more DoFs, different strategies must be taken to make the grasping and in-hand manipulation feasible. Chapter 8, inspired by the form-closure grasping, proposes one controller for four-pin hands to grasp an unknown object. A DexNet object set is used as the training library, and a convolutional neural network is used to model the mapping from the grasp posture and the observed image to the grasp score. Using the predicted grasp score as the cost function, PPO is used in an RL framework to compute the optimized grasping action. Instead of exploiting the powerful but "black-box" neural network, the authors in Chapter 10 propose a hierarchical learning strategy. The whole in-hand manipulation skill is divided into three primitives: finger gating, sliding, and repositioning. A low-level controller is based on the dynamics model, and a high-level controller is parametrized and learned by DRL. This approach can be generalized to unseen-object in-hand manipulation.

A.4 Why adaptive control

Adaptive control is the lowest implementation component for manipulation tasks. With a well-designed adaptive controller, a robot can resist external disturbances and implement learned skills robustly. In a lot of modern skill learning work, adaptive controllers are also used as basic skill primitives to support more complex skill learning. Usually, an "adaptive" algorithm runs online to tune the parameters needed for the final desired actuator motion.

In Chapter 13, the authors present an adaptive controller of a robotic arm for medical minimally invasive surgery. The "adaptivity" is mainly shown by using a neural network to observe the external disturbance. After the disturbance is compensated, model predictive control is used for the surgery tool's tracking. In Chapter 12, the authors tackle a unique adaptive control. Instead of using the measured data to tune the parameters, this work uses human-in-loop (teleoperation) to guarantee the robustness of manipulation. A human uses his or her eyes to closed-loop control, so the main work is to map the motion of the human's arm and hand to a robotic arm and hand. In order to tackle the challenging anthropomorphic hand control, the authors propose an end-to-end mapping approach. A neural network is proposed that maps the human hand's image directly to the desired robotic hand joint command. This work also considers the coordination between arm and hand. To this end, an IMU located in the human arm provides the desired pose of the human wrist, which then maps to the robotic arm's end-effector. Using a human in the loop and monitoring the progress of manipulation will be helpful for safe manipulation, but pure visual feedback brings a heavy control burden for the human, and it is impossible to perfectly map the human motion to the robot for nontrivial in-hand manipulation tasks. In this situation, one good solution is for the robot to have a partially autonomous capability using tactile sensing for contact-based fine manipulation. This capability will allow it to deal well with losing contacts, grasp force deviation, motion deviation, and others which are unavoidable drawbacks in teleoperation. In Chapter 14, the authors present a sensor-based grasp and an in-hand manipulation framework. Exploiting vision and tactile feedback, the robot can reactively grasp and in-hand manipulate an object even with the knowledge of its geometry. A tactile-based exploration mechanism is proposed to extract geometrical features of the unknown object, and this provides the possibility for an "adaptive" component to improve the quality of grasping and in-hand manipulation. In Chapter 15, the author extends this idea to the dynamics domain of grasping and in-hand manipulation. Thanks to the modeling procedure, it is theoretically possible to quantify the robustness to external parameters. Comparing with Chapter 14, the author in Chapter 15 formalizes tactile exploration as one optimization problem to find good "grasping points" by employing a continuous grasp quality function. For the discontinuous quality area, e.g., when edge contacts happen, manually designed processing is presented. To this end, it is necessary to use tactile sensing to recognize contact geometry.

A.5 Conclusion

As we described in the previous sections and all chapters, tactile sensing, skill learning, and adaptive control are essential components for dexterous manipulation tasks, especially when the tasks are in the context of contacts.

Tactile sensing can provide high-quality contact measurements which cannot be obtained from other modalities like vision. The raw tactile measurements are processed further to compute task-relevant information, e.g., contact-level information and object-level information. Tactile sensing can be independent or integrated with other modalities like vision to learn manipulation skills and adaptive control. Many studies have shown the fusion's robustness for object recognition or manipulation.

Skill learning requires measuring and storing sensor-motor data and finding human neurophysiological mechanisms. This provides the possibility of deploying and adapting the mechanisms in robotic controllers for innovative tasks that cannot be performed well with traditional feedback control. Another way towards skill learning is to let the robot explore its state and action spaces to find the best control policy. In order to improve the efficiency of exploration, manually designing the learner's hierarchy or using the control policy initialized by imitation are two effective strategies presented in these chapters. Some studies even focus on the autonomous learning of hierarchy. If this concept can be proved and generalized to different tasks, it can lead to fully autonomous learning approaches.

Adaptive control is used for implementing the learned skills and learning new higher-level skills based on the available primitive controllers. It takes the tactile sensing as input and implements the control policy (manually designed or via learning) robustly. This requires the external states to be well estimated and the robot's position, force, and stiffness to be fine-tuned according to the estimated parameters. The fine-tuning laws are designed based on adaptive control theory, and the manipulation stability can also be guaranteed theoretically. Another interesting direction is to use an available adaptive controller to learn the skills. Taking the action of grasping glass–lifting glass–pouring liquid–placing glass as an example, the robot can learn the sequenced complex skills by observing the following: visual and tactile sensing, generalizing action decisions to unseen scenarios e.g. the glass position, COM as well as liquid density, the volume in the glass, the required poured liquid, and the placed position are not the same like the demonstrated ones. Without the manually designed sequence and implementable grasping, lifting, pulling, and placing the primitive controller as input, learning the whole manipulation is not resolvable. Scientifically, the manually designed sequence should be learned by observing human manipulation.

Index

D

K

L

M

N

U

V

W